Telecommunication Circuit Design

WILEY SERIES IN TELECOMMUNICATIONS

Donald L. Schilling, Editor
City College of New York

Worldwide Telecommunications Guide for the Business Manager
Walter L. Vignault

Expert System Applications to Telecommunications
Jay Liebowitz

Business Earth Stations for Telecommunications
Walter L. Morgan and Denis Rouffet

Introduction to Communications Engineering, 2nd Edition
Robert M. Gagliardi

Satellite Communications: The First Quarter Century of Service
David W. E. Rees

Synchronization in Digital Communications, Volume 1
Heinrich Meyr et al.

Telecommunication System Engineering, 2nd Edition
Roger L. Freeman

Digital Telephony, 2nd Edition
John Bellamy

Elements of Information Theory
Thomas M. Cover and Joy A. Thomas

Telecommunication Transmission Handbook, 3rd Edition
Roger L. Freeman

Digital Signal Estimation
Robert J. Mammone

Telecommunication Circuit Design
Patrick D. van der Puije

Telecommunication Circuit Design

Patrick D. van der Puije
Carleton University
Ottawa, Ontario, Canada

A Wiley-Interscience Publication
JOHN WILEY & SONS, INC.
New York • Chichester • Brisbane • Toronto • Singapore

Library of Congress Cataloging in Publication Data:

Van der Puije, Patrick D.

Telecommunication circuit design/Patrick D. van der Puije.

p. cm.

"A Wiley-Interscience publication."

Includes bibliographical references and index.

ISBN 0-471-50777-6

1. Telecommunication—Apparatus and supplies—Design and construction. 2. Electronic circuit design. 3. Radio circuits—Design and construction. 4. Television circuits—Design and construction. 5. Telephone—Equipment and supplies—Design and construction. I. Title.

TK5103.V365 1992
621.382—dc20 91-34097
CIP

Printed in the United States of America

10 9 8 7 6 5 4 3 2 1

CONTENTS

PREFACE

This book is about telecommunications: the basic concepts, the design of subsystems, and the practical realization of the electronic circuits that go to make up telecommunication systems. The aim of this book is to fill a gap that exists in the teaching of telecommunications and electronic circuit design to electrical engineering students. Frequently, courses on electronic circuits are taught to students without a clear indication of where these circuits may be used. Later in their career, the students may take a course in communication theory where the usual approach is to treat subjects like modulation, frequency changing, and detection as mathematical concepts and to represent them in terms of "black boxes." Thus the connection between the functional "black boxes" and the design of an electronic circuit that will perform the function is glossed over or is completely missing.

The approach followed in this book is to take a specific communication system, for example the amplitude modulated (AM) radio system, and describe in mathematical terms how and why the system is designed the way it is. The system is then broken down into functional blocks. The design of each functional block is examined in terms of the electronic devices to be used, the circuit components, and the requirements for power. The effectiveness of each functional block is determined. In most cases, more than one circuit is presented starting from the very elementary, which usually illustrates the principles of operation best, to more sophisticated and practical varieties. The order in which the signal encounters the functional blocks determines the order of the presentation so that new information is presented at an opportune moment when the interest of the student is optimal. Examples are provided to emphaisze the link from concept to design and realization of the circuits. Systems examined include commercial radio broadcasting, television, and telephone, with a section devoted to satellite communications and data transmission circuits.

This book was written primarily for final-year engineering undergraduate students. Illustrative examples are given whenever possible to promote the active participation of the student in the learning process. However, the practical approach to electronic circuit design will no doubt be useful to people involved in the telecommunications industry for updating, review, and reference.

Prerequisites to the course in telecommunication circuit design are second-year university mathematics, basic electronics, and some familiarity with communication theory (although the latter is not strictly necessary). In every case where communication theory has a direct impact on the design, enough background has been given to gain an understanding of the topic.

A number of specialized topics have been excluded or treated qualitatively in the interest of brevity, including antennas, filters, and loudspeakers. Many books are available on these subjects. Rather than presenting a cursory treatment of these very important subjects, I opted for a qualitative description of the operation and design of antennas, filters, and loudspeakers in the hope that the reader can develop an appreciation of the important features of these devices, which can be built upon if they are of special interest. A list of available reading material has been given in the appropriate chapters.

Although most of the circuits discussed in this book can be found in integrated-circuit form, detailed discussion of integrated circuit design has generally been avoided. This is because the "rules" of integration aim to reduce the area of the chip to a minimum, tending to increase the number of transistors or active components that take up little area at the expense of passive components, such as resistors and capacitors, that take up relatively large areas. Furthermore, because the integrated circuit process can produce very closely matched transistors, the integrated circuit designer often uses symmetry to achieve circuit functions not possible with discrete devices. An explanation of how an integrated circuit works is therefore more complicated than its discrete counterpart. In every case, where the simple and the modern have clashed, I have chosen the simple. However, integrated circuit design techniques have been discussed whenever they are relevant and do not distract the reader from a good understanding of the basic principles of circuit design.

In Chapter 1, a brief history of telecommunications is given. The last 150 years has been a time of tremendous growth and change in telecommunications, with more than enough change to qualify as a "revolution," perhaps the greatest revolution in the history of mankind—the information revolution.

Chapters 2 and 3 describe the AM radio system and the electronic circuits that make it possible, including the design of the crystal-controlled oscillator in the transmitter and the loudspeaker in the receiver. Chapters 4 and 5 repeat the process with the frequency modulated (FM) radio and include sections on stereophonic commercial broadcast and reception.

Television—the transmission and reception of images—is discussed in Chapters 6 and 7. The design of the circuits involved in the acquisition of the video signals, the processing, transmission, coding, broadcasting, reception, and decoding are described.

In Chapters 8, 9, and 10, the growth of the telephone system is traced from its humble beginnings to the world-wide network that it is today. The need to open up the system to an increasing number of subscribers has led to the development of sophisticated signal processing techniques and circuits with which to implement them. The development of channels in new transmission media, such as satellites and fiber optics, as well as improvements to hard wire connections have been made possible because circuit designers have produced the hardware at the right time and at the right price. The growing traffic of "conversations" between machines of various descriptions has accelerated the trend toward "digitization" of signals in the telephone network. The design of circuits capable of accepting data corrupted by noise, restoring the data, and transmtting them is discussed.

This book started off as lecture notes for a senior college course in electrical engineering "Telecommunication Circuits." At the time that I proposed the course, it was becoming increasingly clear that our graduating students did not know enough about communication systems. The students thought that anything analog (including radio, television, and telephone) was *passé*. Digital circuits (computers and software development) on the other hand were considered "cutting edge." It is necessary to balance this situation, and I hope that this book helps to restore some semblance of symmetry. The seemingly simple task of changing a set of lecture notes into a textbook turned out not to be quite as simple as I had imagined. However, I have learned a lot from it and I hope the reader does too.

The material presented in this book is more than can be presented in a normal 13-week term. However, the organization of the chapters is on the basis of the three major telecommunication networks, namely radio, television, and the telephone. It is therefore convenient to organize such a course around a group of chapters with minimal rearrangement of the material and still maintain coherence.

PATRICK D. VAN DER PUIJE

Ottawa, Ontario, Canada
April 1992

Telecommunication Circuit Design

1

THE HISTORY OF TELECOMMUNICATIONS

1.1 INTRODUCTION

According to UNESCO statistics, in 1983, there were about 50,000 radio transmitters broadcasting to 1.5 billion receivers in nearly 200 countries. The figures for television were 72,000 transmitters and 623.7 million receivers [1]. During the same year, it was reported that there were 486.6 million telephones in use world-wide [2]. In addition to this, the military in every country has its own communication network, which is usually more technically sophisticated than the civilian network. These numbers look very impressive when one recalls that electrical telecommunication is barely 150 years old. One can well imagine the number of people employed in the design, manufacture, maintenance, and operation of this vast telecommunication system.

1.2 TELECOMMUNICATION BEFORE THE ELECTRIC TELEGRAPH

The need to send information from one geographic location to another with the minimum of delay has been a quest as old as human history. Galloping horses, carrier pigeons, and other animals have been recruited to speed up the rate of information delivery. The world's navies used semafore for ship-to-ship as well as from ship-to-shore communication. This could be done only in clear daylight and over a distance of only a few kilometers. The preferred method for sending messages over land was the use of beacons: lighting a fire on a hill for example. The content of the message was severly restricted since the sender and receiver had to have previously agreed on the meaning of the signal. For example, the lighting of a beacon on a particular hill may inform one's allies that the enemy was approaching from the north, say. In 1792, the French Legislative Assembly approved funding for the demonstration of a 35-km visual telegraphic system. This was essentially semafore on land. By 1794, Lille was connected to Paris by a visual telegraph [3]. In England, in 1795, messages were being transmitted over a visual telegraph between London and Plymouth, a return distance of 800 km in 3 min [4].

North American Indians are reputed to have communicated by creating puffs of smoke using a blanket held over a smoking fire. Such a system would require clear daylight as well as the absence of wind, not to mention a number of highly skilled operators.

A method of telecommunication used in the rainforests of Africa was the "talking drum." By beating on the drum, a skilled operator could send messages from one village to the next. This system of communication had the advantage of remaining operational in daylight as well as at night. However it would be subject to operator error especially when the message had to be relayed from village to village.

1.3 ELECTRIC TELEGRAPH

The first practical use of electricity for communication was in 1833 by two professors from the University of Goettingen, Carl Friedrich Gauss (1777–1855) and Wilhelm Weber (1804–1891). Their system connected the Physics Institute and the Astronomical Observatory, a distance of 1 km, and used an induction coil and a mirror galvanometer [4].

In 1837, Charles Wheatstone (1802–1875) (of Wheastone Bridge fame) and William Cooke (1806–1879) patented a communication system that used five electrical circuits consisting of coils and magnetic needles to indicate a letter of the alphabet painted on a board [5]. The first practical use of this system was along the railway track from Euston to Chalk Farm stations in London, a distance of 2.5 km. Several improvements were made, the major one being the use of a coding scheme which reduced the system to a single coil and a single needle. The improvement of the performance, reliability, and cost of communication has since kept many generations of engineers busy.

At about this time, Samuel Morse (1791–1872) was busy working on similar ideas. His major contribution to the hardware was the relay also called a repeater. By connecting a series of relays as shown in Figure 1.1, it was possible to increase the distance over which the system could operate [5]. Morse also replaced the visual display of Wheatstone and Cooke with an audible signal, which reduced the fatigue of the operators. However, he is better known for his efficient coding scheme, which is based on the frequency of occurrence of the characters in the English language so that the most frequently used character has the shortest code (E: dot) and the least

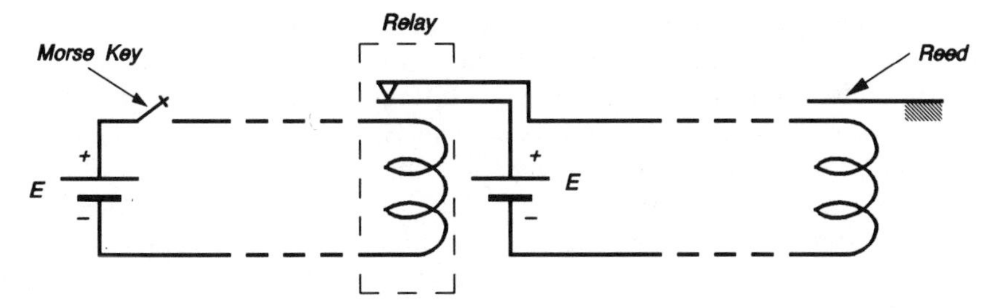

Figure 1.1. *Morse's relay.* With a given battery voltage and minimum current to operate the relay, the maximum distance was determined by the resistance of the connecting wire. Morse's relay removed this barrier at the cost of more relays and batteries.

frequently used character has the longest code ('-apostrophe: dot-dash-dash-dash-dash-dot). This code was in general use until the 1950s and it is still used by amateur radio operators today.

In 1843, Morse persuaded the United States Congress to spend $30,000 to build a telegraph line between Washington and Baltimore. The success of this enterprise made it attractive to private investors, and Morse and his partner Alfred Vail (1807–1859) were able to extend the line to Philadelphia and New York [6]. A number of companies were formed to provide telegraphic services in the East and Midwest of the United States. By 1851, most of these had joined together to form the Western Union Telegraph Company.

Several improvements had been made to the Wheatstone invention by the partnership of Werner Siemens (1816–1892) and Johann Halske (1814–1890) in Berlin by 1847. This was the foundation of the Siemens telecommunication company in Germany.

The next major advance came in 1855 when David Hughes (1831–1900) invented the printing telegraph, the ancestor of the modern teletype. This must have put a lot of telegraph operators out of work (a pattern which was to be repeated over and over again) since the machine could print messages much faster than a person could write. Another improvement that occurred at about this time was the simultaneous transmission of messages in two directions on the same circuit. Various schemes were used, but the basic principle was the balanced bridge.

In 1851, the first marine telegraphic line between France and England was laid, followed in 1866 by the first transatlantic cable. The laying of this cable was a major feat of engineering and a monument to perseverance. A total of 3200 km of cable was made and stored on an old wooden British warship, the HMS Agamemnon. The laying of the cable started in Valentia Bay in western Ireland; but, in 2000 fathoms of water, the cable broke and the project had to be abandoned. A second attempt the following year was also a failure. A third attempt in 1858 involved two ships starting in midocean; it was a success. Telegraphic messages could be sent across the Atlantic. The celebration of success lasted less than a month when the cable insulation broke down under excessively high voltage. Interest in transatlantic cables was temporarily suspended while the American Civil War was fought and it was not until 1865 that the next attempt was made. This time a new ship, the Great Eastern, started from Ireland but after 1900 km the cable broke. Several attempts were made to lift the cable from the ocean-bed but the cable kept breaking off so the project was abandoned until the following year. At last, in 1866, the Great Eastern succeeded in laying a sound cable and messages could once more traverse the Atlantic. By 1880, there were nine cables crossing the ocean [6].

The telegraph was and remained a communication system for business and in most European countries it became a government monopoly. Even in its modern form (telex) it is essentially a cheap long-distance communication network for business.

1.4 TELEPHONE

In 1876, Alexander Graham Bell (1847–1922) was conducting experiments on a "harmonic telegraph" system when he discovered that he could vary the electric current flowing in a circuit by vibrating a magnetic reed held in close proximity to an electromagnet which formed part of the loop. By connecting a second electromagnet together with its own magnetic reed in the circuit, he could reproduce the vibration of the first reed. Using a human voice to excite the magnetic reed led to the first telephone, for which he was granted a patent later that year. He went on to demonstrate his invention at the International Centennial Exhibition in Philadelphia and before the year ended, he transmitted messages between Boston, MA and North Conway, NH, a distance of 230 km. Few people realized the potential of the new invention, and in 1878 when Bell tried to sell his patent to the Western Union Telegraph Company, he was turned down [7].

The early telephone system consisted of two of Bell's magnetic reed–electromagnet instruments in series with a battery and a bell. Bell's instrument worked well as a receiver; in fact, it worked so well that it has survived almost unchanged to this day. As a transmitter however, it left a lot to be desired. It was soon replaced by the carbon microphone (one of the many inventions of Thomas Edison (1847–1931)), which is today the most widely used microphone in the telephone system.

In the early telephone system, each subscriber was connected to a central office by a single wire with an earth return. This led to cross-talk between subscribers. At about this time, electric traction had become very popular, which resulted in increased interference from the noise generated by the electric motors. The earth-return system was gradually replaced by two-wire circuits, which are much less susceptible to cross-talk and electrical noise.

The rapid growth of the telephone system was based almost entirely on the fact that the subscriber could use the system with a minimal amount of training. This outweighed the disadvantages of having no written record of conversations and the requirement that both parties have to be available for the call at the same time. The basic central office is required to respond to a signal from the subscriber (calling party) indicating that he wants service. A buzzer excited by current from a hand-cranked magneto was the standard. The telephone operator would answer and find out to whom (the called party) the calling party wants to talk to. The operator then signals to the called party by connecting his own hand-cranked magneto to the line and cranking it to ring a bell on the called party's premises. When and if the called party responds, the operator connects the lines of the two parties together and withdraws until the conversation is over, at which point the operator disconnects the lines. In order to carry out this function, the operator had to have access to all the lines connected to the exchange. This was not a problem in an exchange with less than 50 lines, but, as the system grew, more operators were required for each group of 50 subscribers. If the calling party and called party belong to the same group of 50, the above sequence was followed. If they belong to two different groups, it was necessary for the two operators to have a verbal consultation before the connection could be made. The errors, delays, and misunderstandings in large central offices led to a reorganization whereby each operator responded to only 50 in-coming lines but had access to all the out-going lines. Another improvement in the system was to replace all the batteries on the subscribers' premises with one battery in the central office.

The motivation for the changeover from manual switching to the automatic telephone exchange was not, as one would expect, the inability of the central office to cope with the increasing volume of traffic. Rather, it was because the operators could listen to the conversations. The inventor of the automatic exchange, Almon B. Strowger, after whom the system was named was an undertaker in Kansas City in the 1890s. There was another undertaker in the city whose wife worked in the local telephone exchange; whenever someone died in the city the telephone operators were the first to know, and the wife would pass a message to her husband giving him a head-start on his competitor [8]. The automatic exchange certainly improved the security of telephone conversations; it was also one more example of machines replacing people.

The success of the telephone system led to a large number of small telephone companies being formed to service the local urban communities. Pressure to interconnect the various urban centers soon grew and techniques for transmission over longer distances had to be developed. These included amplification and inductive loading. Since these transmission lines (trunks) were expensive to construct and maintain, techniques for transmission of more than one message (multiplex) over the trunk at any one time became a matter of great concern and an area of rapid advancement.

1.5 RADIO

In 1864, James Maxwell (1831–1879), a Scottish physicist, produced his theory of the electromagnetic field, which predicted that electromagnetic waves can propagate in free space at a velocity equal to that of light [9]. Experimental confirmation of this theory had to wait until 1887 when Heinrich Hertz (1857–1894) constructed the first high-frequency oscillator. When a voltage was induced in an induction coil connected across a spark gap, a discharge would occur across the gap setting up a damped sinusoidal high-frequency oscillation. The frequency of the oscillation could be changed by varying the capacitance of the gap by connecting metal plates to it. The detector that he used consisted of a second coil connected to a much shorter spark gap. The observation of sparks across the detector gap when the induction coil was excited showed that the electromagnetic energy from the first coil was reaching the second through space. These experiments were in many ways similar to those carried out in 1839 by Joseph Henry (1797–1878). Several scientists made valuable contributions to the subject, such as Edouard Branly, (1844–1940) who invented the "coherer" for wave detection, Aleksandr Popov (1859–1906) and Oliver Lodge (1851–1940) who discovered the phenomenon of resonance.

In 1896, Guglielmo Marconi (1874–1937) left Italy for England where he worked in cooperation with the British Post Office on "wireless telegraphy." A year later, he registered his "Wireless Telegraphy and Signal Co., Ltd. in London to exploit the new technology of radio. On December 12, 1901, Marconi received the letter "S" in Morse code at St. Johns, Newfoundland on his receiver whose antenna was held up by a kite; the antenna, which he had constructed for the purpose, had been destroyed by heavy winds. He had confounded the many skeptics who thought that the curvature of the earth would make radio transmission impossible [10].

Up to this point, no use had been made of "electronics" in telecommunication; high-frequency signals for radio were generated mechanically. The first electronic device, the diode, was invented by Sir John Ambrose Fleming (1849–1945) in 1904. He was investigating the "Edison effect" that is, the accumulation of dark deposits on the inside wall of the glass envelope of the electric light bulb. This phenomenon was evidently undesirable, because it reduced the brightness of the lamp. He was convinced that the dark patches were formed by charged particles of carbon given off by the hot carbon filament. He inserted a probe into the bulb to prevent the charged particles from accumulating by applying a voltage to the probe. He soon realized that, when the probe was held at a positive potential with respect to the filament, there was a current in the probe; however, when it had a negative potential, no current would flow. He had invented the diode. He was granted the first patent in electronics for his effort. Fleming went on to use his diode in the detection of radio signals—a practice which has survived to this day.

The next major contribution to the development of radio was made by Lee DeForest (1873–1961). He got into legal trouble with Marconi, the owner of the Fleming diode patent, when he obtained a patent of his own on a device very similar to Fleming's. He went on to introduce a piece of platinum formed into a zig-zag around the filament, and soon realized that by applying a voltage to what he called the "grid" he could control the current flowing through the diode. This was, of course, the triode—a vital element in the development of amplifiers and oscillators.

1.6 TELEVISION

Shortly after the establishment of the telegraph, the transmission of images by electrical means was attempted by Abbe Caselli in France. His technique was to break up the picture into little pieces and send a coded signal for each piece over a telegraph line. The picture was then reconstituted at the receiving end. The system was slow even for static images, but it established the basic principles for image transmission, that is, the break-up of the picture into some elemental form (scanning), the quantization of each element in terms of how bright it is (coding), and the need for some kind of synchronization between the transmitter and the receiver. Subsequent practical image transmission schemes, whether mechanical or electronic, had these basic units.

The discovery in 1873 by Joseph May, a telegraph operator at the Irish end of the transatlantic cable, that when a selenium resistor was exposed to sunlight, its resistance decreased, led to the development of a light-to-current transducer. Various schemes for image transmission based on this discovery were devised by George Carey, William Ayrton, John Perry and others. None of these was successful, because they lacked an adequate scanning system and each element of the picture had to be sent on a separate circuit, making them quite impractical.

In 1884, Paul Nipkow (1860–1940) was granted a patent in Germany for what became known as the Nipkow disc, which consisted of a series of holes drilled in the form of spirals in a disc. When an image is viewed through a second disc with similar holes driven in synchronism with the first, the observed effect scans point-to-point to form a complete line and line-by-line to cover the complete picture. This was a practical scheme, as the point-to-point brightness of the picture could be transmitted and received serially on a single circuit. The persistence of an image to the human eye could be relied on to create the impression of a complete scene when in fact the information is presented point-by-point. Nipkow's scheme could not be exploited until 1927 when photosensitive cells, photomultipliers, electron tube amplifiers, and the cathode ray tube had been invented and had attained sufficient maturity to process the signals at an acceptable speed for television. Several people made significant contributions to the development of the components as well as to the system. However, two people, Charles Jenkins (1867–1934) and John Baird (1888–1946) are credited with the successful transmission of images at about the same time, both using the Nipkow disc. Mechanical scanning methods of various forms were used with reasonable success until about 1930, when Vladimir Zworykin (1889–) invented the "iconoscope" and Philo Farnsworth (1906–1971) the "dissector." These inventions finally removed all the moving parts from television scanning systems and replaced them with electronic scanning [11]. The application of very-high-frequency carriers and the use of coaxial cables have contributed significantly to the quality of the pictures. The use of color in television was shown to be feasible in 1930 but did not become available to the general public until the mid 1960s. By the 1980s, satellite communication systems brought a large number of television programs to viewers who could afford the cost of the dish antenna.

1.7 GROWTH OF BANDWIDTH AND THE DIGITAL REVOLUTION

Electrical telecommunication started with a single wire with a ground return; but, as the system grew, the common ground return had to be replaced with a return wire—hence the advent of the open-wire telephone line. The open-wire system with its forests of telegraph poles along city streets strung with an endless array of wires eventually gave way to the buried twisted pair cable. The twisted pair cable owes its existence to improved insulating materials, mainly plastics, which reduced the space requirements of the cable. The bandwidth of an unloaded twisted pair is about 4 kHz and it decreases rapidly with length. This can be improved by connecting inductors (loading coils) in series with the line at specific distances and by various equalization schemes to about 1 MHz. The twisted pair has found a niche in the modern telephone system where its bandwidth approximately matches that required for analog audio communication. This is still the dominant mode of telephone communication to the central office. Beyond the central office the network of interoffice trunks uses a variety of conduits for the transmission of the signal.

telecommunication traffic. High-frequency carriers had to be developed to exploit fully the bandwidth capability of new telecommunication media, such as coaxial cables, terrestrial microwave networks, and fiber optics. The development of the coaxial cable, which confines the electromagnetic wave to the annular space between the two concentric conductors, reduced significantly the radiation losses that would otherwise occur. As a result, the bandwidth was increased to approximately 1 GHz and attenuation was reduced. Terrestrial as well as satellite microwave communication systems have further expanded the bandwidth to the terra-hertz range; for those who can afford the dish antenna and its associated equipment, it has increased the number of television channels available to over 60. The application of fiber optics to telecommunication has extended the channel bandwidth to that of visible light (1×10^{12} Hz). It is now possible for one optical fiber to carry as many as 300×10^9 telephone channels at the same time.

An increasingly dominant factor in telecommunication is the enormous popularity of digital techniques. The information is reduced to a train of pulses (1s and 0s) and sent over the channel. The limited bandwidth and the noise in the channel cause the signal to deteriorate, so it is necessary to "refresh" or regenerate the signal at various points along the channel. This is accomplished by using repeaters, whose function is to distinguish between the 1s and 0s and to generate new 1s and 0s and transmit them on. At the receiving end, the 1s and 0s are converted back into an analog signal. The compact disc music recording system is a common example of this technique. Although the need for information transfer between computers spurred the development of digital communication, speech signals increasingly are being converted into digital form for telephone transmission.

REFERENCES

1. *Statistical Yearbook*, UNESCO, Paris, 1986.
2. *The World's Telephones*, AT&T, Indianapolis, 1983.
3. Berto, C., *Telegraphes et Telephones de Valmy au Microprocesseurs*, Le Livre de Poche, 1981.
4. Stumpers, F. L. H. M., "The History, Development and Future of Telecommunications in Europe," *IEEE. Comm. Mag*., 22(5), 1984.
5. Fraser, W., *Telecommunications*, Macdonald, London, 1957.
6. Tebo, J. D., "The Early History of Telecommunications," *IEEE Comm. Soc. Dig*., 14(4), 12–21, 1976.
7. Osborne, H. S., "Alexander Graham Bell, Biographical Memoirs," *Nat. Acad. Sci*., 23, 1–29, 1945.
8. Smith, S. F., *Telephony and Telegraphy*, 2nd ed., Oxford University Press, London, 1974.
9. Bernal, J. D., *Science in History*, Vol. 2, Penguin, Middlesex, 1965.
10. Carassa, F., "On the 80th Anniversary of the First Transatlantic Radio Signal," *IEEE Antennas Propt. Newsl*., 11–19, Dec. 1982.
11. Knapp, J. G., and Tebo, J. D., "The History of Television," *IEEE Comm. Soc. Dig*., 16(3), 8–21, 1978.

PART 1

RADIO COMMUNICATION CIRCUITS

2

AMPLITUDE-MODULATED RADIO COMMUNICATION

2.1 INTRODUCTION

A radio signal can be generated by causing an electromagnetic disturbance and making suitable arrangements for this disturbance to be propagated in free space. The equipment normally used for creating the disturbance is the transmitter and the transmitter antenna ensures the efficient propagation of the disturbance in free space. To detect the disturbance, one needs to capture some finite portion of the electromagnetic energy and convert it into a form that is meaningful to one of the human senses. The equipment used for this purpose is, of course, a receiver. The energy of the disturbance is captured using an antenna, and an electrical circuit then converts the disturbance into an audible signal.

Assume for a moment that our transmitter propagated a completely arbitrary signal (i.e., the signal contained all frequencies and all amplitudes). Then no other transmitter could operate in free space without severe interference, because free space is a common medium for the propagation of all electromagnetic waves. However, if each transmitter is restricted to one specific frequency (i.e., continuous sinusoidal waveforms), then interference can be avoided by incorporating a narrow-band filter at the receiver to eliminate all other frequencies except the desired one. Such a communication channel would work quite well, except that its signal cannot convey information, because a sinusoid is completely predictable and information, by definition, must be unpredictable.

Human beings communicate primarily through speech and hearing. Normal speech contains frequencies from approximately 100 Hz to approximately 5 kHz and a range of amplitudes starting from a whisper to very loud shouting. An attempt to propagate speech in free space faces two severe obstacles. The first is similar to that of the transmitters discussed earlier, in which they interfere with each other because they share the same medium of propagation. The second obstacle is that low frequencies, such as speech, are not propagated efficiently in free space, whereas high frequencies are. Unfortunately, human beings do not hear frequencies above 20 kHz, which is in fact not high enough for free space transmission. However, if we can arrange to change some property of a continuous sinusoidal high-frequency source in accordance with speech, then the prospects for effective communication through free space become a distinct possibility. Changing some property of a (high-frequency) sinusoid in accordance with another signal (e.g., speech) is called *modulation*. It is possible to change the amplitude of the high-frequency signal, called the *carrier*, in accordance with speech and/or music. The modulation is then called *amplitude modulation* or AM for short. It is also possible to change the phase angle of the carrier, in which case we have *phase modulation* (PM), or the frequency, in which case we have *frequency modulation* (FM).

2.2 AMPLITUDE MODULATION THEORY

To simplify the derivation of the equation for an amplitude-modulated wave, we make the simplification that the modulating signal is a sinusoid of angular frequency ω_s and that the carrier signal to be modulated (also sinusoidal) has an angular frequency ω_c.

Let the instantaneous carrier current

$$i = A \sin \omega_c t \tag{2.2.1}$$

where A is the amplitude. The amplitude-modulated carrier must have the form

$$i = [A + g(t)] \sin \omega_c t \tag{2.2.2}$$

where

$$g(t) = B \sin \omega_s t \tag{2.2.3}$$

is the modulating signal. Then

$$i = (A + B \sin \omega_s t) \sin \omega_c t \tag{2.2.4}$$

The waveform is shown in Figure 2.1.

The current may then be expressed as

$$i = (A + kA \sin \omega_s t) \sin \omega_c t \tag{2.2.5}$$

where

$$k = B/A \tag{2.2.6}$$

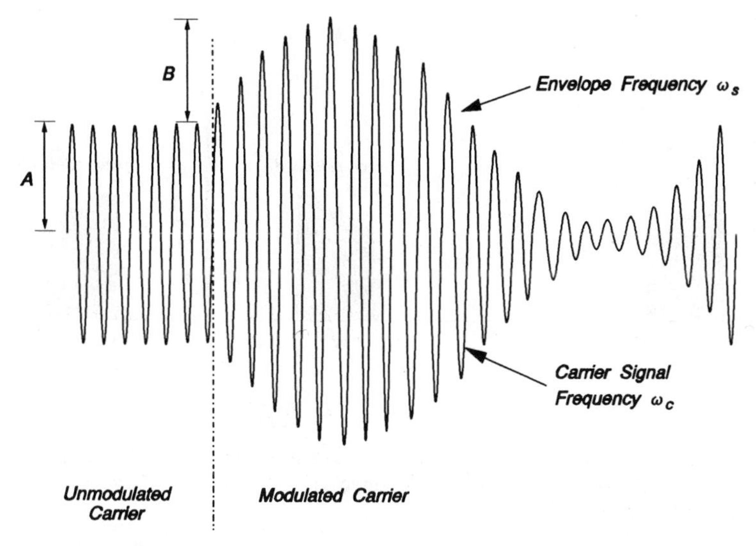

Figure 2.1. Amplitude-modulated wave: the carrier frequency remains sinusoidal at ω_c while the envelope varies at frequency ω_s.

The factor k is called the depth of modulation and may be expressed in percentage. Simplification of Equation (2.2.5) gives

$$i = A \sin \omega_c t + \frac{kA}{2}[\cos(\omega_c - \omega_s)t - \cos(\omega_c + \omega_s)t] \qquad (2.2.7)$$

The frequency spectrum is shown in Figure 2.2.

From Equation (2.2.7), it is evident that the modulated carrier current has three distinct frequencies present: the carrier frequency ω_c, the frequency

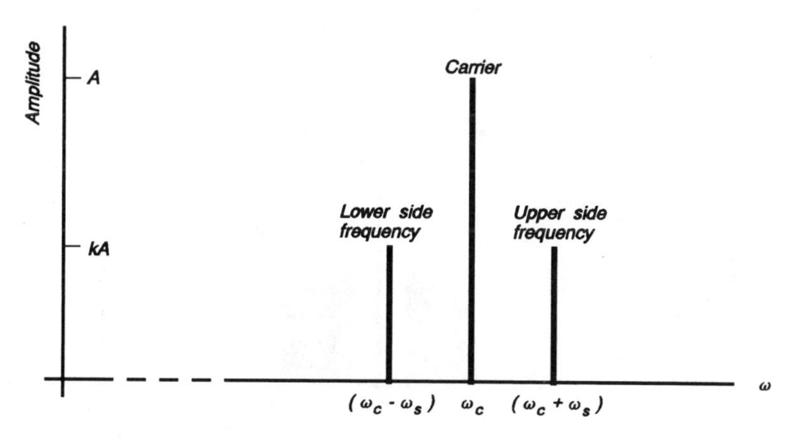

Figure 2.2. Frequency spectrum of the AM wave of Figure 2.1. Note that there are three distinct frequencies present.

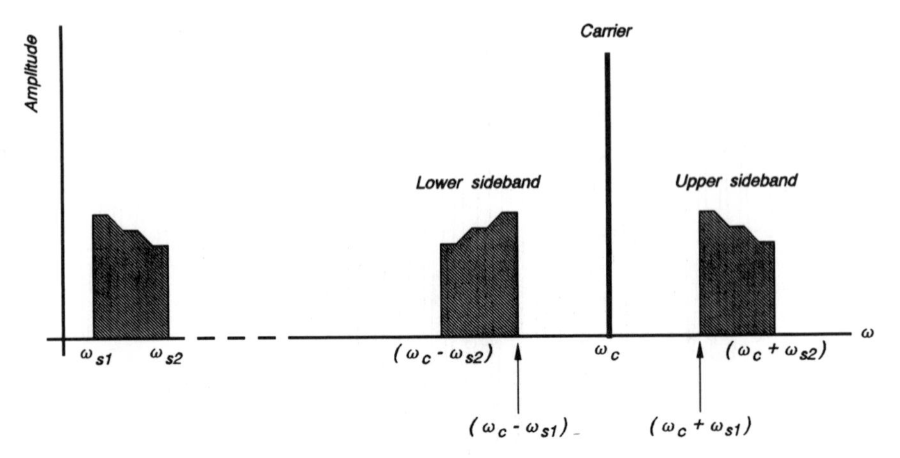

Figure 2.3. Frequency spectrum of the AM wave when the single frequency-modulating signal is replaced by a band of audio frequencies. Note that the information in the signal resides in the sidebands only.

equal to the difference between the carrier frequency and the modulating signal frequency ($\omega_c - \omega_s$), and the frequency equal to the sum of the carrier frequency and the modulating signal frequency ($\omega_c + \omega_s$). The difference and sum frequencies are called the "lower" and "upper" side-frequency, respectively.

To make the situation more realistic, let us assume that the modulating signal is speech, which contains frequencies between ω_{s1} and ω_{s2}. Then it follows from Equation (2.2.7) that the sum and difference terms will yield a band of frequencies symmetric about the carrier frequency as shown in Figure 2.3.

Figure 2.4 shows how two audio signals that would normally interfere with each other, when transmitted simultaneously through the same medium, can

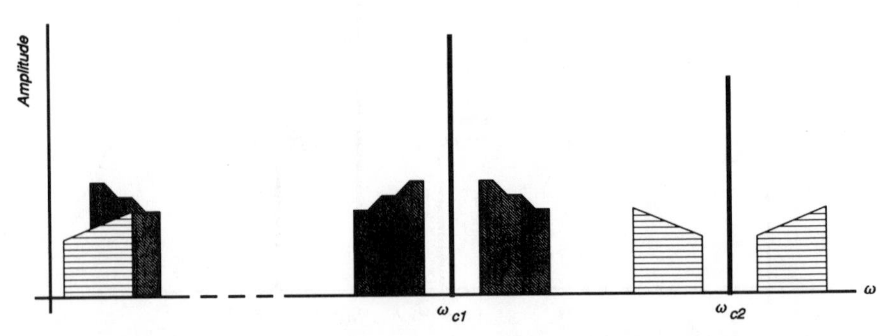

Figure 2.4. The diagram illustrates how two audio frequency sources that would normally interfere with each other can be transmitted over the same channel with no interaction.

be kept separate by choosing suitable carrier frequencies in a modulating scheme. The method of transmitting two or more signals through the same medium simultaneously is sometimes referred to as *frequency division multiplex* and will be discussed in more detail in Chapter 9.

2.3 SYSTEM DESIGN

The choice of a carrier frequency for a radio transmitter is largely determined by government regulations and international agreements. It is evident from Figure 2.4 that, in spite of frequency division multiplex, two stations can interfere with each other if their carrier frequencies are so close that their sidebands overlap. In theory, every transmitter must have a unique frequency of operation and sufficient bandwidth to ensure no interference with others. However, bandwidth is limited by considerations such as cost and the sophistication of the transmission technique used, so that in practice two radio transmitters may operate on frequencies that would normally cause interference so long as they propagate their signals within specified limits of power and are located (geographically) sufficiently far apart. The location and the power transmitted by each transmitter is monitored and controlled by the government.

Once the carrier frequency is assigned to a radio station, it is very important that it maintains that frequency as constant as possible. There are two reasons for this: (1) If the carrier frequency were allowed to drift, then the listeners would have to adjust their radios to keep listening to that station, which would be unacceptable to most listeners. (2) If a station drifts (in frequency) toward the next station, their sidebands would overlap and would cause interference. The carrier signal is usually generated by a crystal oscillator to meet the required precision of the frequency. At the heart of the crystal oscillator is a quartz crystal, cut and polished to very tight specifications, which maintains the frequency of oscillation to within a few hertz of its nominal value. The design of such an oscillator can be found below. Figure 2.5 is a block diagram of a typical transmitter.

2.3.1 Crystal-Controlled Oscillator

The purpose of the crystal oscillator is to generate the carrier signal. To minimize interference with other transmitters, this signal must have extremely low levels of distortion so that the transmitter operates at only one frequency. As discussed earlier, the frequency must be kept within very tight limits, usually within a few hertz in 10^7 Hz. It is difficult to design an ordinary oscillator to satisfy these conditions, so it is common practice to use a quartz crystal to enhance the frequency stability and to reduce the harmonic distortion products.

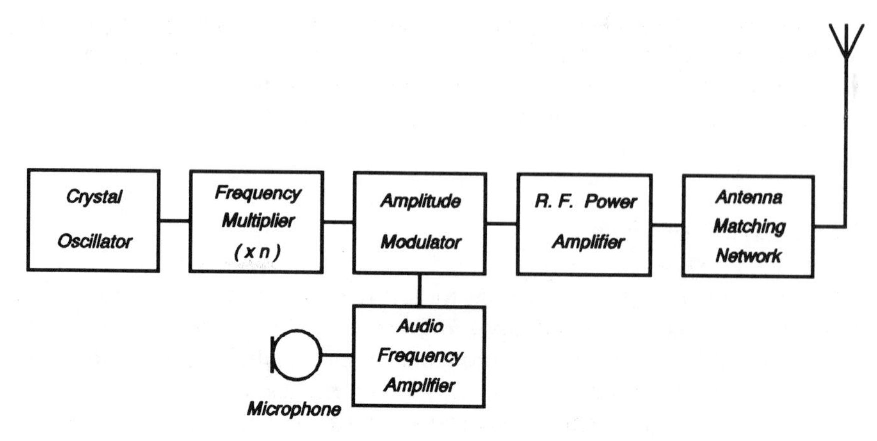

Figure 2.5. Block diagram showing the components that make up the AM transmitter.

The quartz crystal undergoes a change in its physical dimensions when a potential difference is applied across two corresponding faces of the crystal. If the potential difference is an alternating one, the crystal will vibrate and exhibit the phenomenon of resonance. For a crystal, the range of frequency over which resonance is possible is very narrow hence the frequency stability of crystal-controlled oscillators is very high. In general, the larger the physical size of the crystal, the lower the frequency at which it resonates. Thus a high-frequency crystal is necessarily small, fragile, and has low reliability. To generate a high-frequency carrier, it is common practice to use a low-frequency crystal to obtain a signal at a subharmonic of the required frequency and use a frequency multiplier to increase the frequency. Figure 2.5 shows that the crystal oscillator is followed by a frequency multiplier.

2.3.2 Frequency Multiplier

The purpose of the frequency multiplier is to accept an incoming signal of frequency f_c/n and to produce an output at a frequency f_c, where n is an integer. A frequency multiplier can have a single stage of multiplication or it can have several stages. The output of the frequency multiplier goes to the carrier input of the amplitude modulator.

2.3.3 Amplitude Modulator

The amplitude modulator has two inputs, the first being the carrier signal generated by the crystal oscillator and multiplied by a suitable factor and the second being the modulating signal (voice or music) which is represented in Figure 2.5 by the single frequency f_s. In reality, the frequencies present in the modulating signal are in the audio range 20–20,000 Hz. The output from the amplitude modulator consists of the carrier, the lower, and upper sidebands.

2.3.4 Audio Amplifier

The audio amplifier accepts its input from a microphone and supplies the necessary gain to bring the signal to the level required by the amplitude modulator.

2.3.5 Radio Frequency Power Amplifier

The power level at the output of the modulator is usually in the range of watts and the power required to broadcast the signal effectively is in the range of tens of kilowatts. The radio frequency amplifier provides the power gain as well as the necessary impedance matching to the antenna.

2.3.6 Antenna

The antenna is the circuit element responsible for converting the output power from the transmitter amplifier into an electromagnetic wave suitable for efficient radiation in free space. Antennae take many different physical forms determined by the frequency of operation and the radiation pattern desired. For broadcasting purposes, an antenna that radiates its power uniformly to its listeners is desirable, whereas in the transmission of signals where security is important (e.g., telephony), the antenna has to be as directive as possible to reduce the possibility of its reception by unauthorized people.

2.4 RADIO TRANSMITTER OSCILLATOR

Perhaps the simplest way to introduce the phenomenon of oscillation is to describe the common experience of a public address system going unstable and producing an unpleasantly loud whistle. The system consists of a microphone, an amplifier, and a loudspeaker (or loudspeakers) as shown in Figure 2.6. The amplified sound from the loudspeaker may be reflected from walls and other surfaces and reach the microphone. If the reflected sound is louder than the original sound, it in turn produces a louder output at the loudspeaker, which in turn produces an even louder signal at the microphone. It is fairly clear that this state of affairs cannot continue indefinitely; the system reaches a limit and produces the characteristic loud whistle. Immediate steps have to be taken to ensure that the sound level reaching the microphone is less than that required to reach the self-sustained value. If on the other hand we are interested in the generation of an oscillation, then the study of the characteristics of the amplifying element, the conditions under which the

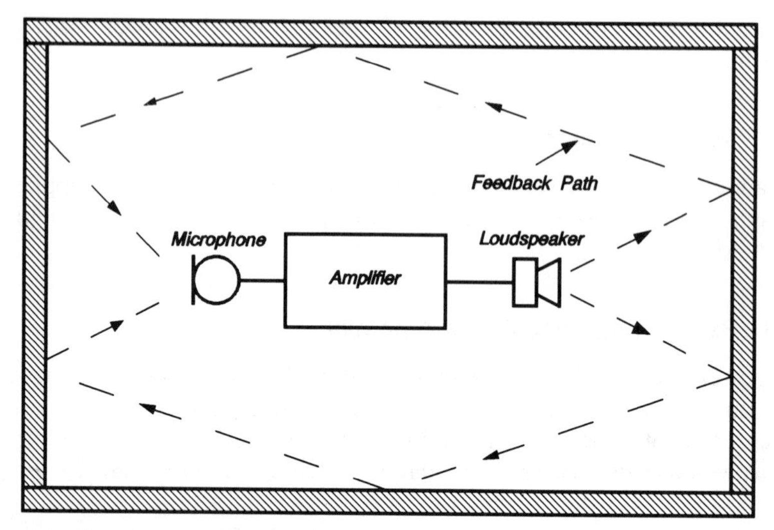

Figure 2.6. The diagrams illustrates how acoustic feedback can cause a public address system to go unstable, making the system an oscillator.

feedback takes place, the frequencies present in the signal, and the optimization of the system to achieve specified performance goals are in order.

The electronic oscillator is a particular example of a more general phenomenon of systems that exhibit a periodic behavior. A mechanical example is the pendulum, which performs simple harmonic motion at a frequency determined by its length and acceleration constant due to gravity, g, if the energy it loses per cycle is replaced from an outside source. In the case of the pendulum used in clocks, the source of energy may be a wound-up spring or a weight whose potential energy is transferred to the pendulum. The solar system with planets performing cyclic motion around the sun is another example of an oscillator, although this time there is no periodic input of energy because the system is virtually lossless.

Three theoretical approaches to oscillator design are presented below. The first is based on the idea of setting up a "lossless" system by cancelling the losses in an LC circuit due to the presence of (positive) resistance by using a negative resistance. The second is based on feedback theory. The third is based on the concept of embedding and the optimization of the power output from the oscillator.

2.4.1 Negative Conductance Oscillator

Consider the circuit shown in Figure 2.7. The externally applied current and the corresponding voltage are related to each other by

$$I = (G_0 + G_n + sC + 1/sL)(V) \tag{2.4.1}$$

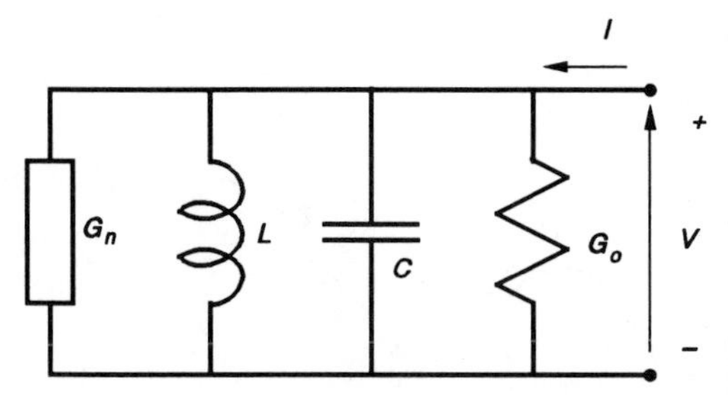

Figure 2.7. The negative conductance oscillator has a negative conductance generating signal power which is dissipated in the (positive) conductance. The components *L* and *C* determine the frequency of the signal. An alternate statement is that the negative conductance cancels all the losses in the circuit. It then oscillates losslessly at a frequency determined by *L* and *C*.

where G_0 is the load conductance, G_n is the negative conductance, I is current, V is voltage, s is the complex frequency, C is capacitance, L is inductance, and G is conductance. If the circuit is that of an oscillator, the external excitation current must be zero since an oscillator does not require an excitation current. Hence,

$$0 = (G_0 + G_n + sC + 1/sL)(V) \tag{2.4.2}$$

For a nontrivial solution, V is nonzero, therefore,

$$G_0 + G_n + sC + 1/sL = 0 \tag{2.4.3}$$

which gives the quadratic equation

$$s^2CL + sL(G_0 + G_n) + 1 = 0 \tag{2.4.4}$$

The solution is then

$$s_1, s_2 = -\frac{(G_0 + G_n)}{2C} \pm \sqrt{\frac{(G_0 + G_n)^2}{4C^2} - \frac{1}{LC}} \tag{2.4.5}$$

when

$$|G_n| = G_0 \tag{2.4.6}$$

that is, the system is lossless, Equation (2.4.5) becomes

$$s_1, s_2 = \sqrt{-\frac{1}{LC}} = j\omega\frac{1}{\sqrt{LC}} \tag{2.4.7}$$

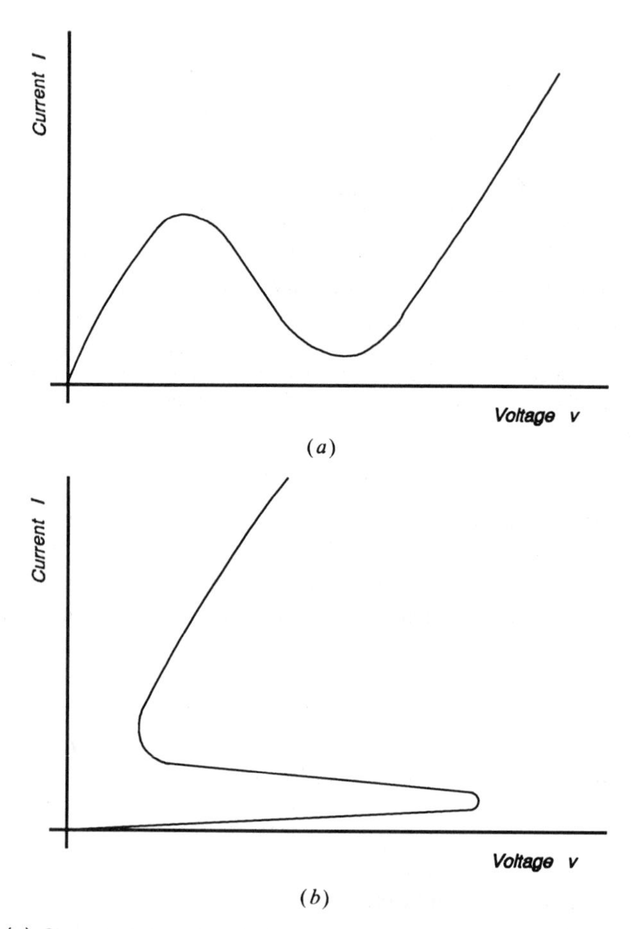

Figure 2.8. (*a*) Characteristics of an N-type negative conductance device. The device is a negative conductance in the region where the slope of the curve is negative. Examples of practical devices that have such characteristics are the tunnel diode and the tetrode. (*b*) Characteristics of an S-type negative conductance device. The device is a negative conductance in the region where the slope of the curve is negative. Examples of practical devices that have such characteristics are the four-layer diode and the silicon controlled rectifier.

which is the resonant frequency for the tuned circuit. The circuit will continue to oscillate at this frequency as if it were in perpetual motion.

A number of devices exhibit negative conductance under appropriate bias conditions and may be used in the design of practical oscillators of this type. These include tunnel diodes and pentodes (N-type negative conductance) and unijunction transistors and silicon-controlled rectifier (S-type). The voltage-current characteristics of N- and S-type negative conductances are shown in Figure 2.8*a* and *b*, respectively.

2.4.2 Classical Feedback Theory

Consider the system shown in Figure 2.9, where A is the gain of an amplifier and β represents the transfer function of the feedback path. E_s is the signal applied to the input and E_o is the output of the system [1]. In the derivation that follows, it is necessary to make the following assumptions:

(1) The input impedances of both the amplifier and the feedback network are infinite and their output impedances are zero.
(2) Both A and β are complex quantities.

The gain of the amplifier alone is

$$A = E_o/E_g \tag{2.4.8}$$

Application of Kirchhoff's Voltage Law (KVL) at the input gives

$$E_g = E_s + \beta E_o \tag{2.4.9}$$

Substituting Equation (2.4.8) into Equation (2.4.9) gives

$$E_o = A(E_s + \beta E_o) \tag{2.4.10}$$

from which we obtain

$$E_o/E_s = A/(1 - \beta A) \tag{2.4.11}$$

Since the E_s and E_o are the input and output, respectively, of the system, as a whole we can define this as A' where

$$A' = E_o/E_s = A/(1 - \beta A) \tag{2.4.12}$$

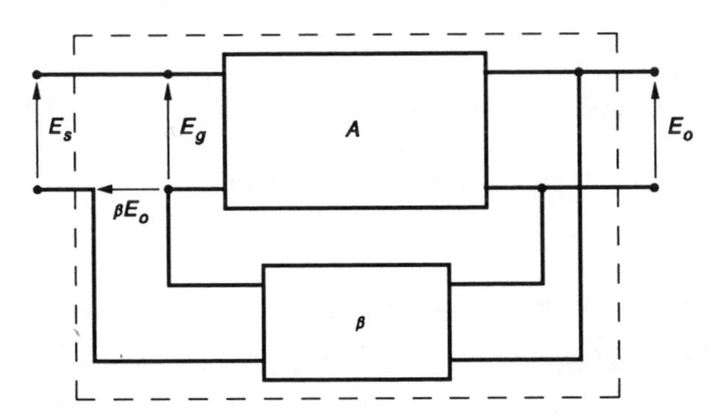

Figure 2.9. Classical feedback system with gain A and feedback factor β.

Three separate conditions must be considered that depend on the value of the denominator of Equation (2.4.12)

(1) *Positive feedback.* If the modulus of $(1 - \beta A)$ is less than unity, then the gain of the system A' is greater than the gain of the amplifier A and therefore the effect of the feedback is said to be positive.
(2) *Negative feedback.* If the modulus of $(1 - \beta A) > 1$, then $A' < A$.
(3) *Oscillation.* If the modulus of $(1 - \beta A) = 0$, then the gain A' is infinite, because with no input ($E_s = 0$) there is still an output. In fact the system is supplying its own input and

$$\beta A = 1 \tag{2.4.13}$$

It must be noted that the waveform of the signal need not be sinusoidal and in fact it can take any form so long as the waveform of the signal that is fed back, βE_o, is identical to the signal E_g. However, the object of this exercise is to generate a carrier for a telecommunication system, and therefore only sinusoidal signals are acceptable—any other waveform will generate other carriers (harmonics of the fundamental) and cause interference with the transmission of other stations.

2.4.3 Sinusoidal Oscillators

Since both A and β are complex quantities, condition (3) implies

$$\text{(a)} \qquad |\beta A| = 1 \tag{2.4.14}$$

Stated in words, the magnitude of the loop-gain, must equal unity, and

$$\text{(b)} \qquad \angle \beta A = 0, 2\pi, 4\pi, \text{etc.} \tag{2.4.15}$$

Again, in words, the loop-gain phase shift must be zero or an integral multiple of 2π radians. The condition given in Equation (2.4.13), which implies Equations (2.4.14) and (2.4.15), is known as the *Barkhausen criterion*.

These two conditions must exist simultaneously for sinusoidal oscillation to occur.

2.4.4 A General Form of the Oscillator

An oscillator circuit is shown in Figure. 2.10 [2] and an equivalent circuit is as shown in Figure 2.11, where the amplifying element is replaced by a voltage-controlled voltage source in series with a resistance R_o to simulate the output resistance of the element. The amplifying element may be a tube, a

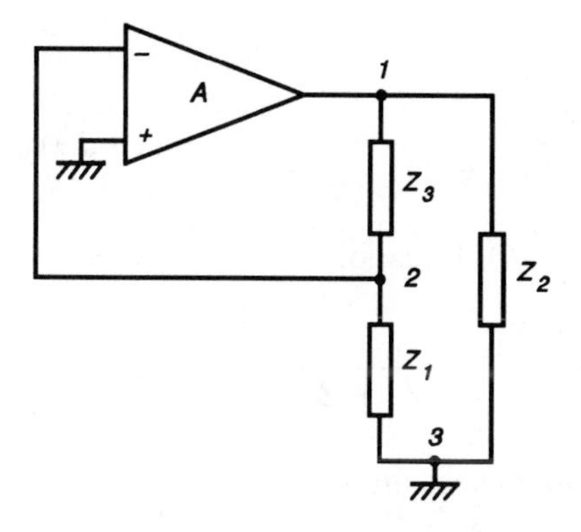

Figure 2.10. Circuit diagram for a more generalized form of the oscillator.

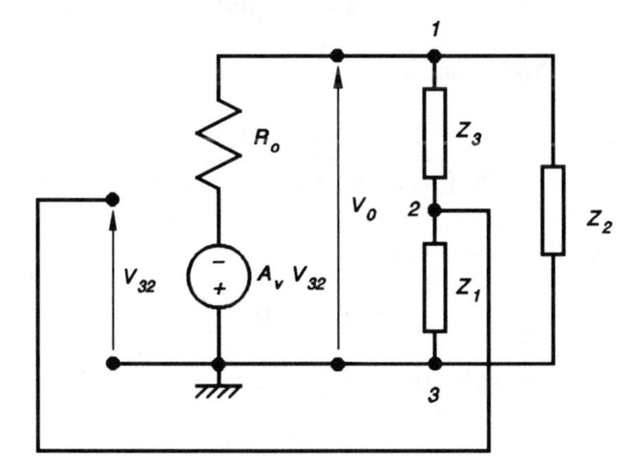

Figure 2.11. The equivalent circuit of the generalized form of the oscillator. R_o represents the output resistance of the amplifier.

transistor, or an operational amplifier. The load seen by the amplifier is

$$Z_L = \frac{Z_2(Z_1 + Z_3)}{(Z_1 + Z_2 + Z_3)} \tag{2.4.16}$$

The amplifier gain without feedback is

$$A = \frac{V_0}{V_{32}} = -\frac{A_v Z_L}{(R_o + Z_L)} \tag{2.4.17}$$

and the feedback constant is

$$\beta = \frac{Z_1}{(Z_1 + Z_3)} \tag{2.4.18}$$

The loop-gain is

$$\beta A = -\frac{A_v Z_1 Z_L}{(R_o + Z_L)(Z_1 + Z_3)} \tag{2.4.19}$$

Substituting Z_L as defined in Equation (2.4.16), Equation (2.4.19) becomes

$$\beta A = -\frac{A_v Z_1 Z_2}{[R_o(Z_1 + Z_2 + Z_3) + Z_2(Z_1 + Z_3)]} \tag{2.4.20}$$

For simplicity, we may assume that the impedances are lossless; hence,

$$Z_1 = jX_1, \qquad Z_2 = jX_2, \qquad \text{and} \qquad Z_3 = jX_3 \tag{2.4.21}$$

Then Equation (2.4.20) becomes

$$\beta A = -\frac{A_v X_1 X_2}{[jR_o(X_1 + X_2 + X_3) - X_2(X_1 + X_3)]} \tag{2.4.22}$$

Recall that for oscillation to occur,

$$1 - \beta A = 0 \tag{2.4.23}$$

This means that βA must be real, and, hence,

$$X_1 + X_2 + X_3 = 0 \tag{2.4.24}$$

that is,

$$X_1 = -(X_2 + X_3) \tag{2.4.25}$$

The expression for the loop gain becomes

$$\beta A = -A_v X_1 / X_2 \tag{2.4.26}$$

Because $\beta A = 1$, it follows that X_1 and X_2 must have opposite signs; that is, if one of them is inductive, the other must be capacitive and X_3 can be capacitive or inductive, depending on the sign of $(X_1 + X_2)$. The two possibilities are shown in Figures 2.12 and 2.13, respectively.

The circuit shown in Figure 2.12 is better known as a *Colpitts oscillator*. The circuit is redrawn in Figure 2.12*b* to emphasize the symmetric structure of the circuit. The circuit shown in Figure 2.13 is better known as a *Hartley oscillator*.

From the point of view of the structure of the circuits, it can be seen that they are the same. It should be noted that the operational amplifier can be replaced by a tube or a transistor.

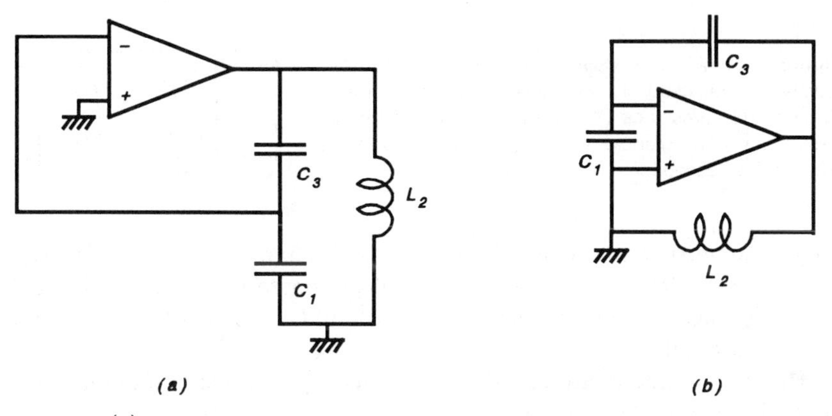

Figure 2.12. (*a*) The generalized form of the oscillator with two of the impedances replaced by capacitors and the third by an inductor to form a Colpitts Oscillator. (*b*) The diagram in (*a*) has been redrawn to emphasize the symmetry of the circuit.

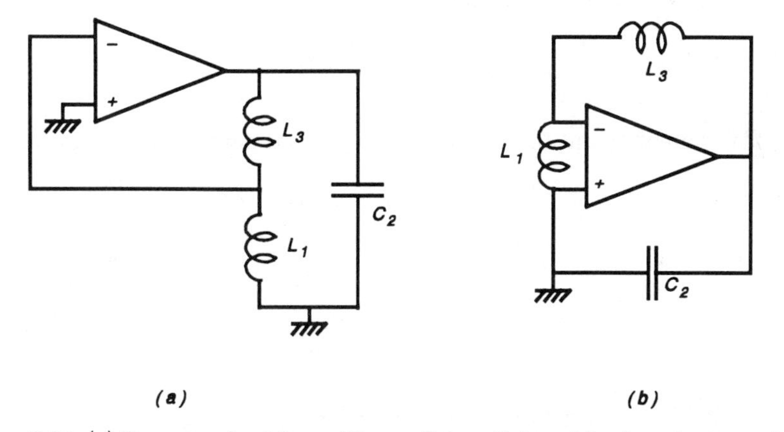

Figure 2.13. (*a*) The generalized form of the oscillator with two of the impedances replaced by inductors and the third by a capacitor to form a Hartley Oscillator. (*b*) The diagram in (*a*) has been redrawn to emphasize the symmetry of the circuit.

2.4.5 Oscillator Design for Maximum Power Output

A major flaw in the above design is that it does not anticipate the necessity for the oscillator to supply power to a load. The theory [3] is based on the characterization of the amplifying element ("active device") as a two-port. A discussion of two-ports is beyond the scope of this book but may be found in any standard text on circuit theory.

A two-port can be described in terms of its terminal voltages and currents by four parameters: impedances, admittances, voltage ratios, and current ratios under constraints of open or short-circuit. Without limiting the gener-

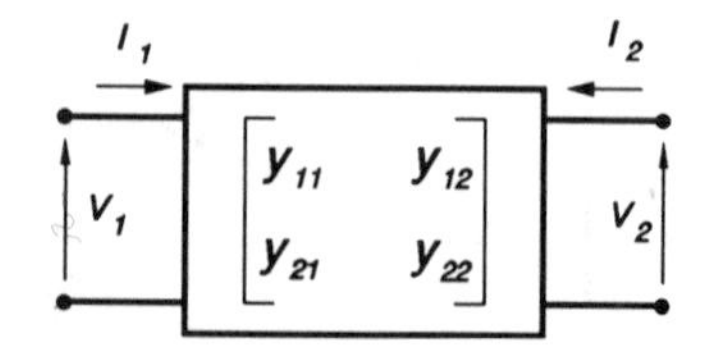

Figure 2.14. A two-port representation of an active device to be used in the design of an oscillator. Short-circuit admittance (Y) parameters are used in the design for convenience. Other parameters could be used in the description.

ality, assume that the active device has been characterized in terms of the short-circuit admittance parameters or Y parameters, for short. Figure 2.14 shows the two-port and its terminal voltages and currents, which are assumed to be sinusoidal.

The Y parameters are functions of frequency and bias conditions, and, in general, are complex, so that

$$Y_{11} = g_{11} + jb_{11} \tag{2.4.27}$$

The total power entering the two-port,

$$P = V_1^* I_1 + V_2^* I_2 \tag{2.4.28}$$

The Y parameters and the terminal voltages and currents are related by

$$I_1 = Y_{11}V_1 + Y_{12}V_2 \tag{2.4.29}$$

and

$$I_2 = Y_{21}V_1 + Y_{22}V_2 \tag{2.4.30}$$

Substituting for I_1 and I_2 in Equation (2.4.28) gives

$$P = Y_{11}|V_1|^2 + Y_{22}|V_2|^2 + Y_{12}V_1^* V_2 + Y_{21}V_1 V_2^* \tag{2.4.31}$$

The ratio of the output voltage, V_2, to the input voltage, V_1, can be defined as

$$\frac{V_2}{V_1} = \mathbf{A} = A_R + jA_I \tag{2.4.32}$$

The real power entering the two-port is

$$P_R = |V_1|^2\left[g_{11} + g_{22}\left(A_R^2 + A_I^2\right) + (g_{12} + g_{21})A_R - (b_{12} + b_{21})A_I\right] \tag{2.4.33}$$

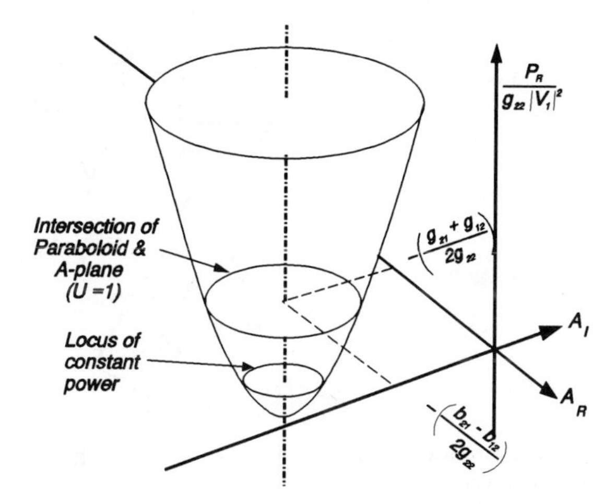

Figure 2.15. Three-dimensional representation of the output power of the oscillator as a function of the complex parameter **A**.

This can be rearranged as follows:

$$\frac{P_R}{g_{22}|V_1|^2} = \left[A_R + \frac{(g_{21}+g_{12})}{2g_{22}}\right]^2 + \left[A_I + \frac{(b_{21}-b_{12})}{2g_{22}}\right]^2 + \frac{4g_{11}g_{22} - (g_{21}+g_{12})^2 - (b_{21}-b_{12})^2}{4g_{22}^2} \tag{2.4.34}$$

This equation is of the form:

$$z = (x-a)^2 + (y-b)^2 + c \tag{2.4.35}$$

and therefore it is that of a paraboloid in space with axes $P_R/(g_{22}|V_1|^2)$, A_R, and A_I as shown in Figure 2.15.

It was assumed that real, positive power was supplied and dissipated in the two-port; therefore, it follows that negative values of power, as shown in Figure 2.15, must represent power generated by the two-port and dissipated in the surrounding circuit; that is, above the A plane real power is supplied to the two-port, and below it the device supplies real power to the embedding circuit. Because the object of the exercise is to generate and supply real power to an external circuit, the most interesting part of Figure 2.15 is the part below the A plane. It is clear that movement toward the apex of the paraboloid represents increasing levels of power supplied by the "active" two-port and that the maximum power supplied occurs at the apex. We shall return to this remark when we consider the optimization of the power output.

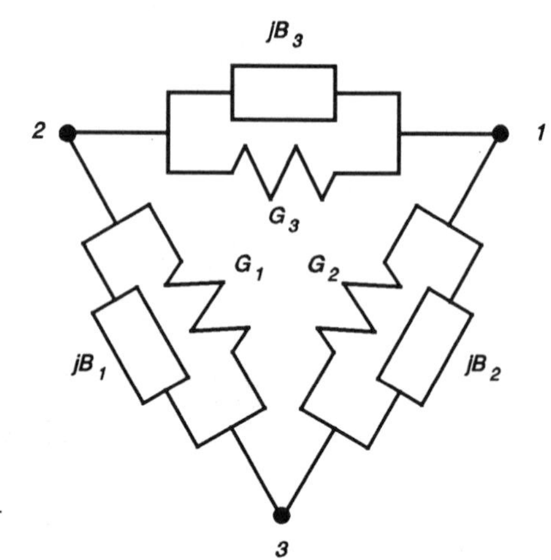

Figure 2.16. The general passive embedding circuit for a two-port device.

The most general embedding circuit for the two-port is shown in Figure 2.16, with each branch made up of a conductance in parallel with a susceptance. The susceptances can be considered as the tuned circuit that will determine the frequency of oscillation and the conductances as the destination of the power generated by the active two-port. The embedding network can also be described in terms of a two-port as follows:

$$I_1' = (Y_2 + Y_3)V_1 - Y_3V_2 \tag{2.4.36}$$

$$I_2' = -Y_3V_1 + (Y_1 + Y_3)V_2 \tag{2.4.37}$$

When the active device and the embedding are connected, as shown in Figure 2.17, the composite circuit can be described by the two-port equations

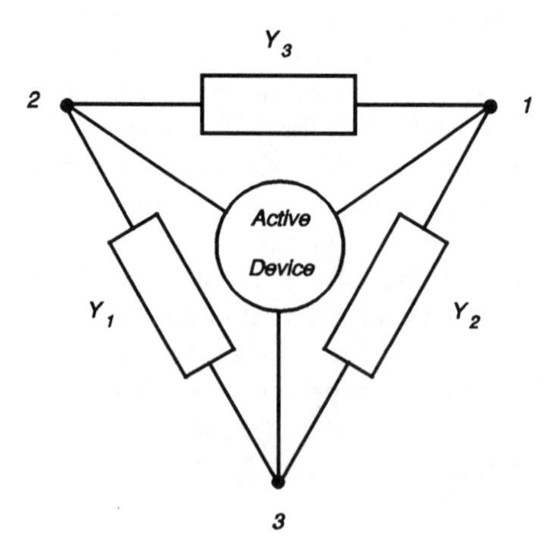

Figure 2.17. The active two-port device is shown with the passive embedding connected.

[Equations (2.4.29) + (2.4.36) and Equations (2.4.30) + (2.4.37)]

$$I_1 + I_1' = (Y_{11} + Y_2 + Y_3)V_1 + (Y_{12} - Y_3)V_2 \tag{2.4.38}$$

$$I_2 + I_2' = (Y_{21} - Y_3)V_1 + (Y_1 + Y_3 + Y_{22})V_2 \tag{2.4.39}$$

For an oscillator, no external signal current is supplied at port 1 and therefore $I_1 + I_1' = 0$. Similarly $I_2 + I_2' = 0$. From Equation (2.4.32) we have

$$V_2 = V_1(A_R + jA_I) \tag{2.4.40}$$

from Equation (2.4.38) we have

$$V_1[Y_{11} + Y_2 + Y_3 + (A_R + jA_I)(Y_{12} - Y_3)] = 0 \tag{2.4.41}$$

and from Equation (2.4.39) we have

$$V_1[Y_{21} - Y_3 + (A_R + jA_I)(Y_1 + Y_3 + Y_{22})] = 0 \tag{2.4.42}$$

For nontrivial values of V_1, real and imaginary parts of Equations (2.4.41) and (2.4.42) are separately equal to zero; that is,

$$g_{11} + G_2 + G_3 + A_R(g_{12} - G_3) - A_I(b_{12} - B_3) = 0 \tag{2.4.43}$$

$$b_{11} + B_2 + B_3 + A_R(b_{12} - B_3) + A_I(g_{12} - G_3) = 0 \tag{2.4.44}$$

$$g_{21} - G_3 + A_R(G_1 + G_3 + g_{22}) - A_I(B_1 + B_3 + b_{22}) = 0 \tag{2.4.45}$$

and

$$b_{21} - B_3 + A_R(B_1 + B_3 + b_{22}) + A_I(G_1 + G_3 + g_{22}) = 0 \tag{2.4.46}$$

Equations (2.4.43) to (2.4.46) can be written in the form of a matrix as follows:

$$\begin{bmatrix} A_R & 0 & (A_R - 1) & -A_I & 0 & -A_I \\ A_I & 0 & A_I & A_R & 0 & (A_R - 1) \\ 0 & 1 & (1 - A_R) & 0 & 0 & A_I \\ 0 & 0 & -A_I & 0 & 1 & (1 - A_R) \end{bmatrix} \begin{bmatrix} G_1 \\ G_2 \\ G_3 \\ B_1 \\ B_2 \\ B_3 \end{bmatrix}$$

$$= \begin{bmatrix} -g_{21} - \mathrm{Re}(Ay_{22}) \\ -b_{21} - \mathrm{Im}(Ay_{22}) \\ -g_{11} - \mathrm{Re}(Ay_{12}) \\ -b_{11} - \mathrm{Im}(Ay_{12}) \end{bmatrix} \tag{2.4.47}$$

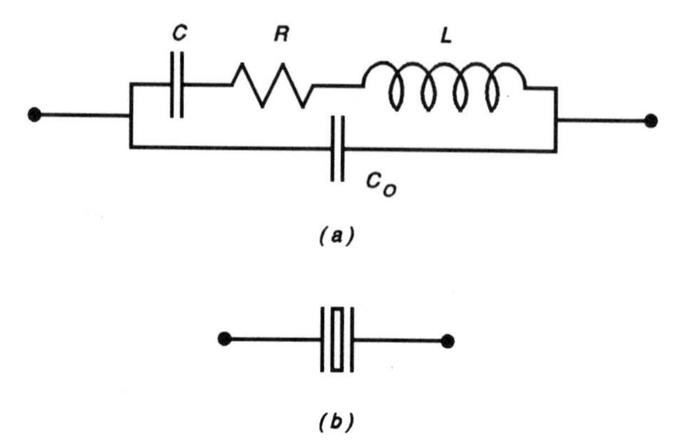

Figure 2.18. (*a*) The equivalent circuit of the crystal and its package. (*b*) The electrical symbol of the crystal.

All the terms in the matrix are known except G_1, G_2, G_3, B_1, B_2, and B_3, that is, there are six unknowns but only four equations so a unique solution cannot be found unless arbitrary values are chosen for at least two of the unknowns. Fortunately, an oscillator normally has only one conductive load and therefore two of the three conductances can be set to zero. The matrix equation can then be solved for one conductance and three susceptances.

2.4.6 Crystal-Controlled Oscillator

The oscillator used in a transmitter must have little tolerance on the stability of its frequency. This is necessary if interference between radio stations is to be avoided. The drift of the frequency of an ordinary *LC* oscillator, for example, makes it unsuitable for this purpose. Greater frequency stability can be achieved by using a crystal as a part of the oscillator circuit. In Section 2.3.1, the behavior of the crystal when it is excited by an ac signal was discussed. It is evident that, since the crystal reacts to electrical excitation, it must be possible to devise an electrical circuit made up of inductors, resistors, and capacitors whose frequency characteristics are approximately those of the crystal. Such a circuit is shown in Figure 2.18. The approximate circuit is reasonably accurate at frequencies close to the resonant frequency. Over a larger frequency range, a more complicated equivalent circuit has to be used.

Typical values of the components of the equivalent circuit are $C = 0.0154$ pF, $R = 8\ \Omega$, $L = 0.0165$ H, $C_o = 4.55$ pF. The capacitance C_o is due largely to the electrodes that are attached to the crystal. The crystal will therefore resonate in the series mode at a frequency ω_s where

$$\omega_s^2 = \frac{1}{LC} \tag{2.4.48}$$

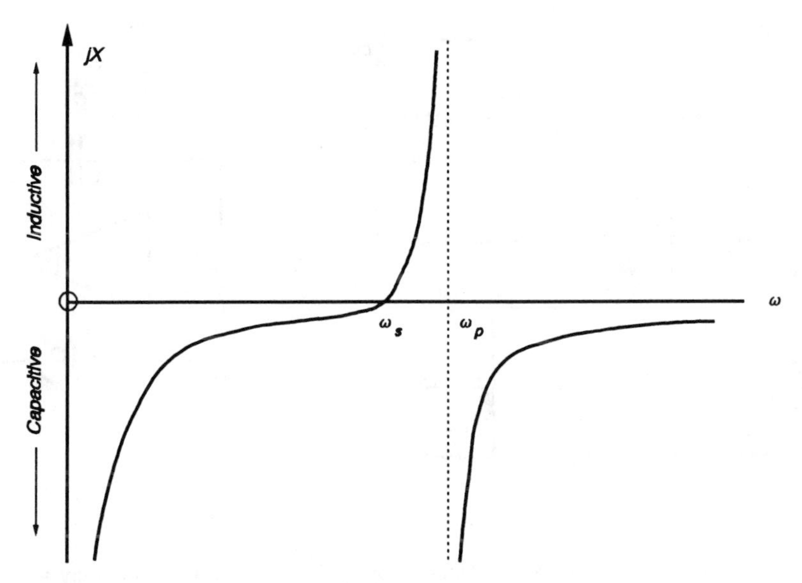

Figure 2.19. The reactance characteristics of the crystal. Note that this is not to scale.

which gives $f_s = 9.984 \times 10^6$ Hz. It will resonate in the parallel mode at an angular frequency given approximately by

$$\omega_p^2 = \frac{1}{L[CC_o/(C + C_o)]} \tag{2.4.49}$$

which gives a resonant frequency of $f_p = 10.001 \times 10^6$ Hz—a change of less than 0.2%. The corresponding quality-factor of the crystal is then $Q_0 =$ 130,000.

Figure 2.19 shows the reactance of the crystal plotted against frequency. It should be noted that the reactance of the crystal is inductive over a narrow band of frequency and also that both reactance and frequency are not to scale. Figure 2.20 shows a typical crystal-controlled oscillator. The crystal is substituted for one of the inductors in what would otherwise be classified as a Hartley oscillator. This type of crystal-controlled oscillator is called the *Pierce oscillator*. Similarly, the crystal-controlled oscillator corresponding to the Colpitts variety is called the *Miller oscillator*. In the circuit shown in Figure 2.20, the active element is a field-effect transistor whose gate-to-drain capacitance plus stray capacitance constitute C_3.

The very high Q_0 of the crystal ensures that the oscillator has an extremely limited range of frequencies in which it can continue to oscillate. Various other measures may be taken to improve the frequency stability, such as placing the crystal in a temperature-controlled environment, and the Q factor can be enhanced by evacuating the glass envelope that protects it.

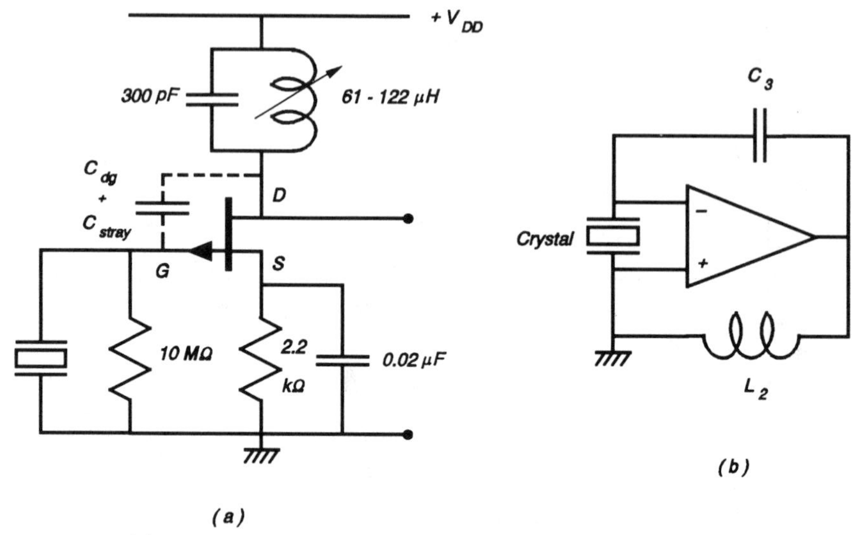

Figure 2.20. (*a*) A Hartley Oscillator with one of the inductors replaced by a crystal. This circuit is called a Pierce Oscillator. The field-effect transistor may be replaced by any suitable active device. (*b*) The equivalent circuit of the Pierce Oscillator demonstrating its symmetrical structure.

High-precision oscillators are invariably connected to their load through a buffer amplifier. This ensures that variations in the load do not affect the operation of the oscillator.

2.5 FREQUENCY MULTIPLIER

The purpose of the frequency multiplier is to raise the frequency generated by the crystal-controlled oscillator to the value required for the transmitter carrier. As explained earlier, it is not possible to obtain physically robust crystals at high frequency since their physical size gets smaller as the frequency of oscillation gets higher. The standard technique is therefore to use a crystal to generate a signal at a frequency that is a subharmonic of the required carrier frequency and then to raise the frequency up to the required value using a cascade of frequency multipliers.

A useful analogy of how a frequency multiplier operates is a child's swing. With a child on the swing, the adult must give it a push to get the swing into operation. Subsequent to that, further supplies of energy must take place at a frequency determined by the length of the swing and the gravitational constant of acceleration, g. It is also necessary to supply the energy at a point in time when it enhances the swinging action rather than opposes it; on average, the adult will have to supply energy equal to that lost during the cycle. If the energy supplied per cycle is less than the energy lost, the

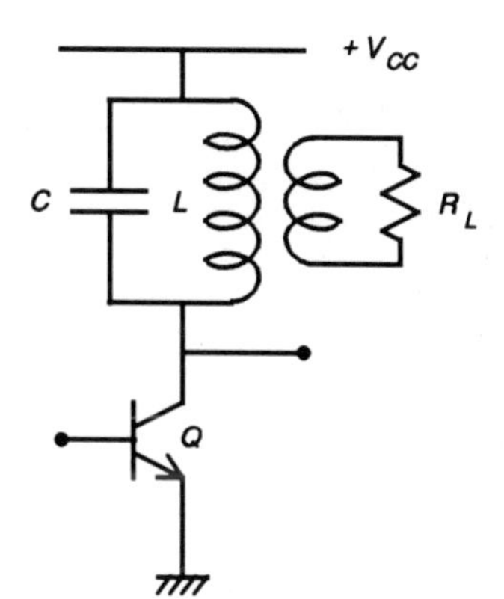

Figure 2.21. A class C amplifier to be used, with minor modifications, as a frequency multiplier.

amplitude of the swing will decrease to a smaller value so as to restore the energy balance. If the energy supplied per cycle is greater than that lost per cycle, the amplitude will grow to a new steady-state value. The motion of the child will be very nearly a simple harmonic one if the total energy stored in the system is large compared to the energy supplied by the adult; that is, the system Q has to be large if the child is to execute a near-sinusoidal motion. The most important point of this analogy is that the energy does not have to be supplied at the same frequency as the swing; it can be supplied at a subharmonic frequency, that is, the push can be given every other cycle of the swing or every third cycle or higher so long as enough energy is supplied to maintain the energy balance. When the push occurs every other cycle, it is clear that the output of the system is at twice the frequency of the input—this is a frequency multiplier with a multiplication factor of two. When the energy is supplied every third cycle, a multiplication factor ofthree is obtained, and so on. Evidently, there is a limit to how high the multiplication factor can be, which is determined by the amount of variation in the amplitude of the swing that can be tolerated. Table 2.1 shows a comparison of the swing and the frequency multiplier.

Figure 2.21 shows a typical frequency multiplier. Energy is fed into it by applying a suitable positive pulse to the base of the transistor. This causes the transistor to conduct momentarily, that is, current flows from the direct current (dc) power supply through the inductor and a finite amount of energy is stored in the inductor. The current flow is shut off when the input pulse ends, and the transistor is essentially an open circuit. The energy stored in the magnetic field of the inductor is transformed into energy stored in the electric field of the capacitor. The transformation of energy from one form to the other and back again would continue indefinitely in a sinusoidal form if the system were lossless and this would take place at a frequency determined by the values of the inductance and capacitance. The resistance R represents the losses in the system—the amplitude of the sinusoid will decay with time. The steady-state amplitude of the voltage or current will be determined by

TABLE 2.1 Swing Analogy for a Frequency Multiplier

Swing, Including Child	Frequency Multiplier
Adult (timing)	Input signal source
Adult (energy transferred to child)	dc power supply
Length of swing, l and gravitational constant, g.	Inductance, L, and capacitance, C.
Air resistance and bearing friction	Energy loss in R
Frequency, $f = \dfrac{1}{2\pi\sqrt{1/g}}$	Frequency, $f = \dfrac{1}{2\pi\sqrt{LC}}$
Amplitude of swing	Amplitude of voltage or current in tank circuit

the equalization of the energy input and the energy output (loss). Subsequent to the initial input pulse, all input pulses must be timed to enhance rather than oppose the stored energy of the system. It is of course not necessary to have an input pulse for every cycle of the output; the input pulse can be supplied every two, three, or more cycles. When the input frequency is the same as the output, the frequency multiplier (multiplication factor = 1) is simply a class C amplifier. Since a class C amplifier represents the simplest frequency multiplier, the design of a class C amplifier will be discussed next.

2.5.1 Class C Amplifier

In a class C amplifier, the current in the active device flows for a period much less than π radians of the output waveform. The active device current waveform is therefore highly nonsinusoidal. In a class A or B amplifier this would give a correspondingly nonsinusoidal output. However, in the class C amplifier, the collector load consists of a parallel LC circuit which is tuned to the frequency of the input signal. The tuned circuit (sometimes referred to as a "tank" circuit) presents a very high impedance at the resonant frequency to the collector of the transistor and hence a high gain is obtained at this frequency. Other frequencies, such as harmonics of the input frequency, are attenuated. Therefore, the output is very nearly sinusoidal.

In a practical class C amplifier, the base of the transistor is held at a voltage such that the transistor is in the off state. The input signal brings the transistor into conduction at its positive peaks and causes enough current to flow in the inductor to store energy equal to that dissipated in the load resistor R and at other sites in the circuit. When the input signal drops below the threshold of the transistor, it switches off, and the LC tank oscillates freely with a sinusoidal waveform at the resonant frequency of the tank circuit. Because of the dissipation in the circuit, the waveform is actually an exponentially damped sinusoid, as shown in Figure 2.22.

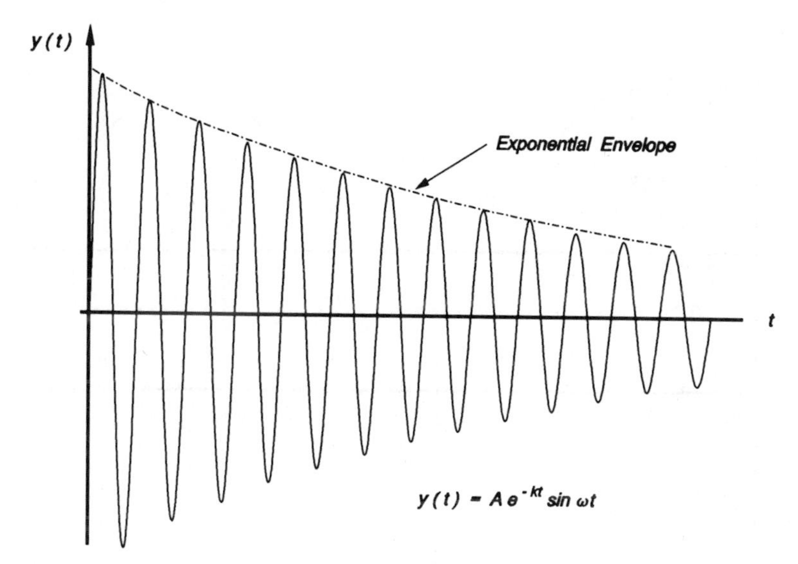

Figure 2.22. The output waveform of a class C amplifier after a single pulse excitation. Note the sinusoidal waveform and the exponential decay of the envelope.

The collector voltage, the base voltage, and collector current waveforms are shown in Figure 2.23. Note that the quiescent value of the collector voltage waveform is V_{cc} and that its maximum amplitude is $2V_{cc}$.

The conversion efficiency of the amplifier is given by

$$\eta = \frac{\text{ac power output}}{\text{dc power input}}$$

A class C amplifier can have a relatively high conversion efficiency, usually near 85%, because the dc current flows for a very short part of the cycle and this happens when the collector voltage is at its lowest value. Thus the power lost in the transistor is minimal.

2.5.2 Converting the Class C Amplifier into a Frequency Multiplier

To convert a class C amplifier into a frequency multiplier with a multiplication factor of two, the L and C of the tank circuit are chosen to resonate at $2\omega_0$ when the input signal frequency is ω_0. Successful operation of the system demands that the Q factor of the circuit is sufficiently high so that the amount of damping at the end of the second cycle is negligible. Higher multiplication factors are possible with the damping problem progressively getting worse as the multiplication factor increases. Figure 2.24 shows the input current and output voltage waveforms of a frequency multiplier with a multiplication factor of three. Note the effect of damping (exaggerated) on the amplitude of the output voltage.

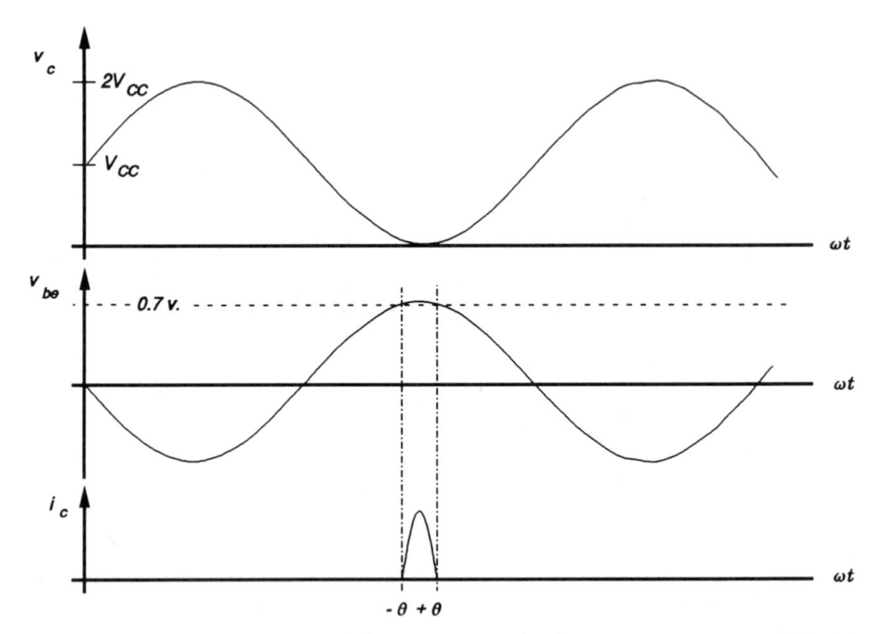

Figure 2.23. The collector voltage (v_c), base voltage (v_{be}), and collector current (i_c) of the class C amplifier. Note that the base voltage need not be sinusoidal for the collector voltage to be sinusoidal.

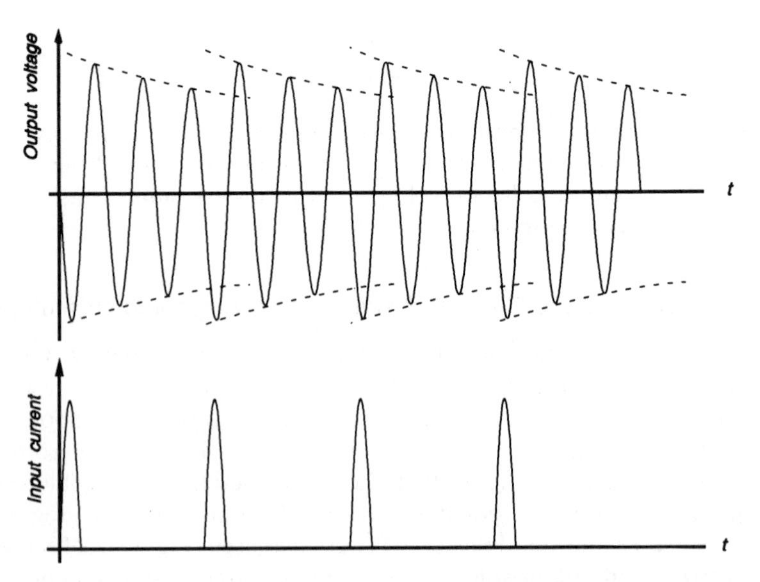

Figure 2.24. The output voltage waveform of a ×3 frequency multiplier and its driving current. The decay is exaggerated to show the effect of low Q factor.

Example 2.5.1 Frequency Multiplier. A frequency multiplier with a multiplication factor of two is driven by an input current whose waveform can be assumed to be a half-sinusoid with a peak value of 200 μA at a frequency of $2/\pi$ MHz. The input current flows for a period corresponding to 5 degrees of one cycle of the output waveform and the average input impedance over this period is 750 Ω resistive. The following data applies to the circuit:

(1) Multiplier load = 50 Ω resistive
(2) Transformer turns ratio = 15 : 1
(3) Transformer coupling, $k = 1$
(4) dc voltage supply = 15 V
(5) Current gain of the transistor = 100
(6) Loaded Q factor of the transformer primary = 50 (min)

Calculate the following:

(a) The value of the primary inductance
(b) The value of the tuning capacitance
(c) The impedance of the collector load at the output frequency
(d) The ac power dissipated in the load
(e) The power gain of the multiplier

Solution. A frequency multiplier is essentially a class C amplifier whose input is driven at a frequency that is a subharmonic of the output frequency. A suitable circuit is shown in Figure 2.25.

$$\text{Input frequency} = 2\pi f = 4 \times 10^6 \text{ rad/sec}$$
$$\text{Output frequency} = 8 \times 10^6 \text{ rad/sec}$$

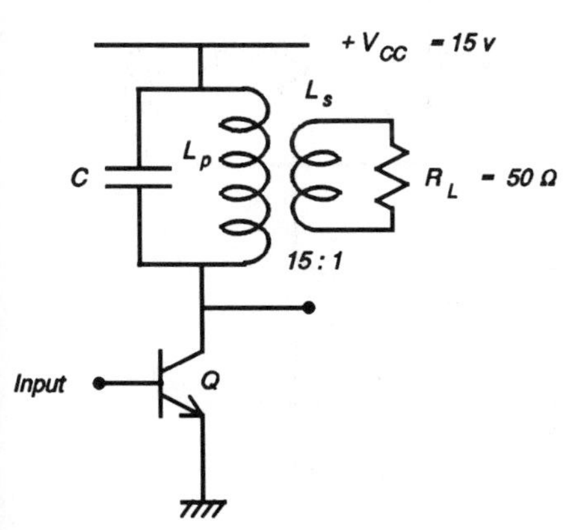

Figure 2.25. Circuit diagram for the frequency multiplier example.

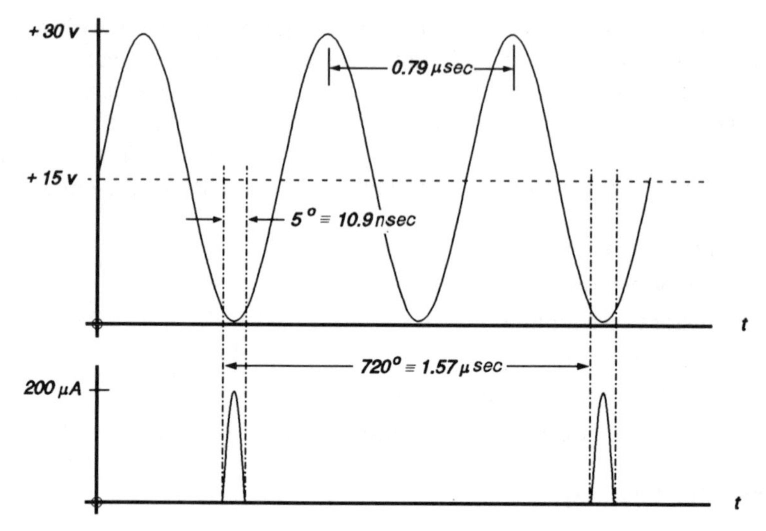

Figure 2.26. Collector and base current waveforms of the frequency multiplier example.

The input current and the output voltage waveforms are shown in Figure 2.26. When the load is transferred to the primary, the collector circuit is as shown in Figure 2.27, with the equivalent resistor having a value $n^2R_L = (15)^2 \times 50 = 11.25$ kΩ.

(a) Assuming that the tank circuit is lossless, we can assign the loss in n^2R_L to the inductor and determine the series equivalent RL circuit (see

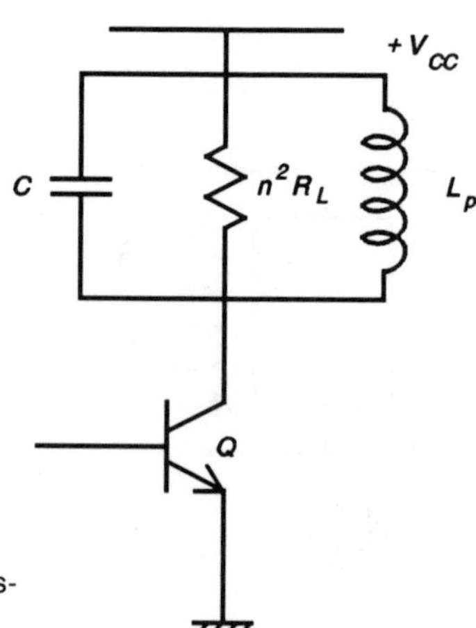

Figure 2.27. The frequency multiplier with the transformer load transferred to the primary.

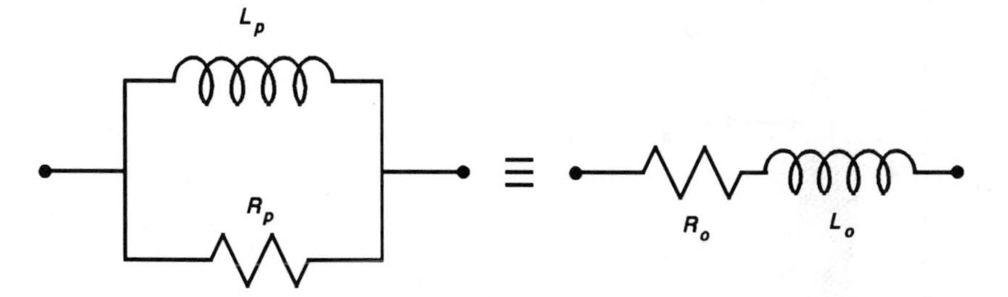

Figure 2.28. Transformation of a parallel *RL* to a series *RL* circuit.

Figure 2.28). Equating the impedances,

$$\frac{j\omega L_p R_p}{R_p + j\omega L_p} = R_0 + j\omega L_0 \tag{2.5.1}$$

Rationalizing the left-hand side and equating real and imaginary parts gives

$$R_0 = \frac{\omega^2 L_p^2 R_p}{R_p^2 + \omega^2 L_p^2} \tag{2.5.2}$$

and

$$L_0 = \frac{L_p R_p^2}{R_p^2 + \omega^2 L_p^2} \tag{2.5.3}$$

The loaded Q factor of the tank circuit has a minimum value of 50; therefore

$$Q_0 = 50 = \frac{\omega L_0}{R_0} = \frac{R_p}{\omega L_p} = \frac{n^2 R_L}{\omega L_p} \tag{2.5.4}$$

The primary inductance

$$L_p = \frac{n^2 R_L}{50\omega} = \frac{11.25 \times 10^3}{50 \times 8 \times 10^6} = 28.13\ \mu\text{H} \tag{2.5.5}$$

(b) For the tank circuit at resonance,

$$\omega^2 = \frac{1}{L_p C} \tag{2.5.6}$$

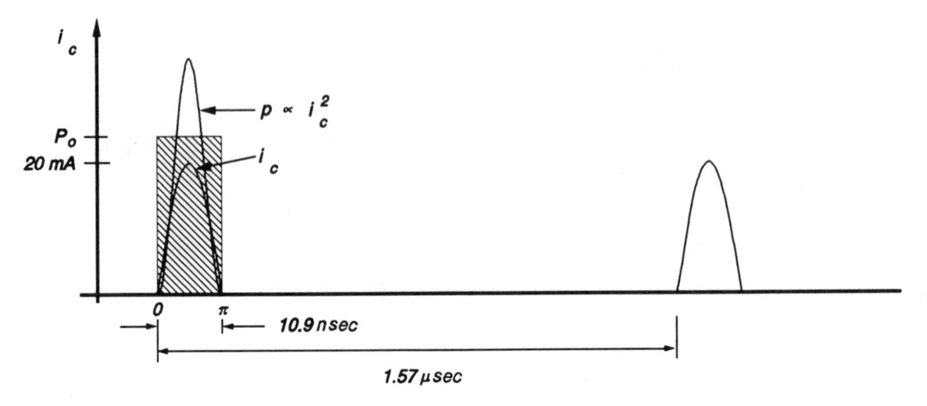

Figure 2.29. This diagram illustrates the various stages involved in the calculation of the output power of the frequency multiplier.

and

$$C = \frac{1}{\omega^2 L_p} = 555 \text{ pF} \tag{2.5.7}$$

(c) Because the parallel LC tank circuit is at resonance, it is actually an open circuit. Therefore the load seen by the collector is $n^2 R_L = 11.25$ kΩ.

(d) The base and collector currents are related by

$$\begin{aligned} i_c &= \beta i_b \\ &= 20 \text{ mA peak} \end{aligned} \tag{2.5.8}$$

Current in the secondary $= n \times 20$ mA peak. Since the system is in steady-state, the energy dissipation associated with the current $n \times 20$ mA peak flowing in the load can be averaged over one cycle of the input signal (2 cycles of the output) and the ac output power calculated on that basis.

The collector current waveform is shown in Figure 2.29. The instantaneous power in the collector circuit is

$$p(\theta) = i_c^2 R_L = (n \times 20 \times 10^{-3} \times \sin\theta)^2 R_L \tag{2.5.9}$$

The average power is

$$P = \frac{1}{\pi} \int_0^{\pi} i_c^2 R_L \, d\theta \tag{2.5.10}$$

The total energy stored in the tank circuit during π rad (10.9 nsec) of the input current is

$$W = (10.9 \text{ nsec}) \times \frac{1}{\pi} \int_0^{\pi} i_c^2 R_L \, d\theta \tag{2.5.11}$$

This amount of energy must be dissipated during a period corresponding to 4π rad (equivalent to 720 degrees or 1.57 μsec in time) of the output waveform (two cycles) since the system is in equilibrium. Therefore,

$$P_0 = \frac{10.9 \times 10^{-9}}{1.57 \times 10^{-6}} \times \frac{1}{\pi} \int_0^{\pi} i_c^2 R_L \, d\theta = 33.75 \text{ mW} \tag{2.5.12}$$

(e) The input power is

$$P_i = \frac{1}{\pi} \int_0^{\pi} i_b R_{in} \, d\theta = 15 \ \mu\text{W} \tag{2.5.13}$$

(f) The power gain is

$$G = 10 \log_{10} \frac{P_0}{P_i} = 33.5 \text{ dB} \tag{2.5.14}$$

2.6 MODULATOR

In Section 2.3, the basic principles of modulation were discussed in general, and amplitude modulation was treated at some length. Briefly then, modulation is a form of multiplexing, that is, providing a common conduit in which several messages can be transmitted at the same time without interference. In terms of amplitude modulation, the message (voice, code, or music) is used to control the amplitude of a high-frequency sinusoidal signal called a carrier. In this section, several circuits that can be used to generate amplitude modulation will be presented with the derivation of the appropriate design equations.

2.6.1 Square-Law Modulator

From Equation (2.2.5), it can be seen that the modulated current has two components:

(1) $A \sin \omega_c t$
(2) $kA \sin \omega_s t \sin \omega_c t$

These suggest that the circuit required to achieve amplitude modulation must

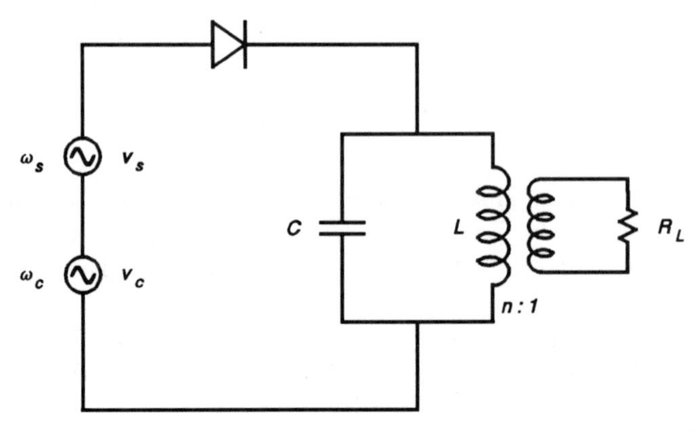

Figure 2.30. Using the nonlinearity of the diode to obtain amplitude modulation. The circuit is tuned to ω_c with sufficient bandwidth to select the carrier and sidebands and attenuate all spurious signals.

generate the product of the carrier and the modulating signal frequency and add to this a suitably scaled carrier signal. The multiplication can be approximated by a nonlinear element, such as a diode, and the addition by connecting the two sources in series. Since the diode has a voltage–current characteristic, which is approximately a square-law relationship, it is commonly used in the square-law modulator. The modulator is sometimes called the diode modulator.

Consider the circuit shown in Figure 2.30. Let

$$v_s = V_s \cos \omega_s t \tag{2.6.1}$$

and

$$v_c = V_c \cos \omega_c t \tag{2.6.2}$$

and assume the following:

(1) Operation is at the resonant frequency of the circuit; hence,

$$\omega_c^2 = \frac{1}{LC} \tag{2.6.3}$$

(2) The diode current is given by

$$i = a_1 v + a_2 v^2 + \cdots \tag{2.6.4}$$

Since the two voltage sources are connected in series,

$$v = V_c \cos \omega_c t + V_s \cos \omega_s t \tag{2.6.5}$$

Substituting Equation (2.6.5) into Equation (2.6.4) gives

$$i = a_1 V_c \cos \omega_c t + a_1 V_s \cos \omega_s t + a_2 V_c^2 \cos^2 \omega_c t$$
$$+ 2a_2 V_c V_s \cos \omega_c t \cos \omega_s t + a_2 V_s^2 \cos^2 \omega_s t + \cdots \tag{2.6.6}$$

Substituting

$$\cos^2 \phi = \tfrac{1}{2}(1 + \cos 2\phi) \tag{2.6.7}$$

into Equation (2.6.6) gives

$$i = a_1 V_c \cos \omega_c t + a_1 V_s \cos \omega_s t + \frac{a_2 V_c^2}{2}(1 + \cos 2\omega_c t)$$
$$+ 2a_2 V_c V_s \cos \omega_c t \cos \omega_s t + \frac{a_2 V_s^2}{2}(1 + \cos 2\omega_s t) \tag{2.6.8}$$

The LC tank circuit is tuned to ω_c and acts as a bandpass filter, and because $\omega_c \gg \omega_s$, signals of frequency ω_s, $2\omega_s$, and $2\omega_c$ are filtered out, leaving

$$i = a_1 V_c \cos \omega_c t + 2a_2 V_c V_s \cos \omega_c t \cos \omega_s t \tag{2.6.9}$$

When the load R_L is reflected into the primary, the voltage across the primary is then given by

$$v_p = a_1 n^2 R_L V_c \left(1 + \frac{2a_2 V_s}{a_1} \cos \omega_s t\right) \cos \omega_c t \tag{2.6.10}$$

This is an amplitude-modulated voltage wave, where the index of modulation is

$$k = \frac{2a_2 V_s}{a_1} \tag{2.6.11}$$

and

$$A = a_1 n^2 R_L V_c \tag{2.6.12}$$

Thus,

$$v = A(1 + k \cos \omega_s t) \cos \omega_c t \tag{2.6.13}$$

2.6.2 Direct Amplitude Modulation Amplifier

Consider an amplifier into which a signal of frequency ω_c is fed [5, 6]. If the dc power supply of the amplifier is varied as a function of another signal of

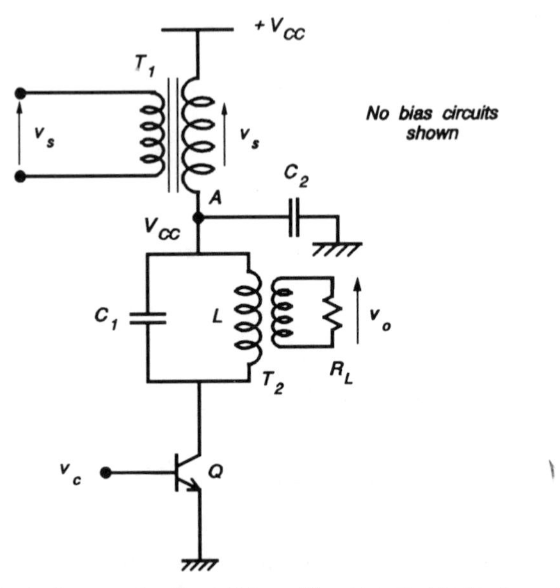

Figure 2.31. The circuit diagram of a direct AM amplifier. Note that in this type of modulator, a high level of audio frequency power is required to drive the circuit.

frequency ω_s, where $\omega_c \gg \omega_s$, then it is evident that the amplitude of the output voltage of the amplifier will be high when the dc supply voltage is high and low when the dc supply voltage is low. The basic circuit for a modulator operating on this principle is shown in Figure 2.31.

The circuit shows a class C amplifier with its power supply connected through the secondary winding of a transformer. Node A is connected to ground by a capacitor whose value is chosen so that, at frequency ω_c, it is a short circuit. From the point of view of the frequency ω_c, node A is at ground potential, as it should be for any amplifier. Assuming that T_1 is a 1:1 transformer, then with v_s applied to the primary, a voltage v_s will be developed in series with the dc supply voltage V_{cc}, such that

$$v_{cc} = V_{cc} + v_s \tag{2.6.14}$$

Let

$$v_s = V_s \cos \omega_s t \tag{2.6.15}$$

and

$$v_c = V_c \cos \omega_c t \tag{2.6.16}$$

Then

$$v_{cc} = V_{cc}(1 + m \cos \omega_s t) \tag{2.6.17}$$

where $m = V_s/V_{cc}$

Since the output voltage is proportional to the input and the supply voltage,

$$v_o = kv_{cc}V_c \cos \omega_c t \tag{2.6.18}$$

where k is a constant

$$v_o = kV_cV_{cc}(1 + m \cos \omega_s t)\cos \omega_c t \tag{2.6.19}$$

This is an amplitude-modulated voltage. It is clear that C_2 must appear to be an open circuit at the frequency ω_s, because the dc supply source V_{cc} is, in fact, an ac short circuit. The two constraints on C_2, namely that it must be a short circuit at ω_c and an open circuit at ω_s, are easily satisfied because, in general, $\omega_c \gg \omega_s$.

2.6.3 Four-Quadrant Analog Multiplier

A modulated signal can be obtained by adding the carrier signal to the product of the modulating (audio) signal and the carrier. Thus,

$$v_o = A \cos \omega_c t + kA \cos \omega_c t \times B \cos \omega_s t \tag{2.6.20}$$

There are a number of techniques that can be used to achieve this. One of the more practical of these is the variable transconductance method [7]. The output current of a common-emitter transistor amplifier depends on the input signal voltage at the base-emitter junction and the magnitude of the emitter resistance that can be controlled by varying the emitter current. The collector current is therefore proportional to the product of the input signal and the emitter current.

Consider the circuit shown in Figure 2.32 Assuming that collector currents are approximately equal to emitter currents, when V_k increases positively, the current I_1 increases at the expense of I_2, because the total current in the

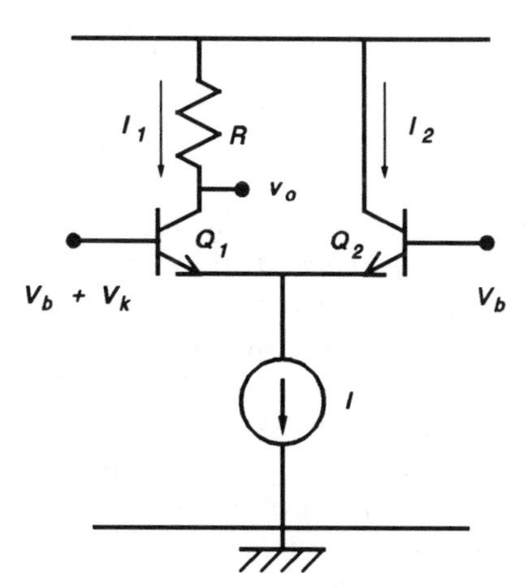

Figure 2.32. The long-tailed pair used as a variable transconductance element in an analog multiplier.

transistors Q_1 and Q_2 are fixed by the ideal current source, I. Let

$$k = I_2/I \tag{2.6.21}$$

then

$$I_1 = (1 - k)I \tag{2.6.22}$$

where k is a current gain and $1 > k > 0$.

Clearly, the ideal current source, I, must have two components: a dc (bias) component and an ac (signal) component

$$I = I_{\text{dc}} + I_c \cos \omega_c t \tag{2.6.23}$$

$$I_2 = kI = k(I_{\text{dc}} + I_c \cos \omega_c t) \tag{2.6.24}$$

Similarly, the voltage source V_k, when it is varied sinusoidally, must have two components: a dc (bias) and an ac (signal) component.

$$k = V_{\text{dc}} + V_s \cos \omega_s t \tag{2.6.25}$$

The output current

$$\begin{aligned} I_2 = m_1 I_{\text{dc}} V_{\text{dc}} + m_2 V_{\text{dc}} I_c \cos \omega_c t + m_3 V_s I_{\text{dc}} \cos \omega_s t \\ + m_4 I_c V_s \cos \omega_c t \cos \omega_s t \end{aligned} \tag{2.6.26}$$

where the ms are constants. To convert I_2 to an amplitude modulated wave, the term

$$m_1 I_{\text{dc}} V_{\text{dc}} + m_3 I_{\text{dc}} V_s \cos \omega_s t \tag{2.6.27}$$

will have to be removed from Equation (2.6.26) to give

$$I_2 = A(1 + B \cos \omega_s t) \cos \omega_c t \tag{2.6.28}$$

where

$$A = m_2 I_c V_{\text{dc}} \tag{2.6.29}$$

and

$$B = (m_4 V_s)/(m_2 V_{\text{dc}}) \tag{2.6.30}$$

Assuming that the circuit is to be realized in integrated circuit form where the transistors characteristics can be very closely matched, the removal of Equation (2.6.27) is most easily achieved by using the concept of "balancing," that is, by duplicating the circuit and interconnecting the two parts as shown in Figure 2.33.

Note that the ideal current source I' has a dc current equal to the dc component of I. A practical realization of the circuit with various bias arrangements, temperature compensation, and a number of other features is shown in Figure 2.34, and it is manufactured by Motorola Semiconductor Products Inc. and designated "MC1595—Four-quadrant Multiplier."

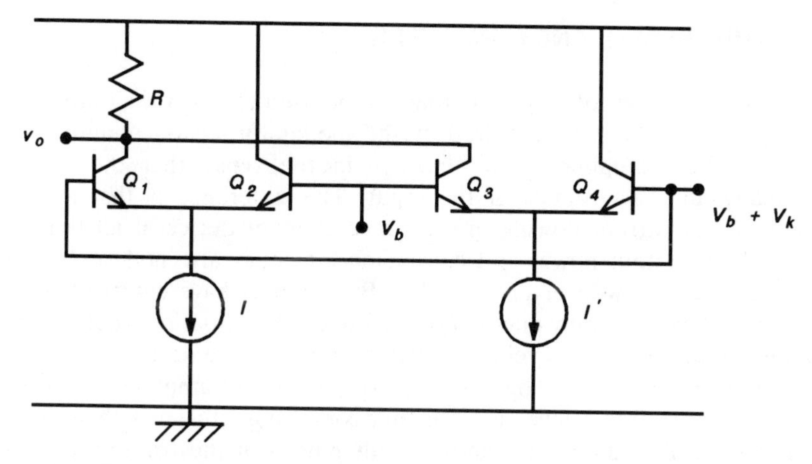

Figure 2.33. The connection of a second long-tailed pair as shown, removes the unwanted signals by using the concept of "balance." This is possible when the circuit is integrated on a single chip, because circuit integration produces matched transistors that can be relied on to track together with temperature.

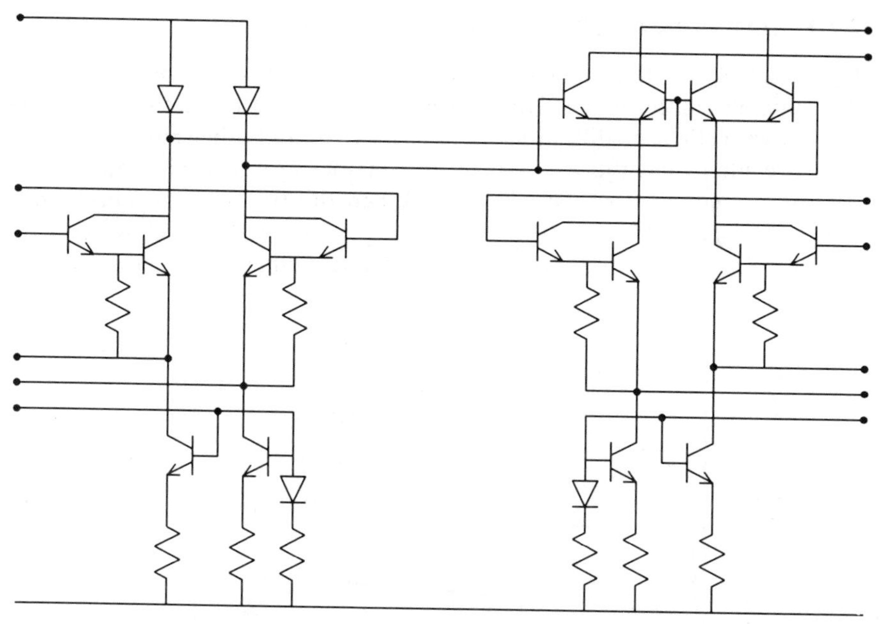

Figure 2.34. The circuit diagram of the four-quadrant Multiplier (MC1595) manufactured by Motorola Semiconductor Products Inc. Reprinted with permission from the *Linear and Interface Integrated Circuits Handbook*, 1983.

2.7 AUDIO FREQUENCY AMPLIFIER

There are two types of amplifiers that can be used to boost the output of the microphone to the level required by the modulator—class A and Class B. Both of these amplifiers are "linear," in the sense that the output is nominally an exact replica of the input. The difference is that a class A amplifier has current flowing in the active element device at all times, and generally its output power and conversion efficiency are modest. It is most often used with low-level signals. Class B amplifiers have current flowing in the active device for only one-half of the cycle. This situation would normally lead to an unacceptable level of distortion, so two active devices are used in a push–pull configuration. The advantage of this type of amplifier is that it has a higher power output level and a higher conversion efficiency than the class A amplifier. The audio frequency amplifier used in the transmitter will be determined by the output of the signal source (microphone) and the signal level required at the input of the modulator. It could be a combination of the two, that is, a class A followed by a class B amplifier. The two types of amplifier are discussed in detail below.

2.7.1 Basic Device Characteristics

Figure 2.35 shows the characteristics of a bipolar junction transistor (BJT) in the common emitter mode. This is used as the basis of the design and can be applied with slight modifications to other devices and other modes.

It is evident that the $V_{CE} - I_C$ characteristics are controlled by the base current, I_B, supplied to the transistor. When the transistor is connected as shown in Figure 2.36 and the value of I_B is reduced, it follows that the collector current I_C will decrease and so will the voltage drop across R_c until

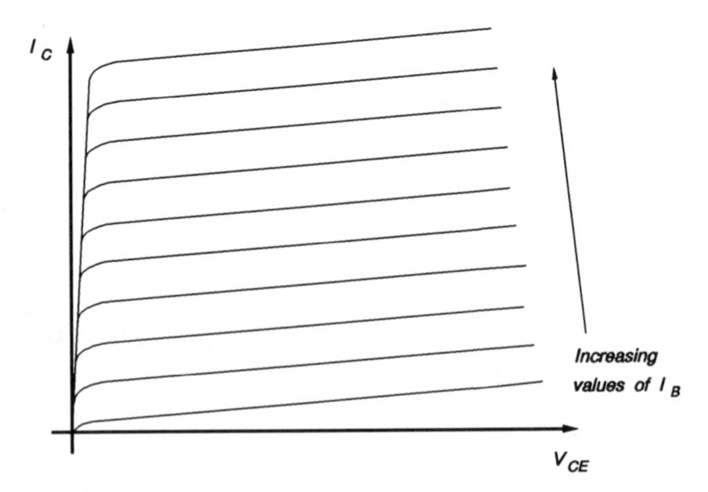

Figure 2.35. The collector characteristics of a junction bipolar transistor showing collector current (I_C) plotted against collector-emitter voltage (V_{CE}) for various values of base current.

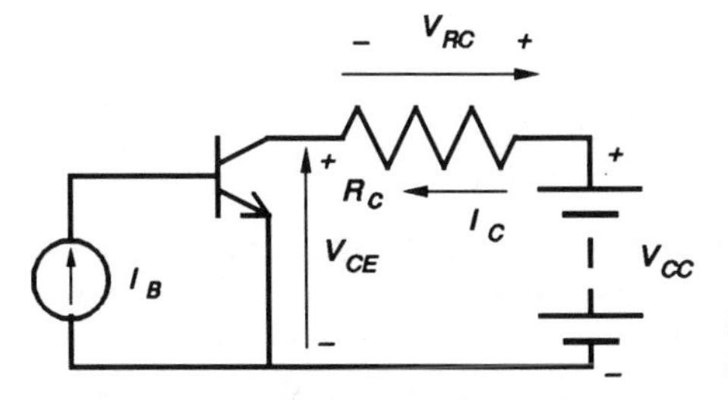

Figure 2.36. The basic common-emitter amplifier with load R_c.

finally when $I_C = 0$, $V_{CE} = V_{cc}$. The transistor is said to be "cut off." On the other hand if I_B is increased, the collector current I_c will increase and so will the voltage drop across R_c until finally $V_{CE} = 0$ and $V_{Rc} = V_{cc}$. The transistor is said to be "in saturation." The collector current is no longer under the control of the base current but determined by the value of R_c. In practice, the value of V_{CE} is not quite zero, and, for most discrete devices, V_{CE} has a value of about 0.5 V.

Applying KVL to the collector current path,

$$V_{cc} = V_{CE} + V_{Rc} \tag{2.7.1}$$

$$= V_{CE} + I_c R_c \tag{2.7.2}$$

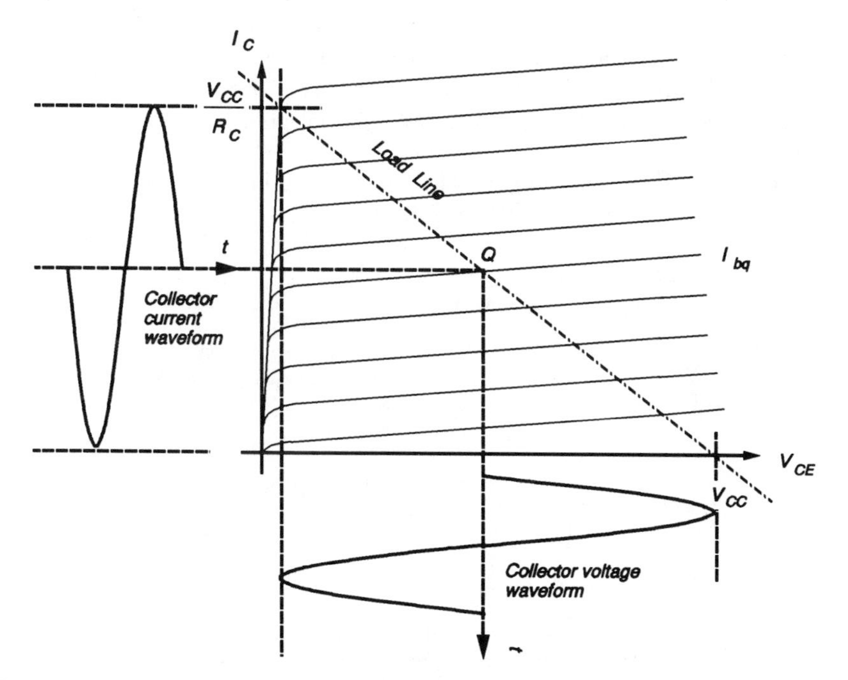

Figure 2.37. The operation of the amplifier is illustrated on the $I_C - V_{CE}$ characteristics by the load line with corresponding current and voltage waveforms.

or

$$I_c = \left(\frac{1}{R_c}\right)V_{cc} + \left(-\frac{1}{R_c}\right)V_{CE} \tag{2.7.3}$$

When Equation (2.7.3) is plotted on the characteristics of the device shown in Figure 2.37, it gives a straight line with a slope equal to $(-1/R_c)$ and intercepts on the y axis equal to V_{cc}/R_c and on the x axis equal to V_{cc}. This line is known as the *load line* and determines the behavior of the device when it is connected to the collector load R_c and the supply voltage V_{cc}.

2.7.2 Class A Amplifier

Because the current in a class A amplifier flows for the entire 2π rad of the input ac cycle, its analysis depends on a simultaneous consideration of the dc as well as the ac conditions of the circuit.

2.7.2a Base-Current-Biased Class A Amplifier. The intersection of the load line and the characteristic representing the value of the base current, I_{bq} establishes the *quiescent* or operating point of the transistor. Figure 2.37 shows how the collector voltage and collector current can be made to change sinusoidally by superimposing a sinusoidal current source, i_b, on the quiescent value. The circuit can be made to provide amplification by a suitable choice of a collector resistor R_c, a base resistor R_b, the supply voltage V_{cc}, and by connecting them as shown in Figure 2.38. The design of an amplifier is a series of judicious choices, including their modification when necessary to meet the design specification. The design is therefore not unique. The technique is best illustrated by a design example.

Example 2.7.1 Base-Current-Biased Class A Amplifier. The following data apply to the circuit shown in Figure 2.38.

Collector supply voltage, $V_{cc} = 10$ V
Base supply voltage, $V_{bb} = 3$ V
Current gain of transistor, $\beta = 100$

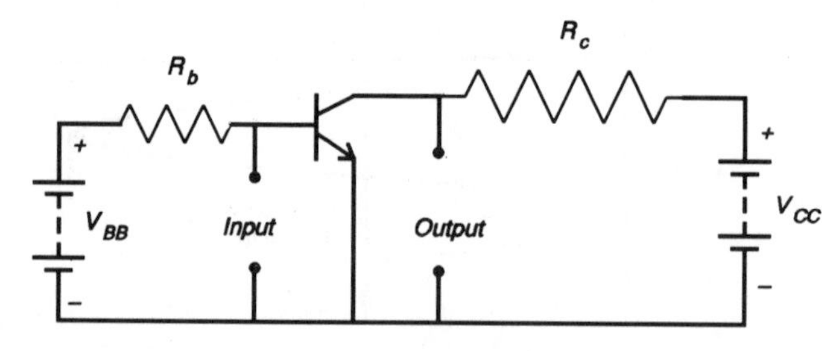

Figure 2.38. The common-emitter amplifier with the quiescent point determined by V_{BB} and R_b.

Base-emitter voltage drop, $V_{be} = 0.7$ V (silicon)
Input voltage = 0.5 V peak-to-peak
Required voltage gain = 10
Amplifier load, $R_L = 10$ kΩ

Solution. The first step is to locate the point corresponding to V_{cc} on the V_{CE} axis of the transistor characteristics. The slope of the load-line is determined by the value of R_c and the choice of R_c depends largely on the current required to drive the load of the amplifier. With an input voltage of 0.5 V peak-to-peak and a voltage gain of 10, the output voltage will be 5.0 volts peak-to-peak. The current required to drive the load is therefore 0.5 mA. peak-to-peak. To ensure that the operation of the transistor is not adversely affected when the load draws off 0.5 mA peak-to-peak from the transistor, we have to allow approximately 10 times the load current in the collector of the transistor *in the form of direct current*, that is $I_c = 5$ mA in the quiescent state.

The next step is to determine the value of R_c. For the amplifier to handle the maximum possible signal, it is necessary to place its quiescent point in such a position that the collector voltage can swing as far in the positive direction as it can swing in the negative direction. For the maximum signal, the transistor goes into saturation at the minimum and is cut off at the maximum value of the collector voltage. It then follows that the required point on the I_c axis is 10 mA. A straight line can now be drawn to join $V_{cc} = 10$ V to $I_c = 10$ mA. From the slope of the load-line, $R_c = 1$ kΩ.

Next, choose the value of R_b to establish the appropriate value of I_b so that the quiescent point is in the middle of the load line. From the diagram, the required base current

$$\begin{aligned} I_b &= I_c/\beta \\ &= 50\ \mu\text{A} \end{aligned}$$

Neglecting the resistance of the base-emitter diode,

$$\begin{aligned} R_b &= (V_{bb} - V_{be})/I_b \\ &= (3.0 - 0.7)/50 \times 10^6\ \Omega \\ &= 46\ \text{k}\Omega \end{aligned}$$

The diode resistance is given approximately by the empirical formula

$$\begin{aligned} R_d &= 26\ \text{mV}/(\text{diode current in amperes}) \\ &= 26 \times 10^{-3}/50 \times 10^{-6}\ \Omega \\ &= 520\ \Omega \end{aligned}$$

R_d is indeed negligible compared to R_b.

The input voltage source has to be adjusted to give the required input voltage at the base of the transistor and the coupling capacitor C_o is chosen so that at the lowest frequency of operation, its reactance is negligibly small compared to the load R_L.

The amplifier, as designed, has the following disadvantages:

(1) It has two dc voltage supplies.
(2) Its input impedance is very low—that of a forward-biased diode.
(3) Its bias point is largely determined by V_{BE}, which is sensitive to changes in temperature.
(4) The voltage gain of the amplifier is sensitive to the current gain of the transistor. The current gain of two transistors fabricated next to each other on an integrated circuit chip can vary by as much as 3 : 1, so it is necessary to design amplifiers that are essentially independent of the current gain of the device.

2.7.2b ***Resistive Divider-Biased Class A Amplifier.*** The amplifier circuit can be modified as shown in Figure 2.39. The base of the transistor is then held at a voltage determined by the resistors R_1 and R_2. A resistor R_e is connected between the emitter and ground. The purpose of this is twofold: it raises the input impedance of the amplifier and redefines the voltage gain as the ratio R_c/R_e.

The amplifier circuit is called a "common emitter," although the emitter is not connected directly to ground. A proper common-emitter amplifier has

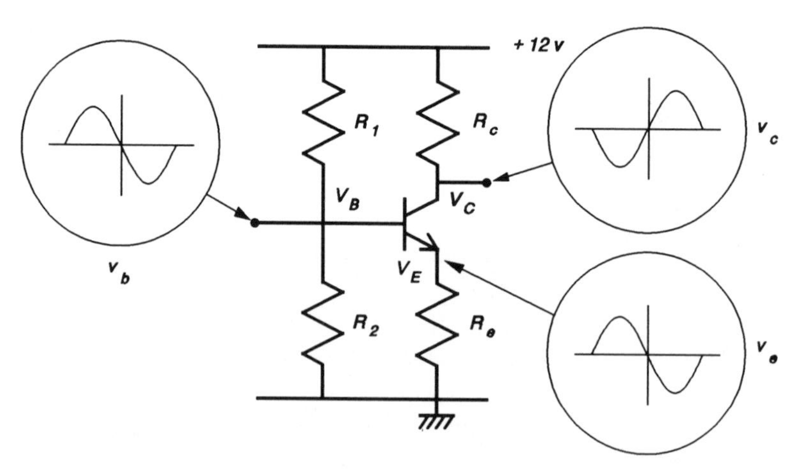

Figure 2.39. The resistive-divider biased "common-emitter" amplifier. The use of the emitter resistor R_e increases the input resistance and reduces the dependence of the amplifier gain on the transistor parameters.

characteristics that are dependent on the parameters of the transistor, the absolute value of the collector load, and other factors, such as the temperature at which it is used. These disadvantages make the use of R_e attractive.

As before, the steps in the design are illustrated by the following example.

Example 2.7.2 Resistive Divider-Biased Class A Amplifier

dc supply voltage, $V_{cc} = 10$ V
Current gain of transistor, $\beta = 100$
Base-emitter diode voltage drop, $V_{be} = 0.7$ V (silicon)
Input voltage = 0.5 V p-p
Required voltage gain = 10
Amplifier load resistance = 10 kΩ

Solution. To keep the transistor in conduction, the base voltage must be kept 0.7 V above the emitter voltage. The point in the cycle where the transistor is most likely to be cut off is at the negative peak (0.5 V peak) of the input voltage. The minimum quiescent base voltage is therefore (0.7 + 0.5 peak) = 1.2 V. The steps leading to the choice of a quiescent collector current remains the same as in Example 2.7.1 and gives the value 5 mA. Assuming that the current gain of the transistor is sufficiently high, then I_c is approximately equal to I_e. Under quiescent conditions, the emitter is held at 0.5 V and the current flowing in R_e is 5 mA. R_e has a value 100 Ω.

The ac voltage at the base of the transistor (0.5 V peak) is the same as what appears at the emitter except for the dc shift of 0.7 V. Since the voltage gain of the amplifier is 10, it follows that the ac signal at the collector must be 10 times as large as what appears across the emitter. Keeping in mind that the emitter and collector currents are approximately equal, it follows that the collector resistance R_c must be $10 \times R_e = 1$ kΩ.

The base current required to maintain the collector current of 5 mA is given by the relationship:

$$I_c = \beta \times I_b$$
$$I_b = 50\ \mu\text{A}$$

The current I_b is drawn from the resistive chain R_1 and R_2, which also define the voltage at the base. To keep the base voltage reasonably constant for changes in I_b due to variations in β, the current in the resistive chain should be approximately $10 \times I_b$, or 500 μA. The chain has 10 V across it, so $(R_1 + R_2) = 20$ kΩ. Since the base voltage must be at 1.2 V, it follows that $R_1 = 17.6$ kΩ and $R_2 = 2.4$ kΩ. The waveforms at the emitter, base, and collector are shown in Figure 2.40. Note that any increase in the amplitude of the input signal will cause the amplifier to clip the output signal in both the cut-off and saturation modes. Coupling capacitors C_i and C_o are chosen so

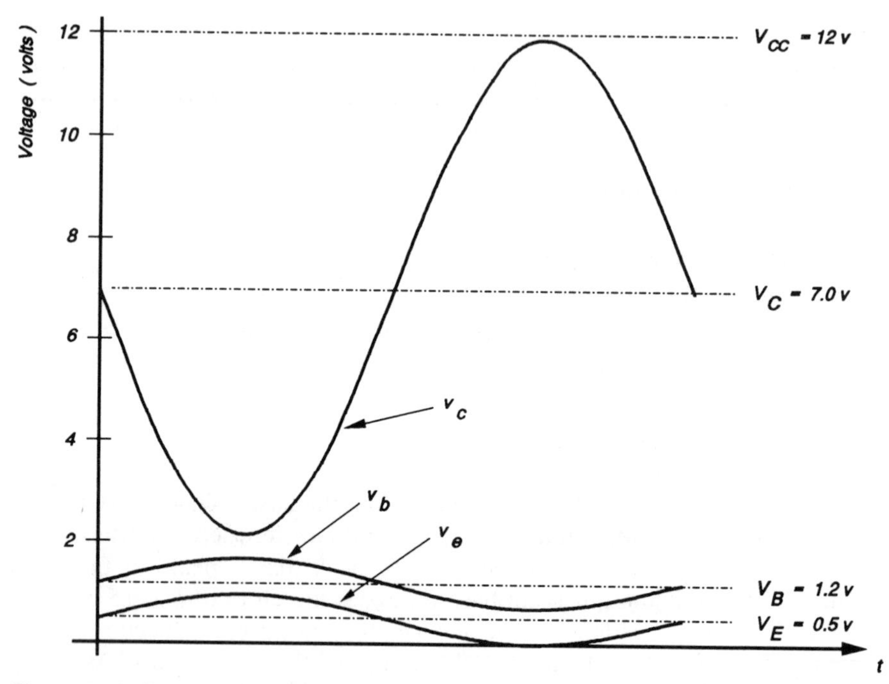

Figure 2.40. The emitter, base, and collector voltage waveforms of the common-emitter amplifier with dc bias voltages shown.

that their reactances at the lowest frequency of operation are less that 10% of the impedance of the load to which they are connected.

The advantage of this configuration over that presented in Example 2.7.1 are as follows:

(1) Only one dc power supply is used.
(2) The input impedance is given by the parallel combination of R_1, R_2, and βR_e, which gives 1.74 kΩ.
(3) The bias point is determined by the ratio of the resistive chain R_1 and R_2, which is independent of temperature.
(4) The voltage gain of the amplifier is determined by the ratio of R_c/R_e and it is not a function of the current gain or any other parameter of the transistor.

2.7.3. Class B Amplifier

2.7.3a Complementary Symmetry Amplifier. The basic circuit for a complementary symmetry amplifier is shown in Figure 2.41. To keep the

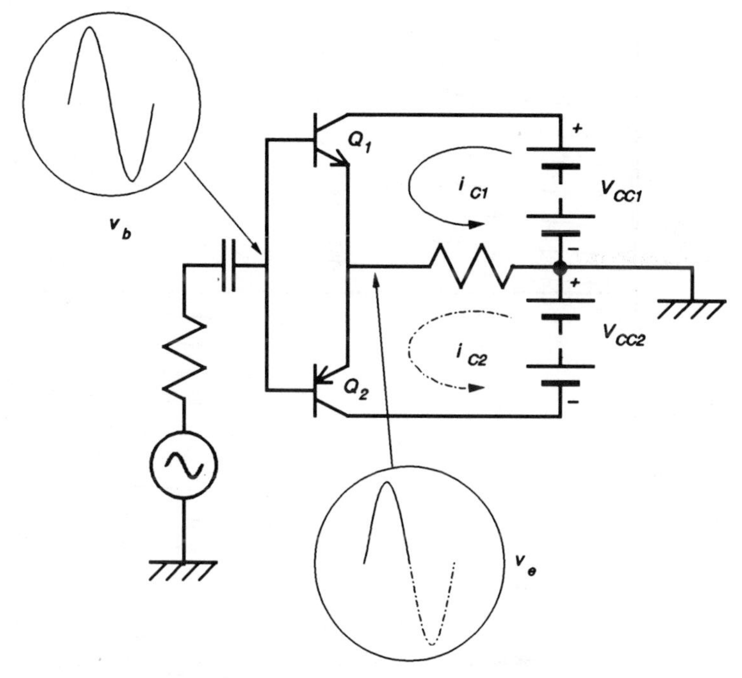

Figure 2.41. Basic class B complementary symmetry amplifier. Note that the circuit is composed of two emitter followers (NPN and PNP) with a common load resistor.

circuit simple, no biasing circuits are shown. It is necessary to make the following assumptions:

(1) The amplitude of the input voltage is much larger than V_{BE}. This assumption ensures that cross-over distortion is kept at a minimum.

(2) The transistors Q_1 and Q_2 are exact complements of each other. This condition (not strictly necessary since we are dealing with emitter followers) ensures that symmetry is maintained.

(3) The dc power supply voltages, V_{cc1} and V_{cc2}, are equal.

When the input voltage becomes positive, the base-emitter junction of Q_1 becomes forward biased and current i_{c1} flows as shown. The base-emitter junction of Q_2 is reverse biased and Q_2 is cut off. Essentially, the circuit acts as an emitter follower.

When the input voltage becomes negative, Q_2 conducts a current i_{c2} in the direction shown. Q_1 is cut off. If i_{c1} and i_{c2} are equal, the voltage developed across the load R_L will be sinusoidal.

The operation of the circuit, in terms of the device characteristics and the load line can be seen in Figure 2.42. Note that the y axis of the PNP transistor, Q_2 has been shifted to make the point V_{cc} coincide with the point $-V_{cc}$. The advantage gained is the simplicity of the diagram.

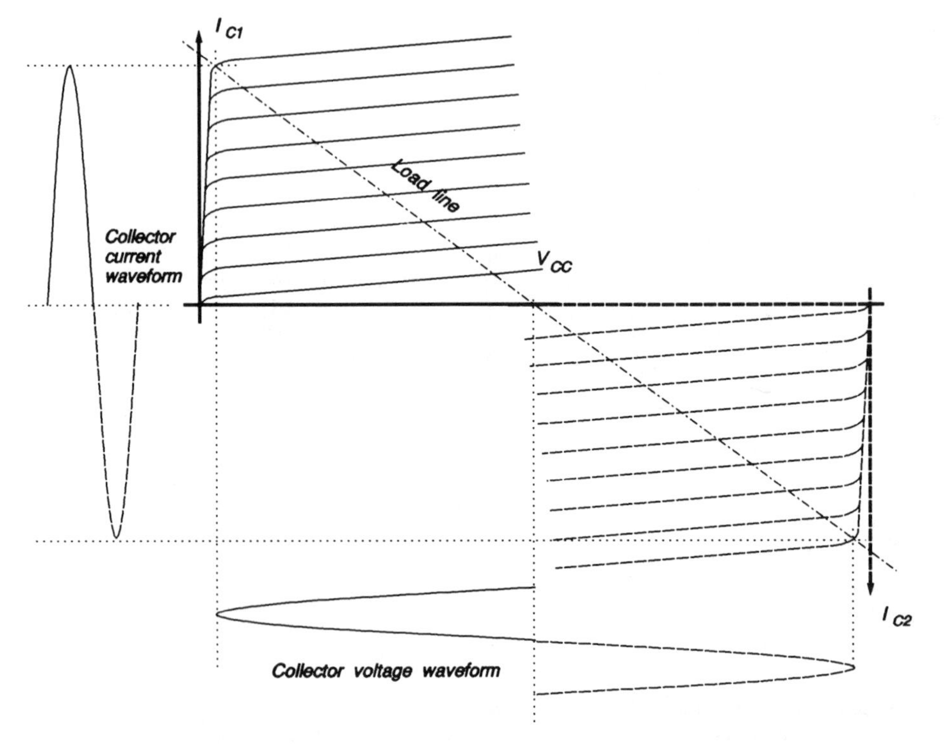

Figure 2.42. Composite characteristics of the complementary symmetry transistor pair with the load line, collector voltage, and current shown. The solid lines refer to Q_1; the dotted to Q_2.

The ac power output is

$$P_o = \frac{\hat{I}_c}{\sqrt{2}} \times \frac{\hat{V}_{cc}}{\sqrt{2}} = \frac{\hat{I}_c \hat{V}_{cc}}{2} \tag{2.7.4}$$

The current flowing in each transistor is a half-sinusoidal whose average value (dc equivalent) is $\hat{I}_c/\pi$. The average current for the two transistors is $(2\hat{I}_c)/\pi$. Because this current flows against a potential difference of V_{cc}, the dc power input is

$$P_{\text{dc}} = \frac{2 I_c V_{cc}}{\pi} \tag{2.7.5}$$

The conversion efficiency is

$$\begin{aligned} \eta &= (\text{ac power})/(\text{dc power}) \\ &= \pi/4 = 78.5\% \end{aligned} \tag{2.7.6}$$

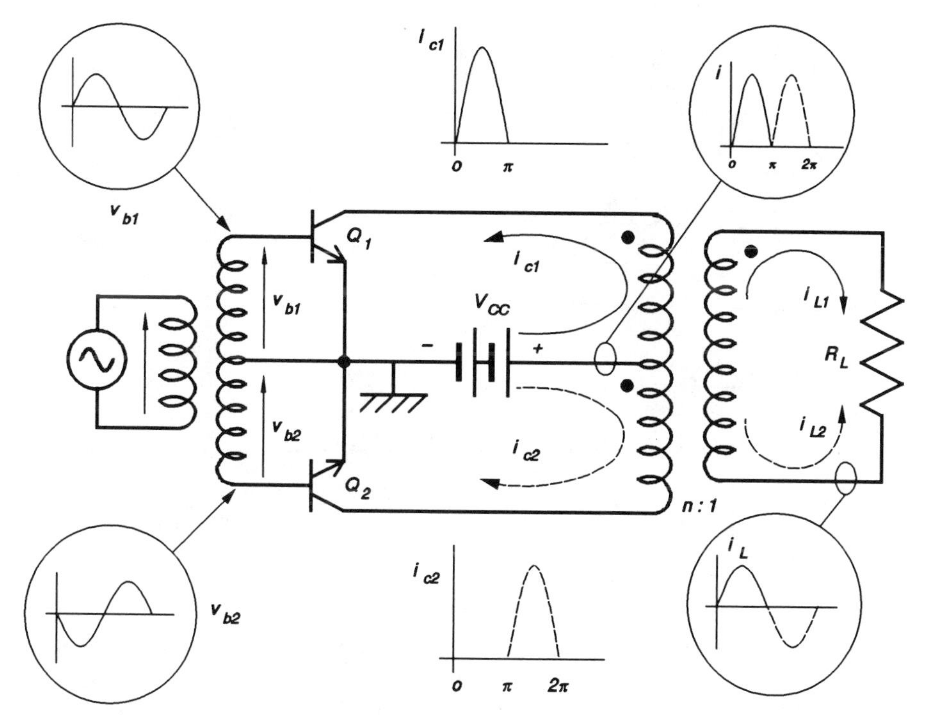

Figure 2.43. Circuit diagram of the transformer-coupled class B amplifier. The waveforms of the currents and voltages at various points are indicated. The solid lines refer to Q_1; the dotted line to Q_2.

The complementary symmetry class B amplifier has the following disadvantages.

(1) It requires two equal dc power supplies.

(2) Complementary pairs of transistors are not easy to manufacture especially as the power level increases.

(3) The circuit cannot have a voltage gain since it is basically two emitter followers with a common load.

To avoid the above disadvantages, the transformer-coupled class B amplifier can be used.

2.7.3b Transformer-Coupled Class B Amplifier. The transformer-coupled class B amplifier uses two center-tapped transformers, as shown in Figure 2.43. The only assumption necessary is that the two transistors are matched. Since they are both NPN (or PNP) this is not, in general unduly restrictive.

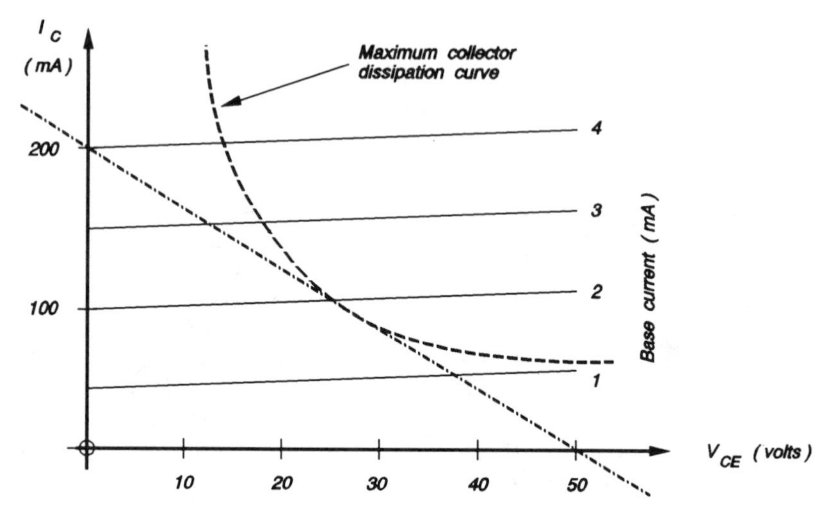

Figure 2.44. The transistor characteristics have been idealized and the maximum collector power dissipation curve superimposed on it.

When the input voltage becomes positive, two voltages v_{b1} and v_{b2} appear across the secondary of the input transformer as shown. The voltage v_{b1} biases Q_1 on and v_{b2} cuts off Q_2. The current i_{c1} that flows in Q_1 and one-half of the primary of the output transformer is a half-sinusoid. The voltage developed across the secondary of the output transformer is therefore a half-sinusoid as shown.

When the input voltage becomes negative, v_{b1} biases Q_1 off and v_{b2} brings Q_2 into conduction. The current i_{c2} that flows in Q_2 and the other half of the primary of the output transformer is a half-sinusoid, but it flows in the opposite direction to i_{c1}. The voltage that is developed across the output transformer is then a negative-going half-sinusoid. The resultant current in the load i_L is a complete sinusoid.

It should be noted that at any point in time only one-half of the primary of the output transformer is coupled to the secondary. Transformers are bulky, heavy, and bandwidth limited, so class B amplifier designs that do not use transformers are available [8, 9].

Example 2.7.3 Class B Amplifier. The idealized output characteristics of a transistor to be used in a class B push–pull amplifier are given in Figure 2.44. The following parameters are used:

(1) Supply voltage $V_{cc} = 50$ V
(2) The amplifier load resistance = 8 Ω
(3) The emitter-base diode characteristics can be approximated by a straight line as shown in Figure 2.45

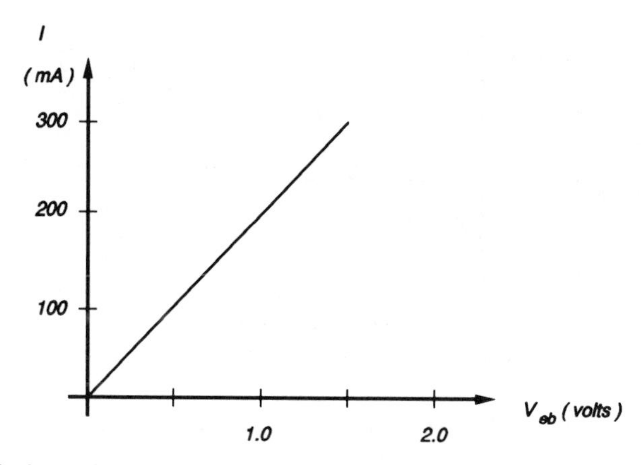

Figure 2.45. To keep the example simple, the nonlinear emitter-base diode has been assumed to be linear.

Calculate the following:

(a) The turns ratio of the output transformer
(b) The conversion efficiency
(c) The input power
(d) The output power
(e) The power gain in decibels

Solution. The collector characteristics has a rectangular hyperbola drawn on it to indicate the maximum dissipation capability of the device. The load line may then touch the locus of the maximum dissipation line, but may not intersect it. In the absence of any requirement on the maximum power output of the amplifier, it may be assumed that it should provide the maximum power. The load line can then be drawn through $V_{CE} = 50$ V and tangential to the maximum dissipation curve. This gives the apparent collector load resistance value of 250 Ω. A suitable circuit for the class B amplifier is shown in Figure 2.46. At any point in time only one half of the transformer primary is linked to the secondary. The turns ratio $a:1$ refers to this link and not to the whole of the primary, which is, in fact, $2a:1$. The load "seen" by the collector of the transistor is to be 250 Ω, but this represents a load R_L transformed through the transformer. Hence,

$$250 = a^2 R_L$$
$$a = 5.59$$

The collector characteristics of the "composite" transistor and the ac load-line together with the collector current and the corresponding collector voltage

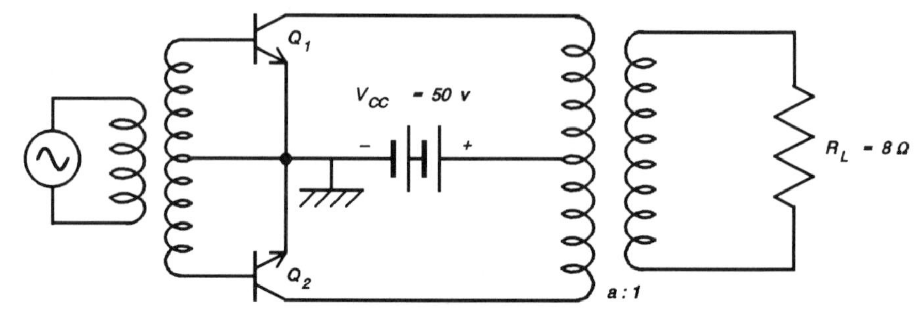

Figure 2.46. The circuit diagram for Example 2.7.3.

are shown in Figure 2.47. The following parameters can be derived:

(a) Current gain of the transistor is approximately equal to 50
(b) Output voltage $= V_{cc}/\sqrt{2}$ V (rms)
(c) Output current $= \hat{I}_c/\sqrt{2}$ A (rms)
(d) Output power, $P_o = VI\cos\phi$, where $\phi = -\pi$ rad

$$
\begin{aligned}
P_o &= -(50 \times 0.2)/2 \text{ W} \\
&= -5 \text{ W}
\end{aligned}
$$

The negative sign can be ignored since it simply indicates that the transistor is generating the power not dissipating it.

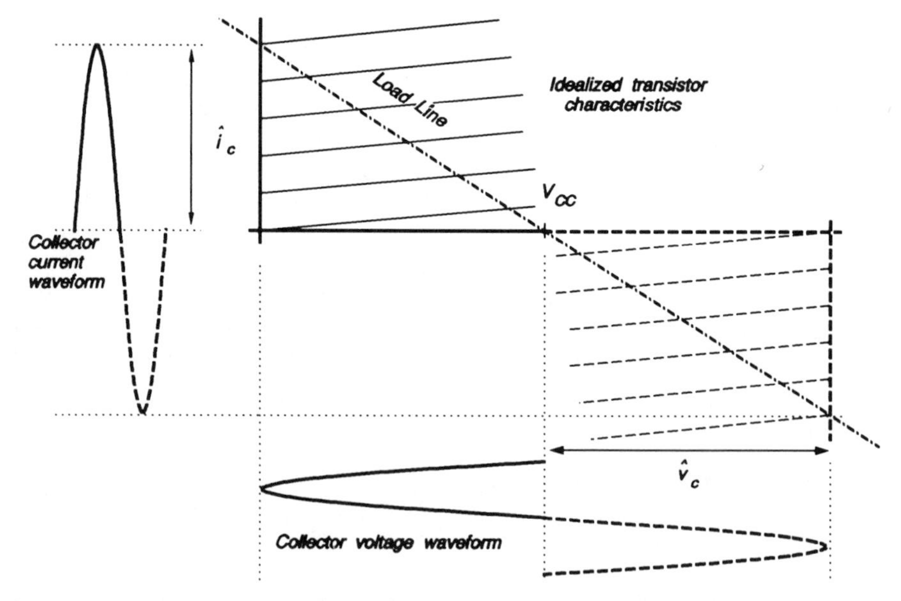

Figure 2.47. The composite transistor collector characteristics have been idealized with the load line, collector current, and voltage waveforms shown. The solid lines refer to Q_1 and the dotted line to Q_2.

(e) The base current required to drive the maximum collector current $\hat{I}_b = 4$ mA.

Using the approximation $I_C = I_E$ and the graph of $I_E - V_{eb}$, for $I_E = 200$ mA,

$$\hat{V}_{eb} = 1.0 \text{ V}$$

$$\text{Input power, } P_i = \left(\hat{V}_{eb}\hat{I}_b\right)/2$$

$$= 0.002 \text{ W}$$

$$\text{Power gain} = 10\log_{10} P_o/P_i = 33.9 \text{ dB}$$

2.8 RADIO FREQUENCY AMPLIFIER

The choice of an amplifier to boost the power coming out of the modulator to the level required to drive the antenna can be made when the following points have been considered:

(1) The amplifier must be linear so as to preserve the nature of the modulated signal.
(2) The output power of the amplifier can vary from a few watts to hundreds of kilowatts.
(3) A high conversion efficiency is necessary when high power is a requirement. The portion of the dc power supplied to the amplifier that is not converted into usable ac signal is dissipated in the amplifier components. Suitable arrangements have to be made to remove the heat. This can add to the complexity and the cost of operation of the transmitter.
(4) The signal to be amplified consists of the carrier and the two sidebands, which normally constitute a narrowband signal.

The usual choice is a class B (actually a class AB) amplifier with a tuned load. The class B is chosen for its linearity, high output power, and high conversion efficiency, and the tuned load ensures that it is a narrowband amplifier. For relatively low power requirements, transistors may be used in the output stage of the amplifier. When the power output required is in excess of a few hundred watts, vacuum tubes are used.

A suitable circuit for the radio frequency power amplifier is shown in Figures 2.48. Its operation is the same as that of the class B audio frequency amplifier discussed in Section 2.7.3b.

Example 2.8.1 Class B Tuned Amplifier. The final stage of a class B radio frequency power amplifier in a radio transmitter has a tuned circuit in which $C = 2500$ pF. The carrier frequency is $\omega_c = 5 \times 10^5$ rad/sec, and the

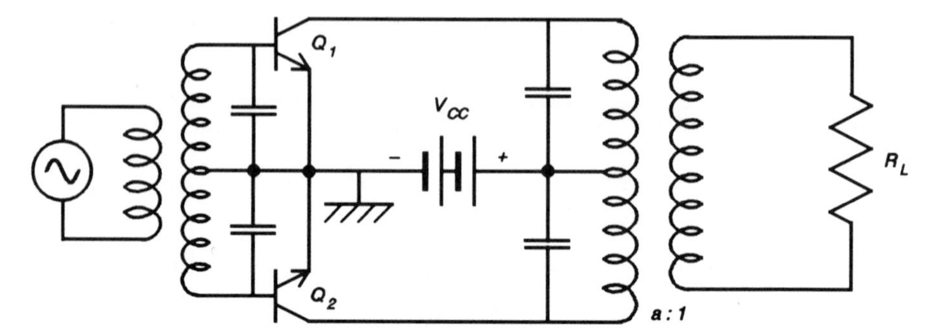

Figure 2.48. Class B amplifier with a tuned input and output. The amplifier is designed to amplify the carrier and sidebands only.

modulating signal has frequencies ranging from $\omega_{s1} = 20$ rad/sec to $\omega_{s2} = 10 \times 10^3$ rad/sec. The power is coupled through a transformer of turns ratio 10 : 1 to an antenna whose impedance is assumed to be 250 Ω resistive.

Calculate the following:

(a) The inductance in the tuned circuit

(b) The Q factor of the tuned circuit if the extreme edges of the sidebands are to be within -3 dB of the carrier

(c) What is the maximum winding resistance that the coil in the tuned circuit can have under the condition in (b)

(d) What is the Q factor of the coil alone?

Solution.

(a)

$$\omega_c^2 = 1/(LC)$$

$$L = 1/(25 \times 10^{10} \times 2500 \times 10^{-12}) \text{ H}$$

$$= 1.6 \text{ mH}$$

(b) Now,

$$Q_0 = \frac{\omega_c}{\omega_2 - \omega_1}$$

where ω_1 and ω_2 are the two -3 dB frequencies

$$Q = 5 \times 10^5 / 20 \times 10^3) = 25$$

(c) There are two sources of loss in the tuned circuit which contribute to give a load Q of 25:

(1) The antenna resistance coupled into the tuned circuit by the transformer ($n^2 R_L$)

(2) The winding resistance of the coil, r.

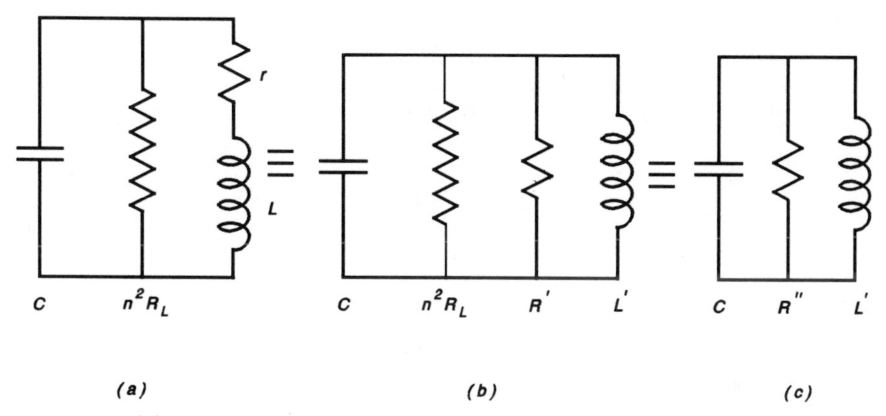

Figure 2.49. (*a*) The tuned collector circuit of the amplifier with the resistive load transferred to the primary of the transformer. (*b*) The collector equivalent circuit when the series *rL* has been replaced with the parallel equivalent. (*c*) The collector circuit when the two resistive elements have been combined.

The equivalent circuit of the load is shown in Figure 2.49*a–c* in three stages. Stage *a* shows the reflected resistance in parallel with the tuned load and the winding resistance in series with the coil. Stage *b* is the equivalent circuit when the winding resistance is converted into a resistance R' in parallel with the inductance. Stage *c* shows the equivalent circuit when the resistances n^2R_L and R' are combined to form R''.

Now, if the two circuits shown in Figure 2.50 are equivalent, then

$$r + j\omega L = \frac{j\omega L_p R_p}{R_p + j\omega L_p} \tag{2.8.1}$$

Rationalizing and equating real and imaginary parts,

$$r = \frac{\omega^2 L_p^2 R_p}{R_p^2 + \omega^2 L_p^2} \tag{2.8.2}$$

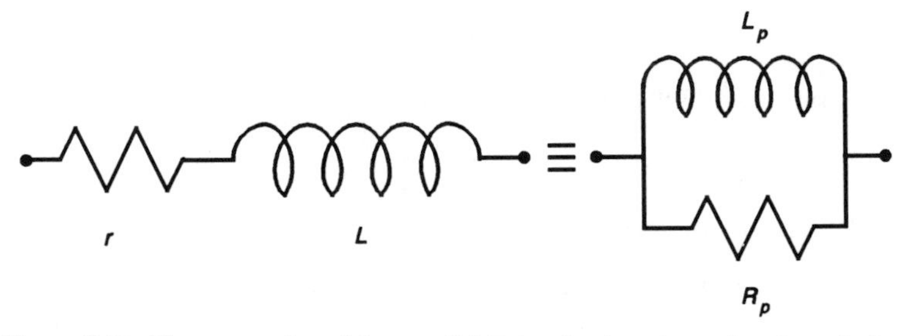

Figure 2.50. The conversion of the parallel R_pL_p circuit to the series *rL* equivalent.

and

$$L = \frac{L_p R_p^2}{R_p^2 + \omega^2 L_p^2} \tag{2.8.3}$$

The Q factor,

$$Q = \frac{R_p}{\omega L_p} \tag{2.8.4}$$

When

$$Q \gg 1, \qquad R_p \gg \omega L_p$$

Then,

$$r \simeq \frac{\omega^2 L_p^2}{R_p} \tag{2.8.5}$$

and

$$L \simeq L_p \tag{2.8.6}$$

From the equivalent circuit c,

$$Q = \frac{R''}{\omega L'} \tag{2.8.7}$$

Therefore

$$R'' = 25 \times \omega L' = 20 \text{ k}\Omega \tag{2.8.8}$$

The resistance, R'' has two components as shown in circuit b, $n^2 R_L$ in parallel with R'

$$n^2 R_L = 10^2 \times 250 = 25 \text{ k}\Omega \tag{2.8.9}$$

Therefore,

$$25R'/(25 + R') = 20 \tag{2.8.10}$$

which gives

$$R' = 100 \text{ k}\Omega \tag{2.8.11}$$

Using Equation (2.8.1) when $R_p = R'$ gives

$$r = 6.4 \ \Omega \tag{2.8.12}$$

(d) The Q factor of the coil only is

$$Q_0 = \frac{\omega L}{r} = 125 \tag{2.8.13}$$

2.9 ANTENNA

Antenna structures can take many physical forms from the "whip" antenna used mostly for in-car radios, through "rabbit ears" for television, to the microwave "dish" used in satellite communication. These are only the most common examples of antennas that are used for reception of broadcast signals. The antenna is the last processor of the signal at the transmitting end and the first at the receiving end. Antennas vary widely in shape, size, and complexity depending on the frequency of operation and the desired field pattern.

Radio communication in free space is possible, because, when an alternating current flows in a conductor, part of the energy is lost in the form of electromagnetic radiation into free space. When the frequency of the current is low, the radiation "loss" is very small, but as the frequency increases, substantial losses can occur. In designing an antenna, the object is to construct a structure that will maximize the radiated energy in a given direction or over a geographic area.

A useful analog of how an antenna radiates energy is a body floating in the middle of a pond on a windless day. For the purpose of this description, assume that the body is a beach ball. If we can get a man suspended from a crane directly above the beach ball to push the ball very slowly into the water and release it equally slowly, then the energy expended in pushing the ball into the water is recovered when it is released and very little will be lost. If the man increases the frequency at which he pushes and releases the ball, an increasing amount of energy will be lost in creating waves that will radiate from the ball outward. The amount of radiated energy at any given point on the pond will bear an inverse relationship to the distance from the ball. It is necessary to bear in mind that the waves on the pond are surface waves (two dimensional), whereas those produced by an antenna are three dimensional.

The speed of propagation of the radio wave is the same as that of light ($c = 3 \times 10^8$ m/sec.). The wavelength and the frequency are related by

$$\lambda f = c \tag{2.9.1}$$

where λ is the wavelength in meters and f is the frequency in Hertz. Electromagnetic radiation consists of two vectors: **E** and **H**, which are orthogonal to each other and to the direction of propagation.

A detailed study of antenna design is beyond the scope of this text. However, the interested reader may consult a suitable textbook on the subject. This discussion will therefore be qualitative.

2.9.1 Radiation Pattern of an Isolated Dipole

A simple way to gain insight into the operation of a dipole is to start from an open-circuit balanced transmission line. Such a transmission line will have a standing-wave distribution of currents in opposite directions, so that the net

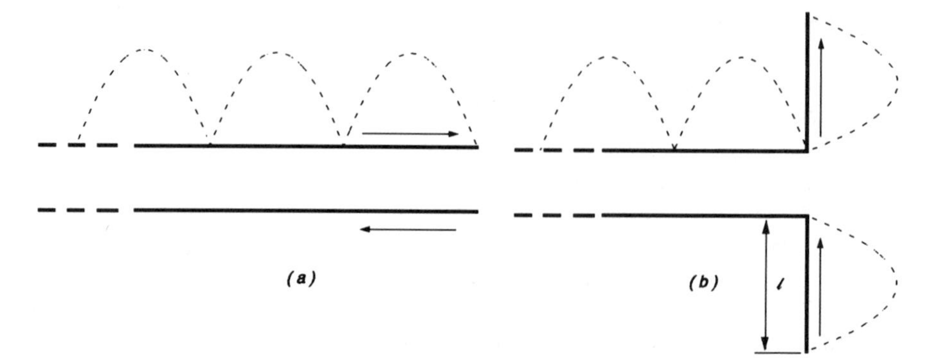

Figure 2.51. (*a*) the current standing wave pattern on an open-circuit balanced transmission line. (*b*) The current standing wave pattern when the line is bent at a point where $l = \lambda / 2$ to form a dipole.

radiation from the structure will be quite small. Figure 2.51*a* shows the transmission line and the standing-wave pattern. It is assumed that the diameter of the conductors is infinitesimally small compared to the wavelength of the signal to be transmitted.

If the transmission line is bent at right angles as shown in Figure 2.51*b*, the structure is called a dipole, which has the current distribution shown. Since the oppositely directed currents are now far apart, they do not counteract each other and hence the dipole is a good radiator of electromagnetic waves.

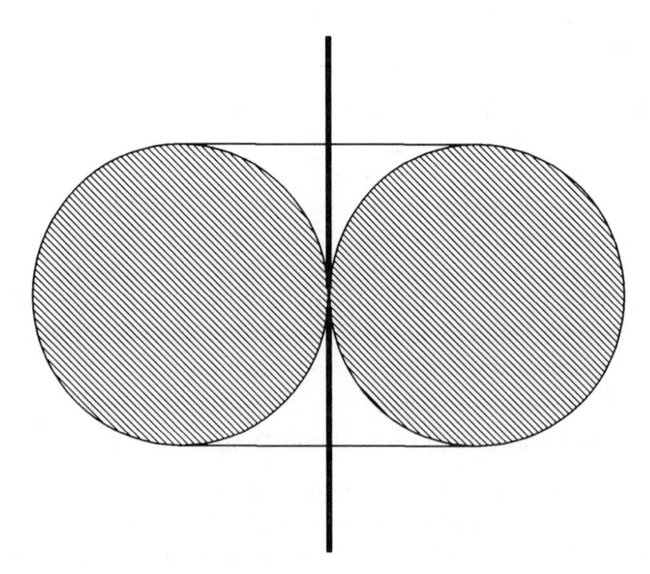

Figure 2.52. The radiation pattern of an isolated half-wavelength dipole.

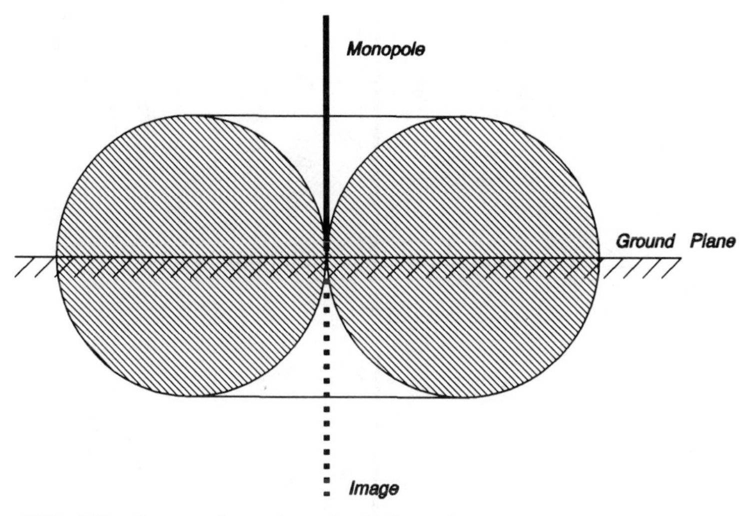

Figure 2.53. This diagram shows how the half-wavelength dipole can be replaced with a monopole and a flat conducting surface (ground plane).

Assume that $2l = \lambda/2$; the dipole is then called a half-wavelength dipole. The **E** field of the half-wavelength dipole when plotted gives a circle tangential to the axis of the dipole touching the axis at the midpoint. In three dimensions, the pattern is a doughnut, as shown in Figure 2.52.

Now assume that $l \ll \lambda/2$; the dipole is said to a "short dipole" and the field pattern is also a doughnut but the cross-section is slightly distorted as if compressed vertically.

2.9.2 Monopole or Half-Dipole

The equations that describe the behavior of the isolated dipole can be used, with slight modification, for the monopole or half-dipole by assuming that the Earth's surface, on which the antenna is to be placed, is a perfectly flat conductor. This permits the monopole and its image on the ground plane to be treated as a dipole. Figure 2.53 shows the monopole and its image. This representation of the antenna is closer to reality than the isolated dipole.

2.9.3 Field Patterns for a Vertical Grounded Antenna

The **E** field pattern generated by a vertical grounded antenna resembles that of the monopole closely, even though the Earth's surface is not exactly a perfectly flat conducting surface. Figures 2.54(a)–(e) are a series of diagrams that present the expected approximate field pattern as the height of the vertical grounded antenna changes. It is important to note that the height is measured in terms of the wavelength of the radiated signal.

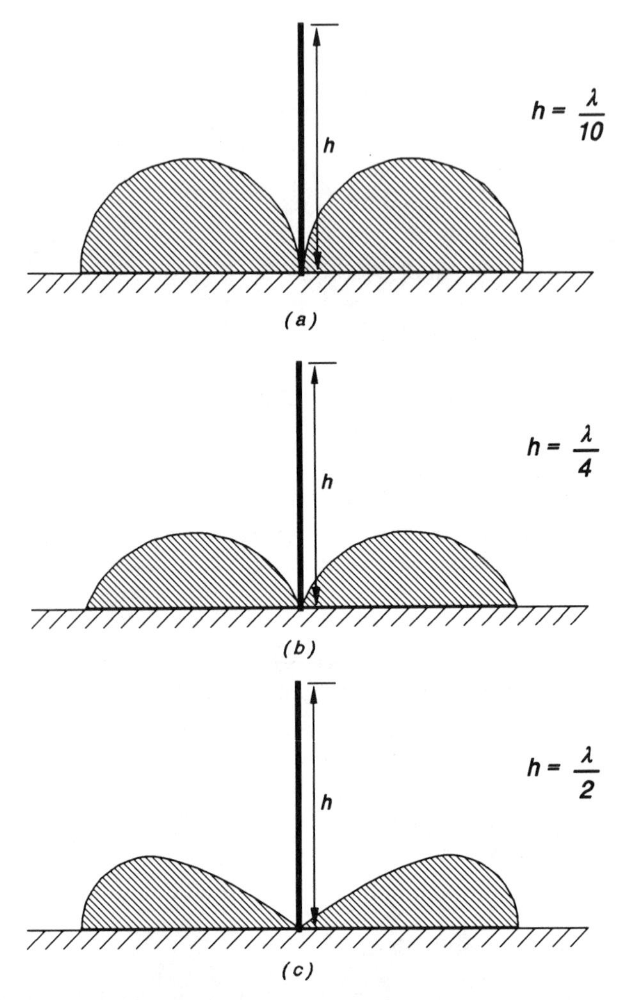

Figure 2.54. (*a*) Radiation pattern when l is approximately $\lambda / 10$. (*b*) Radiation pattern when l is about $\lambda / 4$. Note the slight elongation of the pattern along the ground plane. (*c*) The radiation pattern has undergone further distortion as l increases to about $\lambda / 2$. (*d*) As l approaches $5\lambda / 8$, the ground wave gets more elongated and the sky wave appears. (*e*) With l approximately $3\lambda / 4$, the sky wave has grown considerably at the expense of the ground wave.

2.10 CLASSIFICATION OF AMPLITUDE-MODULATED RADIO FREQUENCY BANDS

Amplitude-modulated radio frequencies are grouped into three bands according to the wavelength of their carrier frequencies. The carrier frequency chosen depends to a large extent on the distance between the broadcasting station and the intended listeners.

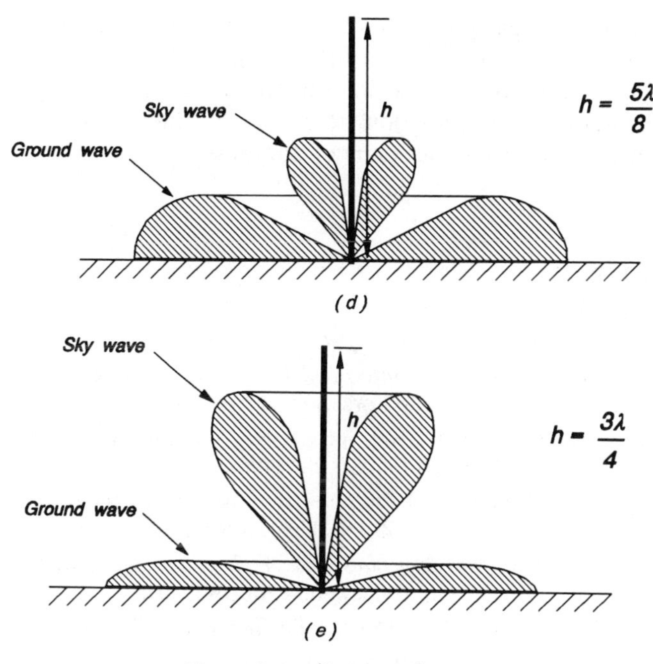

Figure 2.54. (Continued)

1. Long Wave (Low Frequency). All transmission whose carrier frequencies are less than 400 kHz are generally classified as long wave. At a frequency of 100 kHz, a quarter wavelength antenna is 750 m high. Such an antenna poses several problems, such as vulnerability to high winds and danger to low-flying aircraft. Long-wave broadcasting stations therefore use an electromagnetically short antenna, which necessarily limits their reach to a few tens of kilometers, because the short antenna has only the ground wave.

2. Medium Wave. Carrier frequencies in the 300 kHz to 3 MHz range are regarded as medium wave. The height of the antenna becomes more manageable and the possibility of using the sky wave to reach distant audiences is a reality. Generally, it used for local area broadcasting.

3. Shortwave. Shortwave generally refers to carrier frequencies between 3 MHz and 30 MHz. The wavelengths under consideration are between 1 and 100 m. Antenna structures can be constructed to give specified directional properties. Most of the energy can be put into the sky wave and the signal can be bounced off the ionosphere (the layer of ionized gas that surrounds the Earth) to reach receivers half-way round the world. A very severe problem is encountered in shortwave transmission: that is, the signal tends to fade from time to time. This phenomenon is caused by the multiple paths by which the signal can reach the receiver. It is clear that if two signals reach the receiver by different paths such that their phase angles are 180 degrees apart they will cancel each other. The ionosphere sometimes experiences severe turbulence due mainly to radiation from the sun. Shortwave transmission is therefore at its best during the hours of darkness.

REFERENCES

1. Nyquist H., "Regeneration Theory," *Bell System Tech. J.*, 11, 126–147 (1932).
2. Millman, J. and Halkias, C. C. *Integrated Electronics: Analog & Digital Circuits and Systems.*, McGraw-Hill, New York, 1972, Chap. 14.
3. Vehovec, M., Houselander, L., and Spence, R., "On Oscillator Design for Maximum Power," *Trans. IEEE*, CT15, 281–283 (1968).
4. Buchanan, J. P., "Handbook of Piezoelectric Crystals for Radio Equipment Design," WADC Technical Report 56-156, U.S. Airforce, Oct. 1956.
5. Ryder J. D., *Electonic Fundamentals and Applications: Integrated and Discrete Systems*, 5th ed., Prentice Hall, Englewood Cliffs, NJ, 1976.
6. Fraser, W., *Telecommunications*, MacDonald, London, 1957.
7. Kaye, A. R., "A Solid-State Television Fader-Mixer Amplifier," *J. SMPTE*, 74, 602–606, (1965).
8. Bray, D. and Votipka, W., "25-Watt Audio Amplifier with Short-Circuit Protection," Application Bulletin, Fairchild Semiconductor, August 1967.
9. Ruehs, R., "High Power Audio Amplifiers with Short-Circuit Protection," AN-485, Motorola Semiconductors, 1968.

PROBLEMS

2.1 An ideal voltage source has a signal consisting of a sinusoidal carrier of 1-MHz frequency, amplitude-modulated by another sinusoidal signal of frequency 20 kHz to a depth of 100%. This is connected across an LC series tuned circuit. The capacitor has a value 100 pF and may be assumed to be lossless. The Q factor of the circuit is 100. Calculate the following:

(a) The value of the inductance required to tune in the signal.

(b) The value of the coil resistance.

(c) The depth of the modulation of the current that flows in the circuit.

(d) Discuss the reason(s) for the change of the depth of modulation in (c), if any.

2.2 In an AM transmitter, the current flowing in the antenna is 7.5 A when the carrier is not modulated. When a sinusoidal modulation is applied, the current changes to 8.25 A. Calculate the following:

(a) The depth of modulation.

(b) The antenna current when the depth of modulation is 65%.

2.3 A feedback circuit with a $\beta = (8.0 + j4.0) \times 10^{-3}$ is connected to an amplifier and the resulting system oscillates sinusoidally. Calculate the gain and phase shift of the amplifier. Design a circuit that can perform the feedback function and show how it can be connected to the amplifier. The frequency of the oscillation is $100/2\pi$ kHz, and it may be assumed that the amplifier has a high input impedance and a low output impedance.

2.4 Show that when a source of electro-motive force E volts and internal impedance $Z_s = R_s + jX_s$ is connected to a load $Z_L + jX_L$, maximum power is transferred to the load when

(a) $R_s = R_L$

(b) $X_s = -X_L$

A load resistance 1150 Ω is connected across the secondary winding of a transformer. The inductance of the winding is 80 mH and the Q factor is 32. The primary winding has identical characteristics, and it is connected to a voltage source in series with a capacitor C_1. The frequency of the voltage source is 20×10^3 Hz and the internal resistance is 142 Ω. Determine the value of C_1 and the mutual inductance between the two coils for maximum power to be delivered to the load.

2.5 The wiper of a rheostat is moved linearly with respect to time between the 25% and 75% points repetitively at a frequency of 6 Hz. A voltage signal $V_{in} = 10 \cos \omega_0 t$ is connected to the terminals of the rheostat. Sketch the waveform of the voltage across the wiper with respect to one of the terminals when $\omega_0 = 2\pi \times 60$ rad/sec. Derive a general expression for all the frequencies present and evaluate the amplitude of the largest component present other than ω_0.

2.6 A source of sinusoidal signal can be represented by an ideal voltage source of 1-V peak in series with a 600 Ω resistance. This source is to be used to drive a single transistor class A amplifier coupled to a 10-Ω resistive load by means of a transformer. Design a suitable amplifier given that a 12-V dc source is available and the maximum possible power output is required. The β of the transistor is 100 and the lowest frequency of interest is 20 Hz. The dc current in the collector of the transistor is not to exceed 2 mA. Calculate the power gain in decibels assuming that 0 dB refers to the situation when the source is connected directly to the load.

2.7 Show that an amplifier with a voltage gain A and feedback factor β will oscillate when $|\beta A| = 1$. What other condition(s) have to be met for

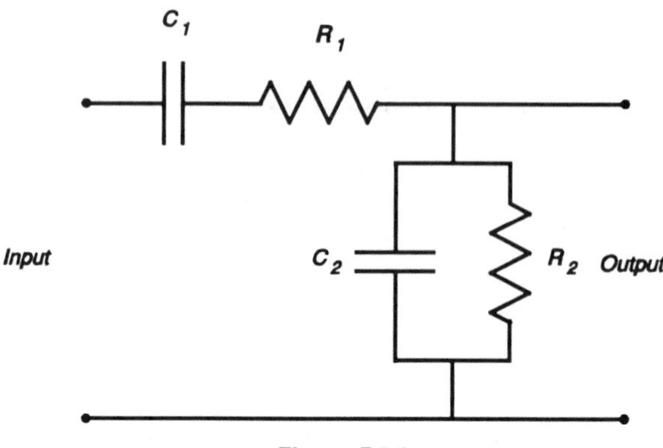

Figure P2.1.

sinusoidal oscillation to occur? The feedback circuit to be used in an oscillator is as shown in Figure P2.1

Assuming an ideal voltage amplifier, derive an expression for:

(a) The frequency of oscillation

(b) The gain of the amplifier

Evaluate (a) and (b) if $C_1 = C_2 = 1000$ pF and $R_1 = R_2 = 20$ kΩ.

2.8 The load of a class C amplifier is coupled through a lossless transformer turns ratio 10 : 1, and it is tuned to a frequency $250/2\pi$ kHz. The average power dissipated in the load at this frequency is 250 mW. The amplifier load is 50 Ω and the loaded Q factor of the primary with the load transferred to the secondary is 25. Calculate the following:

(a) The primary inductance

(b) The secondary inductance

(c) The capacitance required to tune the primary

(d) The frequency at which the output power will equal 125 mW.

(e) The -3-dB bandwidth of the amplifier

(f) The minimum dc power supply voltage required.

Describe how you would convert the amplifier into a frequency multiplier with a multiplication factor of 2. What conditions have to be met to obtain a larger multiplication factor?

3

THE AMPLITUDE-MODULATED RADIO RECEIVER

3.1 INTRODUCTION

The electromagnetic disturbance created by the transmitter is propagated by the transmit antenna and travels at the speed of light as described in Chapter 2. It is evident that if the electromagnetic wave encounters a conductor, a current will be induced in the conductor. How much current is induced will depend on the strength of the electromagnetic field, the size and shape of the conductor, and its orientation to the direction of propagation of the wave. The conductor will then capture some of the power present in the wave, and hence it will be acting as a receiver antenna. However, other electromagnetic waves emanating from all other radio transmitters will also induce some current in the antenna. The two basic functions of the radio receiver are as follows:

(1) To separate the signal induced in the antenna by the transmission, which we wish to receive from all the other signals present.

(2) To recover the "message" signal used to modulate the transmitter carrier.

3.2 THE BASIC RECEIVER

To separate the required signal from all the other signals captured by the antenna, we use a bandpass filter centered on the carrier frequency with sufficient bandwidth to accommodate the upper and lower sidebands but with a sufficiently high Q factor so that all other carriers and their sidebands are attenuated to a level where they will not cause interference. This is most easily achieved by using an LC tuned circuit whose resonant frequency is that of the carrier.

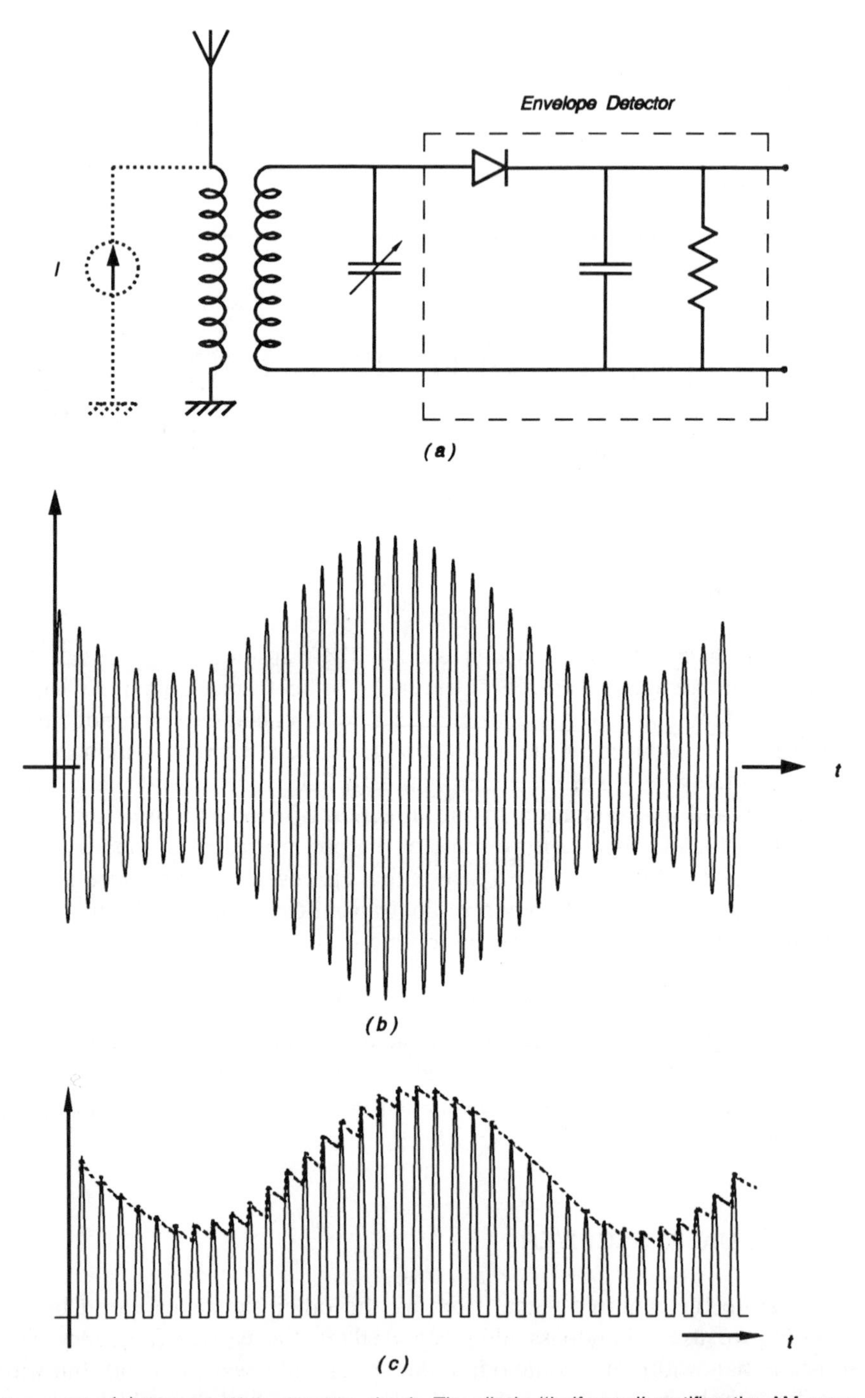

Figure 3.1. (*a*) The envelope detector circuit. The diode "half-wave" rectifies the AM wave and the *RC* time constant "follows" the envelope with a slight ripple. (*b*) The input signal to the envelope detector. (*c*) The output signal of the envelope detector. Note that when the voltage is rising the ripple is larger than when the voltage is falling. A longer time constant will help reduce the ripple, however, it will also increase the likelihood that the output voltage will not follow the envelope when the voltage is falling causing "diagonal clipping."

To recover the "message," we require a circuit that will follow the envelope of the amplitude of the carrier. Such a circuit is called an envelope detector, and it consists of a diode and a parallel *RC* circuit as shown in Figure 3.1*a*. The input signal to the circuit is most appropriately represented by an ideal current source connected to the primary of the transformer. This ideal current source represents all the currents induced in the antenna by all the radio stations broadcasting signals in free space. The signal is coupled to the parallel-tuned *LC* circuit which selectively enhances the amplitude of the signal whose carrier frequency is the same as the resonant frequency of the *LC* circuit. In Figure 3.1*b*, only the enhanced modulated signal is shown at the input of the envelope detector. Because the diode conducts only when the anode has a positive potential compared to the cathode, only the positive half of the signal appears across the output resistor. Because the capacitor is connected in parallel with the resistor, the capacitor must charge up to the peak value of the voltage when the diode conducts. When the input voltage is less than the voltage on the capacitor, the conduction is cut off and the capacitor starts to discharge through the resistor with the voltage falling-off exponentially. With the proper choice of time-constant *RC*, the output voltage waveform will have the form shown in Figure 3.1*c*. This waveform is essentially the envelope of the carrier signal with a ripple at a frequency equal to the carrier frequency. A low-pass filter can be used to remove the ripple.

The circuit shown in Figure 3.1*a* has been used with success as a practical receiver with the resistor *R* replaced by a high-impedance headphone. Needless to say, such a simple circuit has its limitations. The power in the circuit is supplied entirely by the transmitter, and naturally it is at a very low level, especially as the distance between the transmitter and the receiver increases. Secondly, the ability of the *LC* tuned circuit to suppress the signals propagated by all the other transmitters is limited, and therefore such a receiver will be subject to interference from other stations. These limitations can be overcome by using the superheterodyne configuration described below.

3.3 SUPERHETERODYNE RECEIVER SYSTEM DESIGN

The superheterodyne receiver takes the incoming radio-frequency signal whose frequency varies from station to station and transforms it to a fixed frequency (intermediate frequency). It is then easier to filter to eliminate interference and at the same time to provide some power gain or amplification of the desired signal.

A normal AM superheterodyne receiver diagram is shown in Figure 3.2. The antenna has induced in it currents from all the transmitters whose electromagnetic propagation reach it. The first step is to use an LC tuned radio frequency amplifier to enhance the desired carrier and its sidebands.

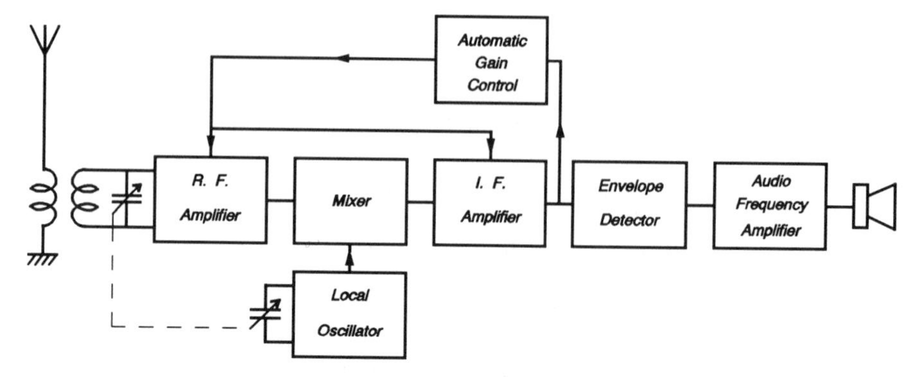

Figure 3.2. The block diagram of the superheterodyne receiver. The capacitor which tunes the radio frequency amplifier is mechanically ganged to the capacitor, which determines the frequency of the oscillator. In the normal AM receiver, the oscillator frequency is always 455 kHz above the resonant frequency of the radio frequency amplifier throughout the range of tuning.

The radio-frequency amplifier is tunable over the frequency for which the receiver is designed by varying the capacitor in the tuned circuit. This capacitor is mechanically coupled or "ganged" to another capacitor, which forms part of the local oscillator circuit. The local oscillator frequency and the frequency to which the radio frequency amplifier is tuned are chosen in such a way that, as the value of the ganged capacitors change, they maintain a fixed frequency difference between them. The outputs from the local oscillator and the radio frequency amplifier are used to drive the frequency changer or mixer. The frequency changer essentially multiplies the two inputs and producers a signal that contains the sum and difference of the input frequencies. Because of the fixed difference between the in-coming radio frequency and the local oscillator frequency, the difference frequency remains constant as the value of the ganged capacitor is changed. The output of the frequency changer is then fed into the intermediate frequency amplifier. The intermediate frequency amplifier is designed to select the difference frequency plus its sidebands and to attenuate all other frequencies present. Since the difference frequency is fixed (for domestic radios the intermediate frequency is 455 kHz) the filters required are relatively easy to design to have sharp cut-off characteristics. The output of the intermediate frequency amplifier, which then goes to the envelope detector, consists of the intermediate frequency and its two sidebands. The envelope detector removes the intermediate-frequency leaving the audio frequency signal which is then amplified by the audio frequency amplifier to a level capable of driving the loudspeaker. It is clear that there will be a very large difference between the signal from a powerful local radio station and a weak distant station. To help reduce the difference an automatic gain control is used to adjust the signal reaching the envelope detector to stay within predetermined values.

The most interesting signal processing step in the system takes place in the frequency changer or frequency mixer or simply the mixer [1]. There are two basic types of mixers: the analog multiplier and the switching types. The analog multiplier frequency changer simply multiplies the radio frequency signal and the local oscillator so that when the modulated carrier current is

$$i_m(t) = A(1 + k \sin \omega_{st})\sin \omega_c t \tag{3.3.1}$$

and the local oscillator signal is:

$$i_o(t) = B \sin \omega_L t \tag{3.3.2}$$

The output of the mixer is essentially the product of the two currents

$$i(t) = A(1 + k \sin \omega_s t)\sin \omega_c t \times B \sin \omega_L t \tag{3.3.3}$$

$$= \tfrac{1}{2}AB(1 + k \sin \omega_s t)\left[\cos(\omega_L - \omega_c)t - \cos(\omega_L + \omega_c)t\right] \tag{3.3.4}$$

$$= \tfrac{1}{2}AB\left[\cos(\omega_L - \omega_c)t - \cos(\omega_L + \omega_c)t \right.$$
$$\left. + k \sin \omega_s t \cos(\omega_L - \omega_c)t - k \sin \omega_s t \cos(\omega_L + \omega_c)t\right] \tag{3.3.5}$$

$$= \tfrac{1}{2}AB\left\{\cos(\omega_L - \omega_c)t - \cos(\omega_L + \omega_c)t \right.$$
$$+ \tfrac{1}{2}k\left[\sin(\omega_L - \omega_c - \omega_s)t + \sin(\omega_L - \omega_c + \omega_s)t\right]$$
$$\left. - \tfrac{1}{2}k\left[\sin(\omega_L + \omega_c - \omega_s)t + \sin(\omega_L + \omega_c + \omega_s)t\right]\right\} \tag{3.3.6}$$

The spectrum of Equation (3.3.6) is shown in Figure 3.3. It should be noted that this has been simplified for clarity. The product formation in Equation (3.3.3) is not a precise process and tends to create a large number of frequencies due to sub- and higher harmonics present in both the radio frequency and the local oscillator signals. The radio frequency and local oscillator signals are usually present in the output as well. It is important to keep all the unwanted signals outside the frequency band of the intermediate frequency or, failing that, to reduce their amplitude to a very low value.

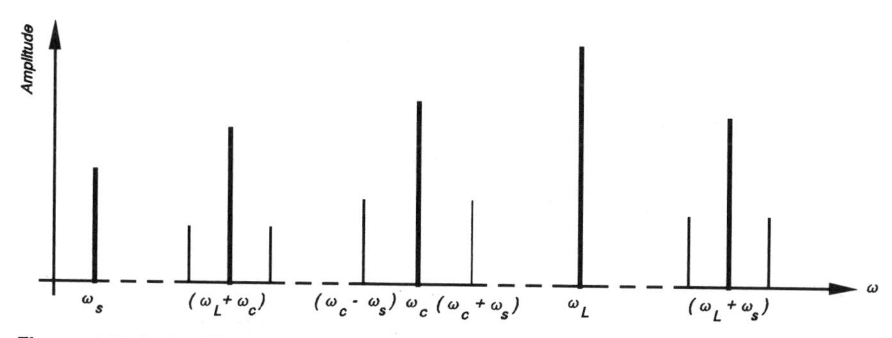

Figure 3.3. A simplified spectrum of the output from a frequency changer that uses a nonlinear device.

TABLE 3.1

	Radio frequency (kHz)	
	Low-frequency end	High-frequency end
In-coming signal, $f_c \pm f_s$	600 ± 5	1600 ± 5
Local oscillator, f_L	$600 + 455 = 1055$	$1600 + 455 = 2055$
Intermediate freq., f_k	455	455
Image frequency,[(a)] f_{im}	$1055 + 455 = 1510$	$2055 + 455 = 2510$
Output, intermediate frequency, A, $f_k \pm f_s$	455 ± 5	455 ± 5
Envelope detector, f_s	0 – 5	0 – 5

[a]The image frequency is the frequency of the unwanted signal which when combined with the local oscillator frequency will give the intermediate frequency. Normally, the radio-frequency amplifier should suppress the image frequency, but this may be difficult if the signal from the desired station is very weak and the image signal is very strong.

It can be seen that the mixing operation gives two additional carriers and their sidebands at frequencies corresponding to the sum $(\omega_L + \omega_c)$ and difference $(\omega_L - \omega_c)$ of the local oscillator and carrier frequencies. The required signal at the difference frequency (intermediate frequency) can now be filtered out by the intermediate frequency stage of the receiver. It should be noted that the mixing operation does not affect the sidebands. To clarify the changes in frequency that takes place as the signal proceeds through the system, the AM broadcast band (600–1600 kHz) is used as an example in Table 3.1.

The frequency changer or mixer presents two immediate problems: the choice of the local oscillator frequency and the design strategy of the mixer itself.

1. It can be seen from Table 3.1 that the local oscillator frequency has been chosen to be higher than the in-coming radio frequency signal. There is a very good reason for this. The ratio of the maximum to the minimum capacitance required to tune the local oscillator across the broadcast band is 3.79 when a higher local oscillator frequency is chosen. If the lower local oscillator frequency had been chosen, the ratio would have been 62.4. Such a variable capacitor would be difficult to manufacture with reasonable tolerance.

2. The mixing operation was treated earlier as an analog multiplication. However, the realization of a precise analog multiplier is a nontrivial problem. A crude analog multiplication can be achieved by using a device whose voltage-current characteristics are nonlinear. An ordinary p–n junction diode can be used to perform the task. The derivation of the output signal is similar to that given in Section 2.6.2 and will therefore not be repeated here.

The switching type of mixer uses a device such as a diode or transistor carrying a current proportional to the radio frequency signal and switches it from one state to another at the local oscillator frequency.

3.4 COMPONENTS OF THE SUPERHETERODYNE RECEIVER

3.4.1 Receiver Antenna

The AM receiver antenna can take many different forms, such as the ferrite bar found in most portable receivers, the whip antenna found on automobiles, and the outdoor wire consisting of several meters of wire strung between two towers. In general, the longer and higher the antenna is, the more likely it is that it will have a strong signal induced in it by the electromagnetic signals propagated by the transmitters. The level of signal induced in the antenna may vary from a few microvolts to a few volts depending on the proximity of the transmitter, its power output, the size of the receiver antenna, and its orientation to the transmitter. Because of the tremendous variation in the input signal, a fixed gain amplifier will very often either not provide enough signal to the frequency changer or will overload it and consequently generate a large number of undesirable frequencies. To ensure a reasonable reception of the largest number of broadcasting stations, the gain of the amplifier is controlled automatically by the in-coming signal—the weaker the signal, the higher the gain of the radio frequency amplifier.

The antenna signal is coupled by a radio-frequency transformer to the input of the radio frequency amplifier. The transformer, is made up of two coils each containing several turns of wire wound on a coil former, which may or may not have a ferrite core. The major consideration in the design of the transformer is that the primary inductance be sufficiently high to ensure that signals at the lowest frequency of interest are not unduly attenuated. Since the signal frequency can vary from 600 to 1600 kHz, the transformer is not tuned.

3.4.2 Low-Power Radio Frequency Amplifier

Since the amplifier input voltage is in microvolts and the signal to be delivered to the demodulator is usually in volts, the amplifier must have a high gain. A multistage amplifier is used to realize the necessary gain. Some of the stages of gain can be placed before the frequency changer, in which case they are referred to as the radio-frequency-amplifier stage, or after the frequency changer, in which case they are called the intermediate-frequency-amplifier stage. It is usual to design the radio frequency stage for a modest gain and the intermediate frequency stage for the high gain. Both the radio and intermediate frequency amplifiers are narrowband amplifiers. This is evident from calculating the Q factor for the two types of amplifiers. Considering that the normal bandwidth of the AM radio is 0–5 kHz, both radio- and intermediate frequency amplifiers have to have a bandwidth of at least 10 kHz. The Q factor of the radio frequency amplifier at the low end of the broadcast band (600 kHz) is 60 and at the high end (1600 kHz) is 160. The Q factor of the intermediate frequency amplifier (center frequency 455 kHz) is 45. However, operation of the radio frequency amplifier with such a high Q factor will cause serious tracking problems with the local oscillator and will also lead to excessive attenuation at the sideband edges. For practical purposes, a Q factor of about 10 is used in the radio frequency amplifier leaving the major part of the filtering problem to the intermediate frequency stage. The design of the intermediate frequency filter about a fixed frequency is a much easier process and can be achieved with greater precision than in the radio frequency stage, where the center frequency of the bandpass filter changes when the tuning capacitor is changed. In spite of the difference in Q factor, the radio and intermediate frequency amplifiers have enough similarities for the same general principles to be used in their design.

The wide variation of the radio frequency input signal level and the need for automatic gain control was discussed earlier. It is usual to amplify the in-coming radio frequency signal by a fixed amount to derive a control signal for the gain of a subsequent variable gain amplifier. A typical fixed gain radio frequency amplifier is shown in Figure 3.4. Although a bipolar transistor is shown, a field-effect transistor can be used. The collector load is an LC tank circuit, in which the capacitance is variable. The variable capacitance is mechanically ganged to the capacitance, which controls the frequency of the

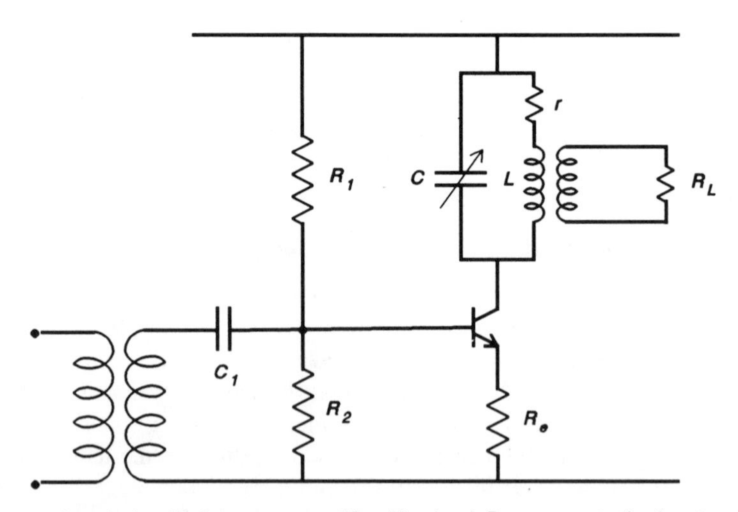

Figure 3.4. A typical radio frequency amplifier. The load R_L represents the input resistance (impedance) of the circuits driven by the amplifier.

local oscillator so that as the capacitance is changed, the resonant frequency of the LC tank circuit tracks the local oscillator frequency with a constant difference equal to the intermediate frequency.

It can be seen that the circuit in Figure 3.4 bears a striking resemblance to the frequency multiplier circuit shown in Figure 2.20. The difference is that the frequency multiplier operates in class C, whereas the radio frequency amplifier operates in class A.

The load driven by the amplifier may be coupled to the collector circuit by a transformer, in which case the inductor in the collector circuit becomes the transformer primary. The load may also be coupled by a capacitor. In both cases, the load can be represented by a resistance R_L in parallel with the tuned circuit. To simplify the analysis of the circuit, the winding resistance r in series with the inductance is transformed into an equivalent shunt resistance (refer to Figure 2.50) R_p, where

$$R_p = \frac{\omega^2 L_p^2}{r} \tag{3.4.1}$$

$$L_p = L \tag{3.4.2}$$

when $Q \gg 1$.

The amplifier load R_L combined in parallel with R_p is now the resistive part of the collector load. The new equivalent circuit is shown in Figure 3.5. where

$$R_q = R_p \parallel n^2 R_L \tag{3.4.3}$$

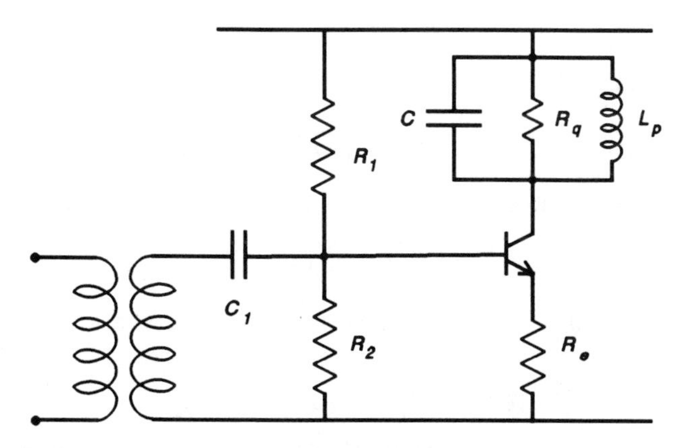

Figure 3.5. The amplifier shown in Figure 3.4 with the transformer load transferred to the primary and combined with the winding resistance r.

It can be seen from Figure 3.5 that:

(1) The emitter resistor R_e has not been bypassed to ground with a capacitor.
(2) At the frequency of resonance, the parallel L and C in the collector circuit will behave like an open circuit. The equivalent collector load is R_q.
(3) Because the inductor is connected directly between $+V_{cc}$ and the collector, the dc voltage on the collector is $+V_{cc}$.

The major advantage of not bypassing R_e is that the gain of the amplifier is determined by the ratio of the collector-to-emitter load impedance, which in this case is R_q/R_e. It is essentially independent of the transistor parameters such as current gain and transconductance. The design steps are illustrated in the following example.

Example 3.4.1 Low-Power Radio Frequency Amplifier. The antenna of an AM radio receiver (600 to 1600 kHz) supplies 100 mV peak to the input of the radio frequency amplifier when the modulation is a sinusoid, the modulation index is unity and the radio frequency frequency is 600 kHz. The dc supply voltage is +6 V and the required gain is 20. The amplifier load represented by the input impedance of the automatic gain control circuit is 10 kΩ resistive and it is capacitively coupled. The variable capacitor used in the tuned circuit (and mechanically coupled to the capacitor used in the local oscillator) has a maximum value of 250 pF and a minimum value of 25 pF. The Q factor of the coil is expected to be approximately 50 and the current gain of the transistor is 100.

Solution. The inductance of the tuning coil is given by

$$L = 1/(\omega^2 C)$$

When $\omega = 2\pi \times 600 \times 10^3$ and $C = 250$ pF,

$$L = 281\ \mu\text{H}$$

When $\omega = 2\pi \times 1600 \times 10^3$ the capacitance required to tune the amplifier

$$C = 35.2\ \text{pF}$$

The combination of L and C can be used to tune the amplifier to any frequency in the AM broadcast band.

The winding resistance of the coil

$$\begin{aligned} r &= \omega L/Q \\ &= 21.2\ \Omega \end{aligned}$$

The equivalent parallel resistance

$$R_p = \omega^2 L_p^2 / r$$
$$= 52.9 \text{ k}\Omega$$

Combining R_p with the load resistance of 10 kΩ gives

$$R_q = 8.41 \text{ k}\Omega$$

The loaded Q of the collector circuit

$$Q_L = R_q/(\omega L)$$
$$= 7.94$$

The relatively low Q should ensure that the sideband "edges" are not subject to severe attenuation. At the resonant frequency, the parallel LC circuit in the collector behaves like an open circuit. The equivalent collector load is therefore R_q. The emitter resistance

$$R_e = R_q/(\text{gain})$$
$$= 420 \ \Omega$$

The output voltage = 100 mV × 20 = 2.0 V (peak). The current drawn by the 10-kΩ load is 0.2 mA (peak).

To ensure that the amplifier is capable of supplying 0.2 mA ac current to the load, the dc current in the collector may be set at ten times the load current, that is $I_c = 2$ mA. The dc voltages at the emitter and the base are then $V_e = 0.84$ and $V_b = 1.54$ V, respectively. The dc voltage on the collector is still 6 V. It is clear that the amplifier will go into saturation when the collector voltage drops to a minimum value of 1.34 V ($V_e + 0.5$) and to cutoff when the collector voltage is 12 V. Because the collector signal is only ± 2 V about a quiescent value of 6 V, there is no danger of clipping at the collector.

The dc base current

$$I_b = I_c/\beta$$
$$= 20 \ \mu\text{A}$$

The values of R_1 and R_2 are chosen so that a dc current of $10 \times I_b$ will flow in the chain but with a voltage of 1.54 V at the base of the transistor. This gives

$$R_1 = 22.3 \text{ k}\Omega$$
$$R_2 = 7.7 \text{ k}\Omega$$

The coupling capacitor is chosen so that, at the lowest frequency of interest, its reactance is negligibly small compared to the load.

3.4.3 Frequency Changer or Mixer

Two distinct approaches can be used in the design of a mixer. The first is based on an analog multiplication of the radio frequency and the local oscillator signals. The second uses the local oscillator signal to switch the radio frequency signal to positive and negative values. In this case, the local oscillator must produce a squarewave.

3.4.3a The Analog Mixer. As discussed earlier, a crude analog multiplication can be achieved by using a nonlinear device such as a p–n junction diode, which can be approximated by

$$i = a_1 v + a_2 v^2 + \cdots \tag{3.4.4}$$

A mixer using diodes produces an output signal with considerable loss. Various schemes exist, some employing several diodes with a single ended or differential output.

If an "active" mixer is used, considerable gain can be obtained in the mixing process [5]. The preferred analog active mixer uses a dual-gate metal-oxide semiconductor field effect transistor (MOSFET). The advantages of this design includes a lower power requirement from the local oscillator and improved isolation between the local oscillator and the receiver antenna. The isolation between the local oscillator and the antenna ensures minimum radiation of the local oscillator signal and hence minimizes interference with other electronic equipment.

To understand the design process, it is necessary to begin with the drain current (i_D) and drain-to-source voltage (v_{DS}) characteristics of the MOSFET. A typical n-channel depletion mode MOSFET is shown in Figure 3.6. It can be seen that the characteristics are similar to those of a BJT except that the drain current is controlled by the gate-to-source voltage. An elementary common-source amplifier is shown in Figure 3.7. Applying KVL to the drain current path,

$$V_{DD} = i_D R_D + v_{DS} \tag{3.4.5}$$

or

$$i_D = (1/R_D)V_{DD} - (1/R_D)v_{DS} \tag{3.4.6}$$

When Equation (3.4.6) is plotted on the field effect transistor (FET) characteristics shown in Figure 3.6, it gives a straight line with a slope of $(-1/R_D)$, an intercept on the x axis given by v_{DS} equal to V_{DD} and on the y axis $i_D = V_{DD}/R_D$. This is the load line, which describes the behavior of the amplifier.

When designing an amplifier, it is necessary to select a bias point, V_{GS}, along the load line and the input signal, v_{gs}, so that the device will remain in

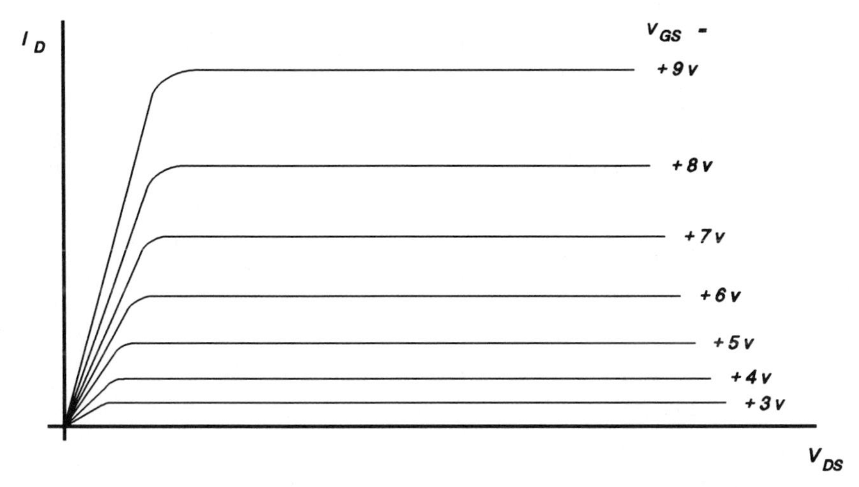

Figure 3.6. Typical characteristics of an n channel, enhancement-mode MOSFET.

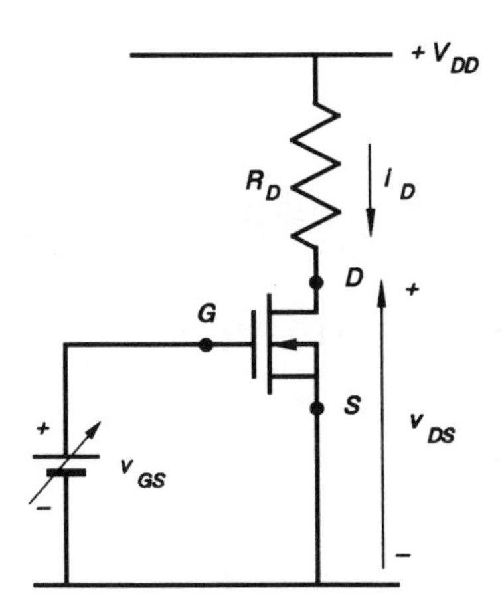

Figure 3.7. Typical biasing arrangement for the common-source MOSFET amplifier.

the "active" region. This is achieved by ensuring that the device is biased above its threshold, V_{th}. Then

$$v_{GS} = V_{GS} + v_{gs} \tag{3.4.7}$$

and the relationship between i_D and v_{GS} can be approximated by

$$i_D = I_{DSS}(1 - v_{GS}/V_{th})^2 \tag{3.4.8}$$

where the V_{th} and I_{DSS} are defined in Figure 3.8. Substituting Equation

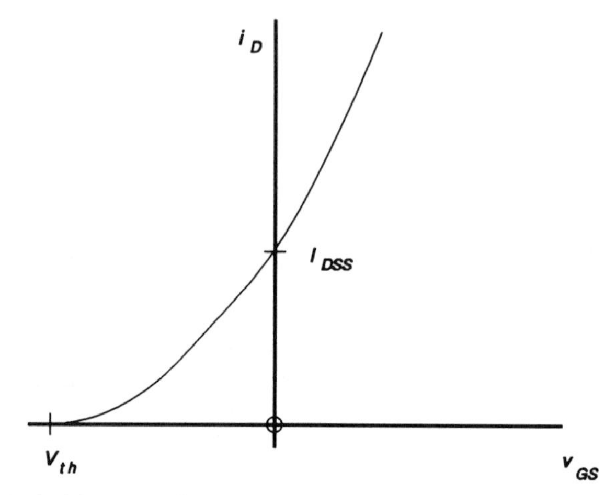

Figure 3.8. A typical $i_D - v_{GS}$ characteristic of an *n*-channel depletion-type MOSFET showing the threshold voltage, V_{th} and the saturated drain-to-source current, I_{DSS}.

(3.4.7) into Equation (3.4.8) gives

$$i_D = I_{DSS}\left[\left(1 - \frac{V_{GS}}{V_{th}}\right)^2 - 2\left(1 - \frac{V_{GS}}{V_{th}}\right)\left(\frac{v_{gs}}{V_{th}}\right) + \left(\frac{v_{gs}}{V_{th}}\right)^2\right] \quad (3.4.9)$$

The first term $(1 - V_{GS}/V_{th})^2$, represents the dc component of the drain current. The second term $2(1 - V_{GS}/V_{th})/(v_{gs}/V_{th})$, is an ac current proportional to the input voltage and represents the normally desired output. The third term $(v_{gs}/V_{th})^2$, represents a nonlinearity, which is normally undesirable. In terms of designing a mixer, however, this is the desired output. The relative value of this term can be increased by making the input signal v_{gs} large. However, since Equation (3.4.8) is an approximation, making v_{gs} too large can produce spurious signals, which may interfere with the required signal. A more practical version of the MOSFET amplifier is shown in Figure 3.9. The MOSFET used in the circuit is an *n*-channel enhancement device, and the resistive chain R_1 and R_2 are chosen to hold the gate at a specific potential above that of the source. R_s is used partly to stabilize the dc bias point and partly to reduce the dependence of the gain on the parameters of the device. In general, semiconductor device parameters vary widely from one device to another, and it is necessary to build some controls into the design. When semiconductor devices are used in the design of circuits whose specifications must be held to very tight tolerances, it is a good idea to use design strategies that rely on ratios of passive components, such as resistances, than on the values of the device parameters. In this case it can be shown that the gain of the amplifier is equal to the ratio of the drain impedance Z_D to the source resistance R_s.

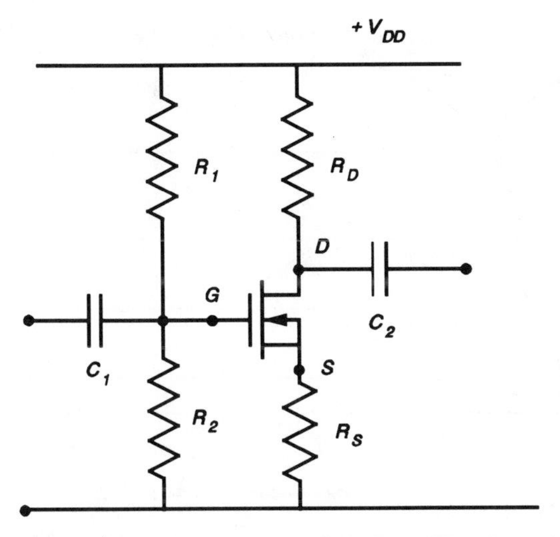

Figure 3.9. A more practical version of the MOSFET amplifier shown in Figure 3.7.

Because the mixer has two input signals of different frequencies, several problems such as frequency "pulling" and local oscillator feedthrough to the antenna can be avoided by ensuring that the two sources are well isolated from each other. It is possible to achieve a high level of isolation by using a dual-gate MOSFET. The design process is best illustrated by an example.

Example 3.4.2 The Mixer. Design a mixer for an AM radio using the dual-gate n-channel depletion MOSFET whose characteristics are given in Figure 3.10. The following are specified:

(1) Supply voltage, $V_{DD} = 12$ V
(2) Drain bias current, $I_D = 5$ mA
(3) Primary inductance of the drain transformer, $L_p = 250\ \mu$H
(4) Center frequency of the output intermediate frequency = 455 kHz
(5) -3-dB bandwidth, 20 kHz
(6) Transformer turns ratio, 10:1

What is the value of the resistive load that the mixer must see, assuming that both the primary and secondary winding resistances are negligibly small?

Solution. A suitable circuit for the mixer is as shown in Figure 3.11. The capacitance required to tune the drain to 455 kHz (intermediate frequency) is

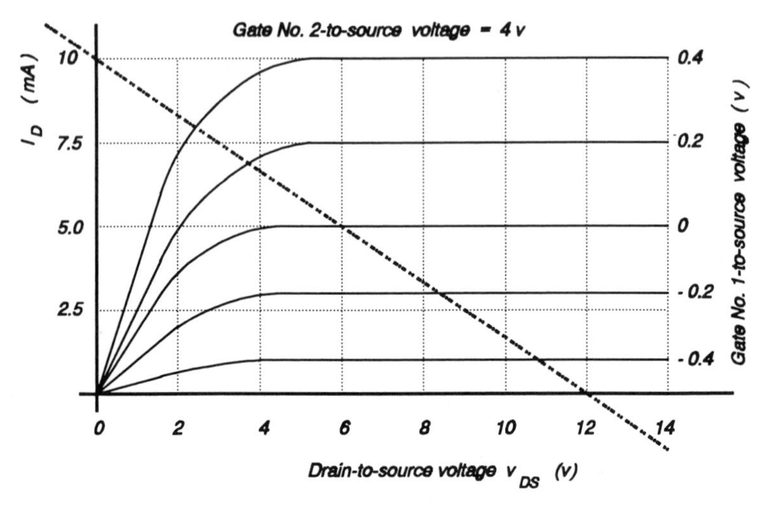

Figure 3.10. The drain characteristics of the dual-gate *n*-channel MOSFET used in Example 3.4.2.

given by

$$\omega^2 = \frac{1}{L_p C_p}$$

$$C_p = \frac{1}{\omega^2 L_p} = \frac{1}{(2\pi \times 455 \times 10^3)^2 \times 250 \times 10^{-6}}$$

$$= 489 \text{ pF}$$

The bandwidth Δf is related to the center frequency f_0 by

$$Q_0 = \frac{f_0}{\Delta f} = \frac{455 \times 10^3}{20 \times 10^3} = 22.75$$

The resistive load transferred to the primary $n^2 R_L$ will be in parallel with L_p. Q_0 for a parallel LR circuit is given by

$$Q_0 = \frac{n^2 R_L}{\omega L_p} = 22.75$$

$$n^2 R_L = 2\pi \times 455 \times 10^3 \times 250 \times 10^{-6} \times 22.75 = 16.26 \times 10^3$$

$$R_L = 162\ \Omega$$

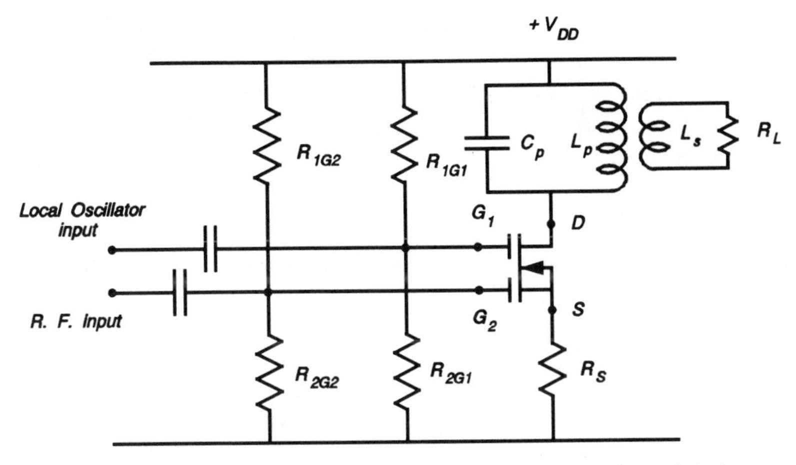

Figure 3.11. A typical MOSFET mixer using the dual-gate *n*-channel device.

From the device characteristics given in Figure 3.10, locate $V_{DS} = V_{DD} =$ 12 V and using a straight edge pivoted at this point determine a point along the line given by $I_D = 5$ mA that will give a wide dynamic range for the drain current. A good point is given by the intersection of $V_{DS} = 6$ V and $I_D =$ 5 mA. The load line can now be drawn in as shown. From the slope, the required load is 1.2 kΩ. In a common-source amplifier, this load would normally be connected in series with the drain. However, from the dc point of view, the drain is connected to V_{DD} by a short-circuit (through the inductor L_p). The device can be correctly biased by connecting the 1.2 kΩ resistor in series with the source, that is, choose

$$R_s = 1.2\ \text{k}\Omega$$

With a drain current of 5 mA, the voltage of the source will be

$$V_s = 1.2 \times 10^3 \times 5 \times 10^{-3} = 6.0\ \text{V}$$

From Figure 3.10, it can be seen that the required gate-source voltage on gate 1 is 0 V. Gate 1 must therefore be biased at 6.0 V by the resistive chain, so that

$$R_{1G1} = R_{2G1}$$

Since no current flows into the gate, the value of the two resistances is arbitrary and can be made as large as practicable. Let

$$R_{1G1} = R_{2G1} = 100\ \text{k}\Omega$$

From Figure 3.10, it can be seen that the required gate-source voltage on gate 2 is 4.0 V. Because the source is biased at 6.0 V, gate 2 must be biased at 10 V. Again no current flows into gate 2, and therefore the resistance can be made as large as practicable. However

$$R_{1G2}:R_{2G2} = 2:10$$

Choosing $R_{2G2} = 100\ \text{k}\Omega$ makes $R_{1G2} = 20\ \text{k}\Omega$. The coupling capacitors, C_1 and C_2, are chosen so that they present negligible impedance to the radio frequency and local oscillator, respectively.

The drain tank circuit is tuned to resonate at the intermediate frequency; therefore, the drain load is

$$n^2R_L = 16.26 \times 10^3\ \Omega$$

The gain of the stage is approximately equal to

$$n^2R_L/R_s = 13.6 = 22.6\ \text{dB}$$

This is not the same as the mixer gain, which is defined as

$$10 \log_{10}(\text{if power output/rf power input})\ \text{dB}$$

In a practical circuit, the relative signal level of both the radio frequency (rf) and the local oscillator will be adjusted to optimize the intermediate frequency (if) signal.

3.4.3b The Switching-Type Mixer. The circuit diagram of one of the simplest switching-type mixers is shown in Figure 3.12 [2]. The radio frequency signal V_s is a sinusoid. The frequency of the local oscillator V_L, which

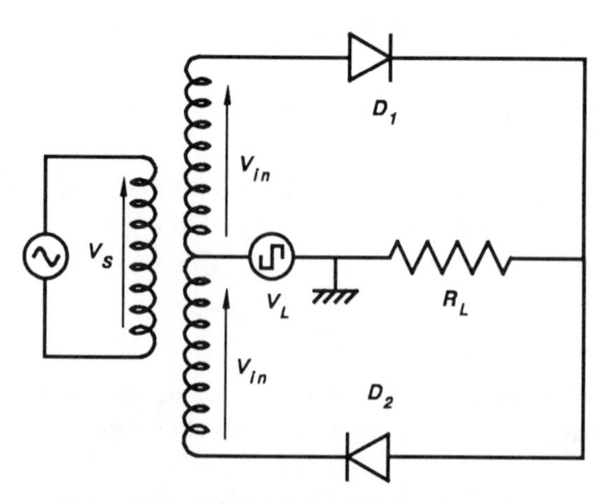

Figure 3.12. The circuit of a switching-type mixer.

is higher than the radio frequency, is a squarewave. The squarewave is defined as

$$g(t) = 1 \text{ for } 0 > t > T/2 \tag{3.4.10}$$

$$g(t) = -1 \text{ for } T/2 > t > T \tag{3.4.11}$$

Assuming ideal diodes and that V_L is larger than V_{in}, then, when $V_L > 0$, D_1 conducts and D_2 is off

$$V_0 = V_L + V_{in} \tag{3.4.12}$$

and, when $V_L < 0$, D_2 conducts and D_1 is off

$$V_o = -(V_L + V_{in}) \tag{3.4.13}$$

The output voltage is then

$$V_o = V_L + V_{in} \times g(t) \tag{3.4.14}$$

This is evident from an examination of Figure 3.13.

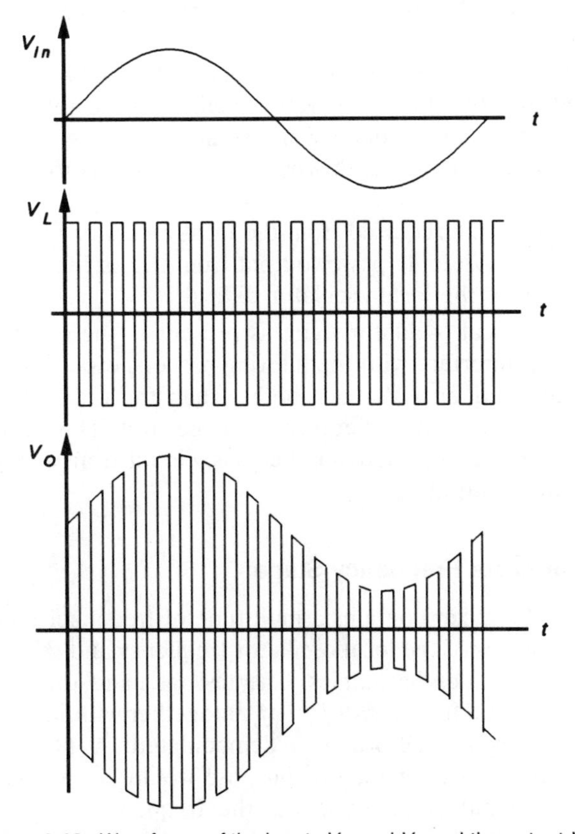

Figure 3.13. Waveforms of the inputs V_{in} and V_L and the output V_o.

The squarewave $g(t)$ can be expressed in terms of its Fourier components as

$$g(t) = \frac{4}{\pi} \sum_{0}^{\infty} \frac{\sin(2n + 1)\omega_L t}{(2n + 1)} \tag{3.4.15}$$

But

$$V_{in} = A \sin \omega_c t \tag{3.4.16}$$

Therefore,

$$V_{in} \cdot g(t) = \frac{2A}{\pi} \sum_{n=0}^{\infty} \frac{\cos[(2n + 1)\omega_L - \omega_c]t - \cos[(2n + 1)\omega_L + \omega_c]t}{(2n + 1)} \tag{3.4.17}$$

The output of the mixer consists of the local oscillator frequency and an infinite number of sums and differences of the local oscillator harmonics and the radio frequency. The desired frequency components can be filtered in the intermediate frequency stage that follows the mixer.

If the AM carrier equation is used in place of Equation (3.4.16), it can be demonstrated that the mixing operation maintains the relationship between the desired intermediate frequency and its sidebands.

The major disadvantages of this type of mixer are as follows

(1) The large signal required from the local oscillator to switch the diodes calls for considerable power output from the oscillator, and this makes the design of the local oscillator difficult.
(2) The large local oscillator signal is present in the output of the mixer, and it can interfere with the filtering process, especially when the local oscillator frequency is much higher than the radio frequency. Hence the desired sum and difference are close to it. This part of the output can be removed by changing the basic circuit from a single-ended to a differential output.

3.4.4 Intermediate Frequency Stage

The output of the mixer contains a multitude of frequencies made up of the sums and differences of the local oscillator frequency and the radio frequency signal and their various harmonics. The task at hand then is to select the signal centered at the frequency $(f_{lo} - f_{rf})$ together with its sidebands and to amplify it if necessary, before it is demodulated. A filter is required to achieve this. An ideal filter for this purpose would be one with rectangular characteristics—a flat response across the frequency band infinitely steep "skirts," with infinite attenuation beyond. A practical filter will, of course, be much less exotic than this.

The only type of frequency-selective circuit discussed so far has been the *LC* tuned circuit. A single tuned *LC* circuit can have steep skirts and high attenuation for out-of-band signals when the Q factor is high, but a high Q factor also means that the in-band frequency is very narrow. The design of a filter that can select the intermediate frequency and its sidebands and suppress to an acceptable level all the other frequency components present in the output of the mixer is beyond the scope of this book. The interested reader will find a short list of sources for more information in the bibliography.

In general, a filter is placed between a resistive source and a resistive load, which may or may not have the same value. For frequencies in the passband, the filter "matches" the source to the load, so that the reflection of the signal from the load is minimal; that is, maximum power is transferred to the load. In the stopband, the filter input presents such a severe mismatch to the source that most of the signal power is reflected with very little reaching the load.

Filters can be classified as follows:

1. Passive LC Filters These filters are made up of only inductors and capacitors, which may be assumed to be lossless. In general, the closer the filter characteristics are to the ideal, the more *L*s and *C*s are required. Discrete *LC* filters are used from frequencies as slow as 20 Hz and as high as 500 MHz. The low-frequency limit is set by the low Q factors of the inductors and at the high-frequency end by the circuit strays of both L and C. Passive *LC* filters can be used at frequencies as high as 40 GHz, but the components have to be considered to have distributed parameters. Inductors and capacitors are then "replaced" by open or short-circuited transmission lines and tuned *LC* circuits by resonators. The physical appearance of the components of these very-high-frequency filters bear no resemblance to *L*s and *C*s.

2. Crystal Filters The very high Q factors that can be obtained from crystals offer the filter designer the possibility of high selectivity and steep skirt gradients. However, crystal parameters are in general not under the direct control of the filter designer and consequently have not been very popular. Crystal filters have an upper frequency limit of a few megahertz.

3. Active *LC* Filters These filters are made up of *LC* sections separated by amplifiers. The advantages gained are that the filter can have an overall gain instead of a loss and the buffering action of the amplifiers between various segments of *L*s and *C*s can reduce the amount of interaction between them, making the tuning of the filter considerably easier. The bandwidth of the amplifier used can limit the range of application of these filters to an upper range of approximately 300 MHz.

4. Active *RC* Filters These filters are made up of resistors, capacitors, and operational amplifiers. The introduction of integrated circuits prompted the development of these filters. Unfortunately, they can be used only at low frequencies with an upper limit of a few megahertz.

5. Digital Filters These filters have become practical since the development of cheap and fast microcomputers with large memories. The signal is sampled by an analog-to-digital converter. The samples are converted into a digital code and multiplied by a function of the desired output characteristics. The resulting signal is fed into a digital-to-analog converter to give an analog output. Even with the fastest microcomputers digital filters are limited to an upper frequency of about 50 kHz.

6. Mechanical Filters In these filters, the electrical signal is converted into a mechanical vibration by a magnetostrictive transducer. A number of mechanical resonators in the form of discs, plates, and rods determine which frequencies will be propagated through the device to reach the output and which will not. At the output, another transducer converts the vibration back into an electrical signal. Mechanical filters normally operate in the frequency range of 50 to 600 kHz. These filters have the attraction of being quite accurate and yet inexpensive.

7. Surface-Acoustic-Wave Filters These filters belong to a more general class called *transversal filters*. In surface-acoustic-wave (SAW) filters, two transducers are fabricated at opposite ends of a highly polished piezoelectric material such as quartz or lithium niobate. When an electrical signal is applied to the input transducer, the material changes its physical shape, and, at the appropriate frequency, will cause a traveling wave to be propagated on the surface of the material. At the output, the traveling wave is converted back into a electrical signal. The mechanical shape, size, and placing of the metallization on the piezoelectric material determine which frequencies are propagated and which are attenuated. SAW filters normally operate in the 20 to 500 MHz range.

8. Switched Capacitor Filters These filters have been made possible by the ease with which metal-oxide semiconductor (MOS) technology can be used to realize capacitors, switches, and operational amplifiers on the same silicon chip. Two switches and a capacitor can be used to simulate a resistor. It is then possible to construct equivalent circuits for most of the filter structures used in active *RC* filters. Switched capacitor filters are generally used where accurate filtering is not required. However, the characteristics of switched capacitors continue to improve as they continue to be the subject of intense research interest. The need to switch the capacitors at a much higher frequency (clock frequencies are 30 to 40 MHz) than the highest frequency present in the signal, limits these filters to operation below 3 MHz.

3.4.5 Automatic Gain Control

The function of the automatic gain control is to ensure that the signal reaching the demodulator is sufficiently high and constant for efficient demodulation. It does this by sensing the level of the signal at the input to the modulator and adjusting the gain of a variable gain amplifier to keep the

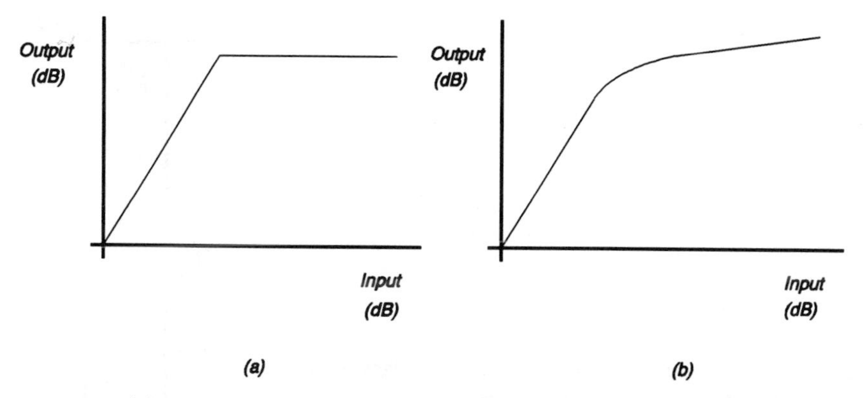

Figure 3.14. (*a*) The ideal input–output characteristics of an AGC circuit. (*b*) The input–output characteristics of a practical AGC circuit.

level constant. In practice it is not possible to boost all signals to a constant level and in any case it is undesirable to amplify noise to the constant level when it is all that is available. The desirable characteristics of an automatic gain control (AGC) circuit is as shown in Figure 3.14*a*. Below the "knee," the signal is amplified by a constant factor. Above the "knee," the output is kept constant. A practical characteristic is shown in Figure 3.14(*b*), where the "knee" is rounded and the output continues to rise, but at a limited rate.

The AGC subsystem consists of a variable gain amplifier whose gain is controlled by a voltage or current derived from the output signal. A block diagram is given in Figure 3.15. The rectifier produces a dc signal which is proportional to the signal that appears at the output of the variable gain amplifier. The dc signal is fed back to the variable gain amplifier and attempts to keep the output signal constant above a predetermined level. It is usual to place the variable gain amplifier before the mixer so that the radio frequency signal reaching the mixer does not vary too widely.

A suitable amplifier for the AGC circuit is the four-quadrant analog multiplier discussed in Section 2.6.3. In the scheme shown in Figure 3.15, the

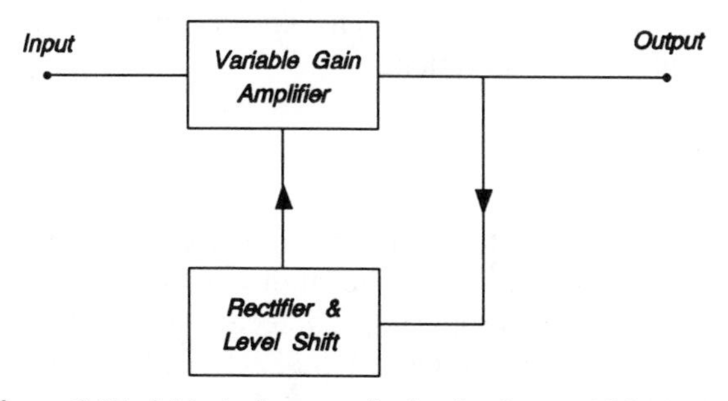

Figure 3.15. A block diagram of a feedback-type AGC circuit.

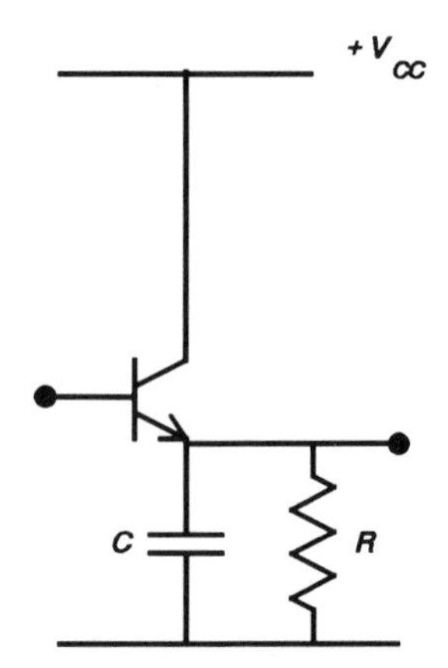

Figure 3.16. A circuit that provides a "fast-attack" and "slow-release" for the AGC control voltage.

four-quadrant amplifier will be part of the radio frequency stage with the radio frequency signal applied to one of its two inputs. The second input is a dc signal derived from the rectified and smoothed intermediate frequency signal, so that, when the intermediate frequency signal is high, the constant of multiplication is low and, when the signal is low, the constant is high.

Because the four-quadrant analog multiplier was discussed at length earlier, only the rectifier and its associated time constant will be discussed. A common characteristic of all AGC circuits is that they have a fast "attack" time and a somewhat slower "release" time. This means that the system can capture sudden increases in signal level but take a longer time to adjust the gain upward for low signals. A suitable rectifier and time constant are shown in Figure 3.16.

The first positive-going signal that appears at the base of the BJT causes it to switch on and very rapidly charge the capacitor C with its emitter current which is much larger than the base current of the BJT. The dc signal required to control a sudden increase in signal is therefore produced very quickly. When the base voltage starts to drop, the charged capacitor holds up the voltage of the emitter, and the falling base voltage ensures that the BJT is cut-off. The discharge of the capacitor is then controlled by the resistor in parallel with it. The attack time can be slowed down by connecting a small resistor in series with the capacitor. The correct attack and release times have to be determined by experiment.

3.4.6 Demodulator

The output to the demodulator is a carrier of frequency 455 kHz with an amplitude envelope determined by the audio signal. The circuit used for demodulation is therefore the envelope detector shown in Figure 3.1. The operation of the circuit was discussed in Section 3.2. An example of an envelope detector follows.

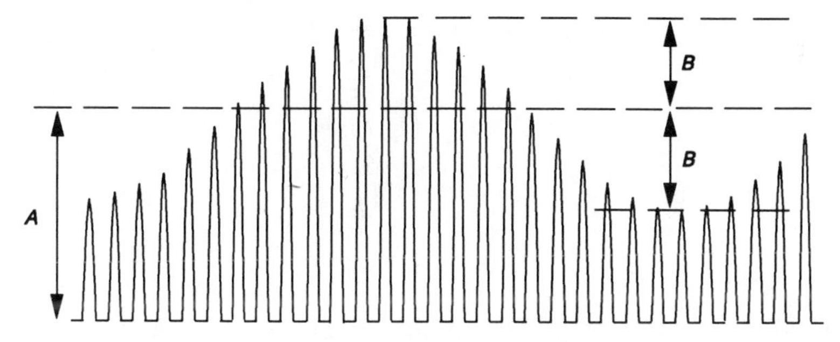

Figure 3.17. The output voltage waveform of the envelope detector used in Example 3.4.3.

Example 3.4.3 Envelope Detector. The amplitude of the signal applied to the input of an envelope detector is 4 V peak when the modulation index is zero and the intermediate frequency is 455 kHz. Choose a suitable time constant for the detector to avoid diagonal clipping when the modulation index is 0.8 and the highest audio frequency component in the input signal is 10 kHz. Assume that the diode is ideal.

Solution. The output waveform with the capacitor removed will be as shown in Figure 3.17. Because $B/A = 0.8$, $B = 3.2$ V peak. The equation for the envelope is

$$v_s(t) = B \cos \omega_s t$$

where $\omega_s = 2\pi \times 10 \times 10^3$ rad/sec. The slope of the envelope

$$dv_s/dt = -B\omega_s \sin \omega_s t$$

From the diagram, it is clear that the maximum slope occurs at $t = T/4$, when $\sin \omega_s t = 1$. Hence,

$$dv_s/dt|_{\max} = -B\omega_s$$

When the capacitor is connected to the circuit and the diode is not conducting, the output voltage decays according to the equation

$$v_c(t) = Ve^{-t/\tau}$$

The slope of the decaying capacitor voltage is

$$dv_c/dt = -(1/\tau)Ve^{-t/\tau}$$

From the diagram, it is clear that when $t = T/4$ or when $\omega_s t = \pi/2$, $V = A$ and

$$dv_c/dt = -(1/\tau)Ae^{-t/\tau}$$

The maximum value occurs at $t = 0$

$$dv_c/dt|_{\max} = -A/\tau$$

To avoid diagonal clipping, the capacitor voltage must decay faster than the envelope voltage, that is,

$$\begin{aligned} |-A/\tau| &\geq |-B\omega_s| \\ \tau &= A/(B\omega_s) \\ &= 4.0/(3.2 \times 2\pi \times 10 \times 10^3) \\ &= 19.89 \times 10^{-6} \text{ sec} \end{aligned}$$

Let $R = 10$ kΩ then $C = 1989$ pF.

3.4.7 Audio Frequency Amplifier

The audio frequency amplifier was discussed in Section 2.7. A few minor changes may be necessary in the design of the final stage of the amplifier so that it can drive a loudspeaker efficiently.

3.4.8 Loudspeaker

So far, the audio frequency signal is in an electrical form. To convert it to an acoustic signal, an electrical-to-acoustic transducer is required. This can take two basic forms: the loudspeaker or the headphone. The more common of the two devices is the loudspeaker [3, 4]. There are two basic types of loudspeakers: the direct radiation and the horn loudspeaker. The diaphragm of the direct radiation loudspeaker is coupled directly to the air, whereas that of the horn loudspeaker is through a horn. The horn loudspeaker is generally more efficient due to the better matching between its diaphragm and the air provided by the horn. However, it is usually large and expensive. A direct radiation loudspeaker by contrast, has a simple construction, occupies a small space, and has a relatively uniform response over a moderate frequency band. The design of loudspeakers is a highly specialized field outside the scope of this book. However, a qualitative description of the behavior of loudspeakers is a subject that should be of interest to radio engineers.

A cross-section of the direct radiation loudspeaker is given in Figure 3.18. It consists of a paper cone, the apex of which is mechanically connected to a

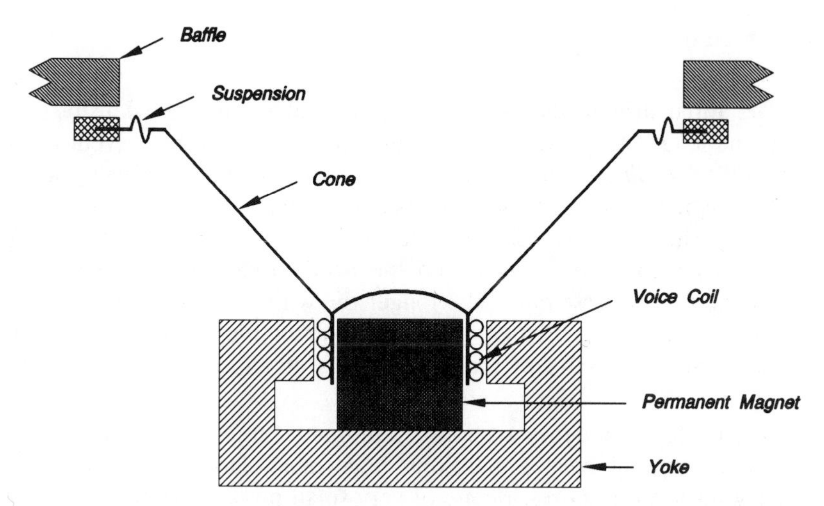

Figure 3.18. A cross-sectional view of a typical loudspeaker.

cylinder on which the voice coil is wound. The voice coil is located in a magnetic field provided by a permanent magnet. The outer rim of the cone is corrugated to permit the cone to move along its axis with the minimum of mechanical resistance. The whole cone may be viewed as the diaphragm of the loudspeaker. The corrugation is referred to as the suspension of the cone. The outer edge of the suspension is fastened to the frame of the loudspeaker and the frame is secured to the baffle.

An ideal loudspeaker should produce a uniform sound pressure at all frequencies with nondirectional characteristics, have a good dynamic range, exhibit high efficiency, and produce no distortion. The characteristics of the practical loudspeaker are quite different, because the properties required to produce a low-frequency vibration are quite different from those required to produce a high-frequency sound. Furthermore, the environment in which the loudspeaker operates has a significant effect on its performance.

When an alternating current is passed through the voice coil, a force is exerted on the voice coil by the interaction of the flux due to the permanent magnet and the flux generated by the current in the voice coil. The result is that the voice coil vibrates and so does the cone to which it is attached. The cone, the voice coil, and the cardboard former on which it is wound have a finite mass, as has the air that the cone must displace as it moves back and forth. Although the suspension is designed to offer the minimum mechanical resistance to motion, this resistance is not zero. When suitable assumptions have been made about the stiffness of the cone and the linearity of the suspension system, and so on, a differential equation describing the motion of the cone can be written and used to study the effects of changing the parameters of the system on its performance. The following broad conclusions can be drawn from both theory and practice:

1. The force driving the cone is proportional to the flux density in the air-gap, the length of the conductor in the voice coil, and the current flowing in the coil ($f = Bli$). A high-flux density can be obtained by using a large powerful magnet in conjunction with a small air-gap. The minimum air-gap is dictated by the thickness of the voice coil former, the diameter of the conductor used to wind the coil, and the need to keep the voice coil from touching the magnet assembly. The longer the wire in the coil the greater its mass, and, because it is part of the cone, a larger force will be required to drive it. A large coil current makes demands on the amplifier required to drive it.

2. A good low frequency response requires a large (area) cone. A large cone will in general have a relatively large mass. However, a good high-frequency response requires a cone of very small mass. It would appear that it is not possible to use one cone for the complete range of frequencies. A reasonable compromise, one that comes with a suitable price tag, is to use a filter to divert the low-frequency component of the signal to a large loudspeaker (woofer) and the high frequency to a small loudspeaker (tweeter).

3. When the frequency is low, the cone moves as one lumped mass, that is, all points on the cone remain in the same position relative to each other as the cone moves back and forth. As the frequency increases, the cone starts to behave like a distributed mass, and a phase difference between different parts occur. The result is that the response becomes irregular with peaks and troughs depending on whether the radiated signals from the different parts of the cone add or subtract from each other at a given point within its field of radiation.

4. The sound wave from the front of the loudspeaker must be separated from the back wave. Otherwise, at low frequency, the high pressure created by the forward movement of the cone is neutralized by the low pressure created at the back of the cone. The energy put into the loudspeaker simply moves the air around the edge of the loudspeaker instead of being converted into radiated energy. This can be prevented by mounting the loudspeaker on a large baffle whose dimensions are at least greater than one-quarter the wavelength of the lowest frequency. Similar results can be obtained by enclosing the back of the loudspeaker in a closed box.

5. The cone, as a mass supported on its suspension, is a system that can have a resonant response when suitably excited. It so happens that the resonant frequency is usually at the low-frequency end of the loudspeaker response. By adjusting the mass of the cone, the characteristics of the suspension and the damping, the peak of the resonant response can be moved to a frequency that improves the low-frequency performance of the loudspeaker.

6. The loudspeaker will generate harmonic distortion when the force-displacement characteristics of the cone suspension are nonlinear. Another source of distortion is the nonuniformity of the flux density in the air-gap especially when the voice coil is at the extremes of its excursion.

3.5 SHORTWAVE RADIO

Frequencies between 3 MHz and 30 MHz are set aside for use in long distance radio communication. This frequency band is generally referred to as the shortwave band. The wavelength that lies between 1 m and 100 m is short compared to that of the medium wave. It was pointed out in Section 2.9 that the increased frequency of the carrier used in shortwave communication gives the transmission increasingly directional properties. Transmission antennas can be designed to concentrate the power of the transmitter in a given direction. The ionosphere and the Earth's surface are used as reflectors to direct the signals from the transmitter to the receiver when very long distances have to be covered.

The superheterodyne arrangement is used in shortwave radio receivers. However, because the radio frequency is much higher, the intermediate frequency used is also high. Intermediate frequencies in use with shortwave radio are usually between 1.5 MHz and 28 MHz. In some special cases, a double superheterodyne system is used. The radio frequency is translated to the first intermediate frequency, a high frequency, and then translated the second time to lower frequency between 455 kHz and 5 MHz. The advantage of this system is improved selectivity and image rejection.

BIBLIOGRAPHY

S. S. Haykin, *Synthesis of RC Active Filter Networks*, McGraw-Hill, London, 1969.

J. L. Herrero and G. Willoner, *Synthesis of Filters*, Prentice-Hall, Englewood Cliffs, NJ, 1966.

L. P. Huelsman (Ed.), *Active Filters: Lumped, Distributed, Integrated, Digital, and Parametric*, McGraw-Hill, New York, 1970.

D. S. Humphreys, *The Analysis and Synthesis of Electrical Filters*, Prentice-Hall, Englewood Cliffs, NJ, 1970.

D. E. Johnson, *Introduction to Filter Theory*, Prentice-Hall, Englewood Cliffs, NJ, 1976.

A. S. Sedra and P. O. Bracket, *Filter Theory and Design: Active and Passive*, Matrix, Champaign, IL, 1978.

M. E. Van Valkenburg, *Analog Filter Design*, Holt, Rinehart & Winston, New York, 1982.

A. I. Zverev, *Handbook of Filter Synthesis*, Wiley, New York, 1967.

REFERENCES

1. W. F. Lovering, *Radio Communication*, Longmans, Green, London, 1966.
2. J. Smith, *Modern Communication Circuits*, McGraw-Hill, New York, 1986.
3. H. F. Olson, *Elements of Acoustical Engineering*, Van Nostrand, New York, 1947.
4. G. A. Briggs, *Loudspeaker*, 5th Ed., Rank Wharfdale Ltd., Bradford, Yorkshire, 1969.
5. C. L. Hutchinson (Ed.), *The ARRL Handbook for the Radio Amateur*, The American Radio Relay League, Newington, CT, 1988.

PROBLEMS

3.1 An AM receiver is tuned to 1000 kHz, but two other stations, the first operating at 1010 kHz and the second operating at 1200 kHz are suspected of causing interference. The intermediate frequency of the receiver is set at 455 kHz. Explain with the aid of suitable calculations which of the two stations is more likely to cause the interference. The intermediate frequency is now set at 100 kHz. Will this improve the situation? Explain your answer fully.

3.2 A diode that has characteristics described approximately by

$$i = 5 + v + 0.05v^2 \text{ mA}$$

is to be used in a demodulator of an AM receiver. The carrier frequency is 1 MHz, with a 100% modulation by a 5 kHz sinusoidal signal. Assuming that both the carrier signal source and the local oscillator have a 5-V output on open circuit and a 1-kΩ internal resistance, design a suitable (single-tuned *LC*) circuit using the diode, transformers (inductors), and capacitors to select the desired intermediate frequency signal and to supply maximum power to a resistive load of 500 Ω. Provide a suitable circuit diagram and specify the values of all the circuit elements used such that the side-frequencies lie within the -3-dB bandwidth. Calculate the mixer gain in decibels.

3.3 A switching-type mixer, as shown in Figure 3.12, is driven from an ideal sinusoidal voltage source of value 5 V peak. The local oscillator voltage (squarewave source) is assumed to be sufficient to cause instantaneous switching of the diodes. Calculate the amplitudes of the output signal for $n = 1$, 2, and 3 and their frequencies, if the sinusoidal source has a frequency 1 MHz and the squarewave is 8 MHz.

3.4 The secondary winding of a transformer has self-inductance $L = 500$ μH and resistance of 15 Ω. The primary current is an amplitude-modulated wave with a carrier frequency of 500 kHz modulated by a single frequency sinusoidal signal of 7.5 kHz to a depth of 50%. Calculate the value of the capacitor required to tune it and the amplitude of the upper or lower side-frequency relative to the carrier in the secondary winding.

3.5 The collector load of a bipolar transistor amplifier consists of a coil of self-inductance $L_1 = 1$ mH whose Q factor is equal to 100 when tuned to resonance at 500 kHz by a parallel capacitor, which may be assumed to be lossless. The amplifier drives a resistive load through a perfectly coupled coil; the mutual inductance being equal to 0.5 mH. If the resistive load has a value of 6.25 kΩ, determine the equivalent impedance of the collector load. What is the -3-dB bandwidth of the circuit? Calculate the value of the emitter resistance for a voltage gain of 25.

3.6 Using the MOSFET whose characteristics are given in Figure P3.1, design a class A amplifier with a voltage gain of 10. Allow for a drain current of 3 mA when the dc supply voltage is 12 Volts. What is the maximum sinusoidal input voltage that the amplifier can take without clipping? Calculate the input impedance of the amplifier.

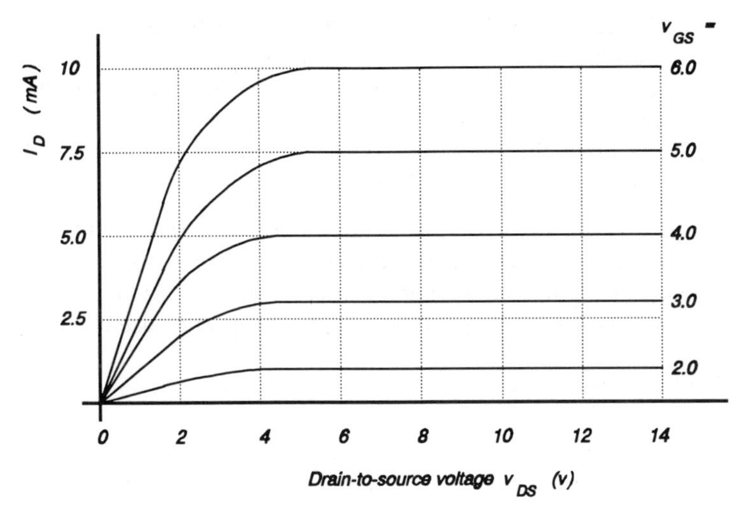

Figure P3.1

3.7 Repeat the design using the device whose characteristics are given in Figure P3.2.

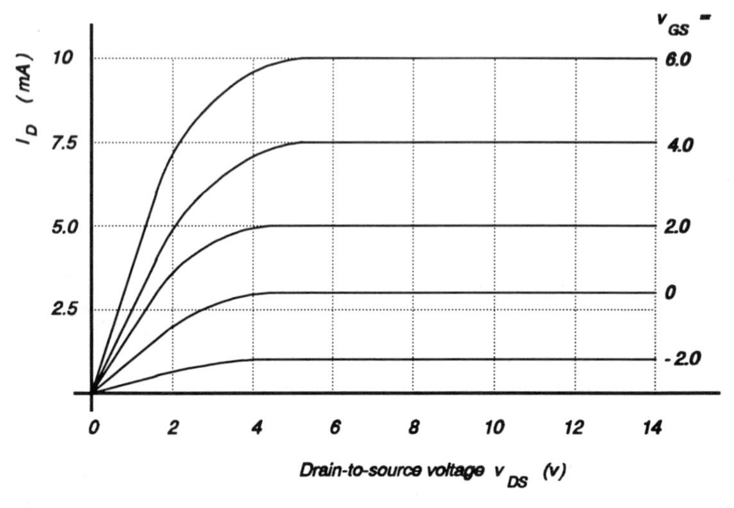

Figure P3.2

3.8 Using the device of Figure P3.1, design a transformer-coupled class A amplifier to operate at $500/2\pi$ kHz with a voltage gain of 10 and -3-dB bandwidth of $20/2\pi$ kHz. The load resistance is 100 Ω, and the dc supply voltage is 15 V. Specify the values of all components used. You may assume that the windings of the transformer have negligible resistances.

3.9 An AM signal applied to an envelope detector has a carrier frequency $\omega_c = 250$ krad/sec. The upper envelope is a triangular waveform that goes from $+10$ V to $+2$ V and back up to $+10$ V at a constant rate in 0.2 msec (the lower envelope is similar!). Assuming that the diode is ideal, calculate the maximum time constant of the detector load to avoid diagonal clipping. Justify any approximation that you make.

3.10 Design a Hartley-type oscillator with a variable capacitor to be used as a local oscillator in a superheterodyne receiver tunable from 600 to 1600 kHz with an intermediate frequency of 465 kHz. The capacitor is to be mechanically ganged to an identical capacitor used to tune a 156 μH inductor in the radio frequency circuit.

4

FREQUENCY-MODULATED RADIO TRANSMITTER

4.1 INTRODUCTION

In Chapter 2, the amplitude of a high-frequency (carrier) signal was varied in accordance with the waveform of an audio frequency (modulating) signal to give an amplitude-modulated (AM) wave that could be transmitted, received, and demodulated to recover the original audio frequency signal.

In frequency-modulated (FM) radio, the frequency of the carrier is varied about a fixed value in accordance with the amplitude of the audio frequency. The amplitude of the carrier is kept constant. The waveform of a sinusoidal carrier modulated by a saw-tooth wave is shown in Figure 4.1.

All signals carried on any transmission system will sooner or later be contaminated by noise, so the susceptibility of the communication system to noise is an important consideration. The noise can be defined as a random variation superimposed on the signal. In AM systems, the information to be transmitted is contained in the envelope of the carrier signal. The noise therefore appears on the envelope and has a direct role in corrupting the signal. In FM systems, the information to be transmitted is contained in the change of frequency of the carrier about a preset value. The amplitude of the FM signal is kept constant, and indeed, if there are changes in the amplitude of the FM signal, they are removed by clipping before demodulation. By comparison, FM systems are less susceptible to degradation by noise.

4.2 FREQUENCY MODULATION THEORY

Although a saw-tooth modulating signal provides a simple picture of the FM signal, a sinusoidal modulating signal is the simplest for the derivation of mathematical equations to describe the FM signal. A sinusoidal voltage can

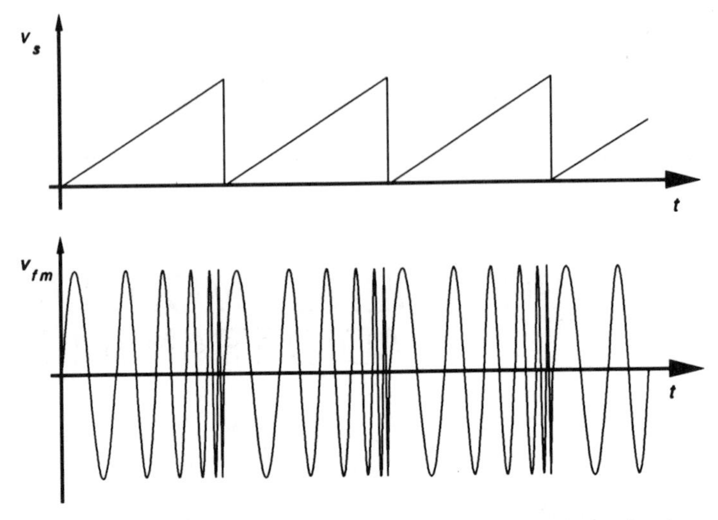

Figure 4.1. The saw-tooth waveform v_s frequency modulates a carrier to give the output v_{fm}. Note that the relative change in frequency has been exaggerated for clarity. In normal FM radio, the change in frequency relative to the carrier is less than 0.15%.

be expressed as

$$v = A \sin \omega t \tag{4.2.1}$$

$$= A \sin \theta(t) \tag{4.2.2}$$

where ω is a constant representing the angular velocity of the sinusoid and θ is a phase angle with respect to an arbitrary datum. In general, the relationship between the phase angle and the angular velocity is given by

$$\frac{d\theta(t)}{dt} = \omega(t) \tag{4.2.3}$$

In a frequency-modulated system, ω is varied about a fixed value ω_c, in accordance with the modulating signal, which is assumed in this case also to be a sinusoid:

$$v_s = B \cos \omega_s t \tag{4.2.4}$$

The instantaneous angular velocity,

$$\omega_i = \omega_c + \Delta\omega_c \cos \omega_s t \tag{4.2.5}$$

where ω_c is the long-term mean angular velocity, $\Delta\omega_c \ll \omega_c$ and $\Delta\omega_c$ is the maximum deviation of the angular velocity about ω_c.

Substituting instantaneous values into Equation (4.2.3),

$$\omega_i(t) = \frac{d\theta_i(t)}{dt} \tag{4.2.6}$$

Substituting Equation (4.2.5) into Equation (4.2.6) and integrating yields

$$\theta_i(t) = \varphi + \int_0^t \omega_c \, dt + \int_0^t \Delta\omega_c \cos \omega_s t \, dt \tag{4.2.7}$$

where φ is the initial value of the phase angle, which without loss of generality can be set equal to zero. Then, substituting Equation (4.2.7) into Equation (4.2.2) gives

$$v_{\text{fm}}(t) = A \sin[\omega_c t + (k\omega_c/\omega_s)\sin \omega_s t] \tag{4.2.8}$$

where $k\,\Delta\omega_c = B$, which is the amplitude of the modulating signal.

The term $k\omega_c/\omega_s$ is called the modulation index, where

$$m_f = \frac{\text{maximum frequency deviation of the carrier}}{\text{modulating signal frequency}} \tag{4.2.9}$$

$$v_{\text{fm}}(t) = A \sin(\omega_c t + m_f \sin \omega_s t) \tag{4.2.10}$$

Expanding,

$$v_{\text{fm}}(t) = A[\sin \omega_c t \cos(m_f \sin \omega_s t) + \cos \omega_c t \sin(m_f \sin \omega_s t)] \tag{4.2.11}$$

The terms $\cos(m_f \sin \omega_s t)$ and $\sin(m_f \sin \omega_s t)$ can be expanded as Fourier series with coefficients which are Bessel functions of the first kind, $J_n(m_f)$ where n is the order and m_f is the argument, to give

$$\cos(m_f \sin \omega_s t) = J_o(m_f) + 2\sum J_{2n}(m_f)\cos 2n\omega_s t \tag{4.2.12}$$

and

$$\sin(m_f \cos \omega_s t) = 2\sum J_{2n+1}(m_f)\sin(2n+1)\omega_s t \tag{4.2.13}$$

Substituting Equations (4.2.12) and (4.2.13) into Equation (4.2.11) and using

$$\cos x \sin y = \tfrac{1}{2}[\cos(x+y) + \cos(x-y)] \tag{4.2.14}$$

and

$$\sin x \sin y = -\tfrac{1}{2}[\cos(x+y) + \cos(x-y)] \tag{4.2.15}$$

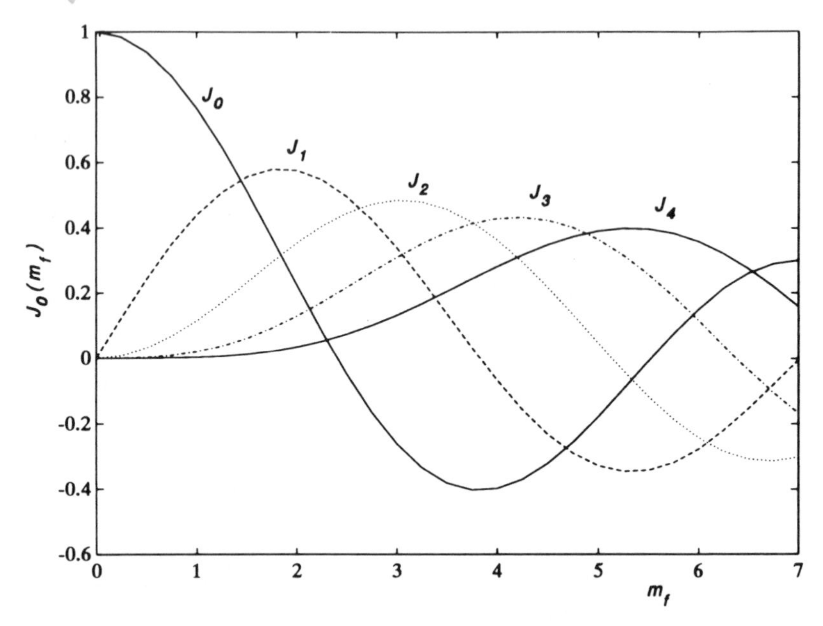

Figure 4.2. A plot of Bessel functions of the first kind, $J_n(m_f)$ against m_f for $n = 0, 1, 2, 3,$ and 4. The values of $J_n(m_f)$ are used to calculate the amplitudes of the side frequencies present in the FM signal.

The result is

$$\begin{aligned} v_{\text{fm}}(t) = A\{&J_o(m_f)\sin \omega_c t \\ &+J_1(m_f)[\sin(\omega_c + \omega_s)t - \sin(\omega_c - \omega_s)t] \\ &+J_2(m_f)[\sin(\omega_c + 2\omega_s)t - \sin(\omega_c - 2\omega_s)t] \\ &+J_3(m_f)[\sin(\omega_c + 3\omega_s)t - \sin(\omega_c - 3\omega_s)t] \\ &+ \dots\} \end{aligned} \tag{4.2.16}$$

where $J_n(m_f)$ is plotted against m_f for various values of n in Figure 4.2.

From Equation (4.2.16), it can be seen that

(1) The carrier frequency ω_c is present and its amplitude is determined by the modulation index m_f.

(2) The next term represent two frequencies, which are the sum $(\omega_c + \omega_s)$ and difference $(\omega_c - \omega_s)$ of the carrier and the modulating frequency with an amplitude $J_1(m_f)$.

(3) The next two terms have an amplitude $J_2(m_f)$ and frequencies $(\omega_c + 2\omega_s)$ and $(\omega_c - 2\omega_s)$.

(4) There are an infinite number of sums and differences of the carrier and integer multiples of the modulating signal.

It would appear that to transmit a simple sine wave in an FM system, an infinite bandwidth is required to accommodate all the multiple sidebands. However, from Figure 4.2 it can be seen that as n increases the amplitudes of the sidebands decrease and their contribution to the signal power falls off rapidly. A second aspect of the bandwidth requirements of the FM system can be seen in Equation (4.2.8). Unlike the AM system where a modulation index greater than unity causes severe distortion, the modulation index in FM does not appear to have an upper limit, except that, for a fixed modulating signal frequency ω_s, increasing the modulation index m_f means a greater deviation of the FM signal from the carrier frequency. This implies a larger bandwidth. With no apparent technical limits to the bandwidth requirements and to permit the maximum number of FM stations to function with minimum distortion and interference, the maximum frequency deviation for commercial FM radio is set at ± 75 kHz. However, since there is substantial signal power beyond the ± 75 kHz limit, the actual bandwidth is set at 200 kHz with the highest modulating frequency limited to 15 kHz.

Frequency-modulated communication channels can be found from approximately 1600 kHz to 4000 MHz. Parts of this spectrum are reserved for the use of police, VHF and UHF television sound channels, VHF mobile communications, and point-to-point communication. The commercial FM radio broadcast band is from 88 MHz to 108 MHz.

4.3 THE PARAMETER VARIATION METHOD

4.3.1 Basic System Design

A simple method for generating an FM signal is to start with any LC oscillator. The frequency of oscillation is determined by the values of C and L. If a variable capacitor, ΔC, is connected in parallel with C and the capacitance variation is proportional to the modulating signal, then an FM signal will be obtained.

Consider a negative conductance oscillator as shown in Figure 4.3. The negative conductance may be obtained from a tunnel diode, a suitably biased pentode, or a bipolar junction transistor with a suitable feedback circuit. The frequency of oscillation is given by

$$\omega_c^2 = \frac{1}{LC} \tag{4.3.1}$$

When a variable capacitor ΔC is connected in parallel with C, the frequency of oscillation will be

$$(\omega_c \pm \Delta\omega_c)^2 = \frac{1}{L(C \mp \Delta C)} \tag{4.3.2}$$

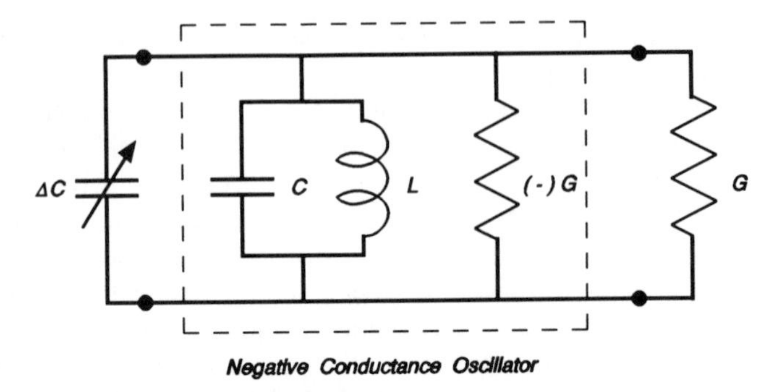

Figure 4.3. The basic FM generator; the value of the variable capacitor ΔC is controlled by the modulating signal, which in turn determines the frequency of the oscillator.

Any *p–n* junction diode in reverse bias can be used to realize ΔC. However the relationship between ΔC and the bias voltage is nonlinear and appropriate steps to obtain the necessary linearity will have to be taken, such as using two diodes in push–pull or specially fabricated diodes (e.g., varactor diodes or diodes with hyperabrupt junctions, both of which come under the classification voltage-controlled capacitors).

The direct generation of FM signals presents further difficulties. Assuming that the transmitter were to operate at 100 MHz, the required inductance and capacitance would be approximately 160 nH and 16 pF, respectively. Stray inductances and capacitances associated with the circuit would render it impractical. A less impractical idea is to generate the FM signal at a lower frequency where larger values of L and C are required and then to use a cascade of frequency multipliers to raise the frequency to the required value. Supposing that the low-frequency FM signal is generated at 200 kHz. Suitable values for L and C are 1.27 mH and 500 pF. Since the frequency has to be multiplied by 500 (actually $2^9 = 512$) to place it within the commercial FM frequency band, the maximum frequency deviation has to be 75 kHz/500 = 150 Hz, so that the frequency deviation will not exceed the legal limit. The required change in capacitance $\Delta C = 2.5$ pF. This is within the range of

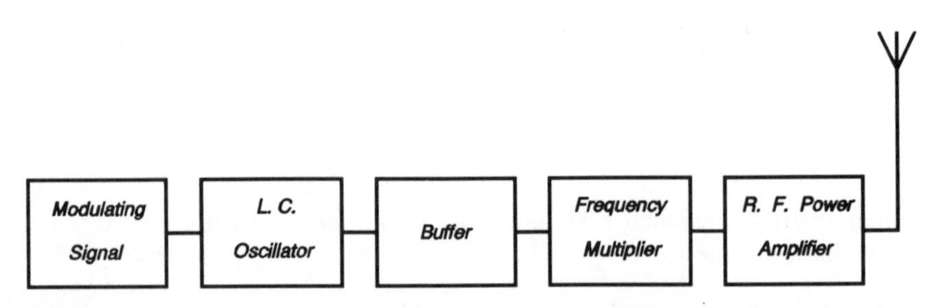

Figure 4.4. An improved FM generator design in which the FM signal is generated at a lower frequency and then multiplied by a suitable factor to give the required frequency.

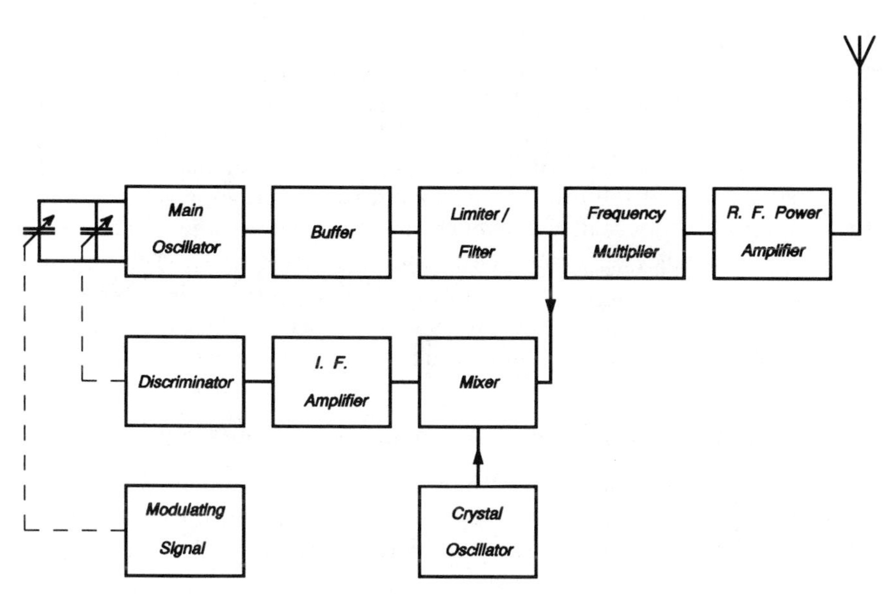

Figure 4.5. The circuit in Figure 4.4 has been improved by the addition of an automatic frequency control (AFC) circuit to stabilize the carrier frequency.

capacitance change that can be obtained from a p–n junction diode. A block diagram of the system is given in Figure 4.4.

The design of all the blocks shown in Figure 4.4 were discussed earlier. Moreover, frequency stability requirements of the system make the approach shown in Figure 4.4 impractical. One of the requirements of a transmitter is that its frequency remain at the assigned value at all times; that is, its frequency does not drift and cause interference with other channels of communication. In the scheme shown in Figure 4.4, there is no built-in mechanism to ensure that the long-term average frequency will be constant. A modified scheme is shown in Figure 4.5, in which the carrier frequency of the FM signal is compared with that of a crystal-stabilized oscillator and the resulting error signal is used to correct the carrier frequency.

4.3.2 Automatic Frequency Control of the FM Generator

The main oscillator is an LC-type similar to the negative conductance oscillator discussed earlier. The variation ΔC generates the FM signal. The output is fed into a buffer amplifier, which provides isolation between the oscillator and its load so that a changing load will have minimal effect on the oscillator operation. The signal proceeds to the *limiter*, which removes any amplitude modulation that may have occurred and also filters out the harmonics generated by the limiter. Part of the amplitude-limited FM signal is then fed into a mixer whose other input is from a crystal-controlled oscillator. The difference frequency is suitably amplified, filtered, and then fed into a *discriminator* to produce a dc signal proportional to the difference between the required and the crystal-controlled oscillator frequency. The dc or error signal is used to control the main oscillator to keep it at the required frequency. Using this scheme, it is possible to impose the stability of the crystal-controlled oscillator on the main oscillator. It should be noted that the time-constant of the error signal should be chosen so it can correct for the long-term frequency deviation of the main oscillator without affecting the short-term frequency deviation due to the modulating signal. The output of the limiter/filter is fed into a chain of multipliers to bring the frequency to its operating value. A power amplifier then boosts the signal power to the desired value and drives the antenna that radiates the signal.

4.3.3 Component Design with Automatic Frequency Control

Two new component blocks were introduced into the system to realize the parameter variation method. These were the limiter and the discriminator. The circuit design of these component blocks follow.

4.3.3a The Amplitude Limiter. An FM signal, by definition, must have a constant amplitude. In practice, circuit nonlinearities cause variations in the envelope of the FM signal related to the modulating signal, that is, some amplitude modulation takes place. It is evident that any amplitude modulation present in an FM system will interfere with the signal during the demodulation process, because most FM demodulators convert the variation of frequency to a variation of amplitude before detection. A second reason for limiting the amplitude of the FM signal is that the noise present in the communication channel generally rides on the envelope of the signal, and therefore, by clipping the amplitude of the signal, some of the noise can be removed.

The ideal amplitude limiter can accept an input signal of any amplitude and convert it to one of constant amplitude. A practical approximation to the performance of the ideal limiter is obtained by amplifying the input by a large factor and clipping a small portion symmetrically about the time axis. The output waveform will then be a squarewave. To get back the sinusoid, the signal must be passed through a bandpass filter, which removes all the harmonics leaving only the fundamental. The amplitude limiter has three identifiable parts: the preamplifier, the symmetrical clipper, and the bandpass filter. The subsystem block diagram is shown in Figure 4.6.

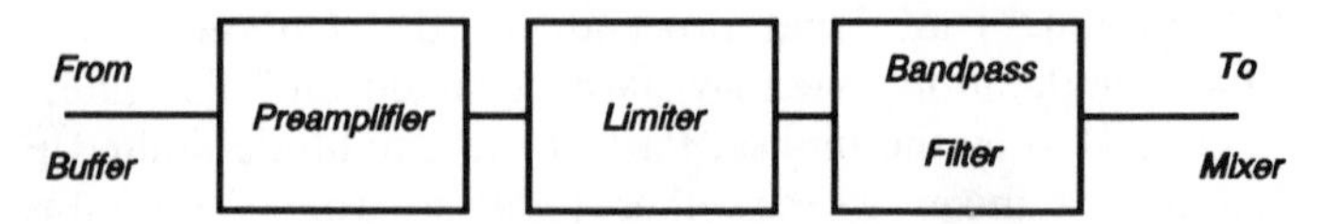

Figure 4.6. A block diagram showing the details of the limiter/filter shown in Figure 4.5.

In Section 4.2, it was shown that an FM signal has an infinite number of sidebands but that it was not necessary to preserve all the sidebands to maintain a high level of fidelity. Working with the permitted modulation index, $m_f = 5$, and retaining sidebands with coefficients $J_n(m_f)$ greater than 0.01 (i.e., 1% of the unmodulated carrier) it can be shown that the first eight sidebands must be preserved. Since the sidebands are spaced ω_s apart, and ω_s has a maximum value of 15 kHz, the required bandwidth is 240 kHz. The -3-dB Q factor of the preamplifier with the required bandwidth will be just over 400. However, this would mean that the outermost sidebands will be subjected to a 3 dB attenuation compared to those nearer the carrier causing further variation in the amplitude of the signal. In a practical situation, the modulating signal is made up of a large number of discrete frequencies, and this causes further complications in the calculation of the required bandwidth. An amplifier with a flat frequency response could be used, but it will have the disadvantage of amplifying the noise below and above the spectrum of interest. A compromise would be to use a tuned preamplifier with a much lower Q factor. It will have the advantage of a minimal variation in its response over the spectrum occupied by the significant sidebands and yet attenuate the noise present in the rest of the spectrum. A loaded Q factor of approximately 20 should be adequate.

It is accepted practice to generate the FM at a lower frequency and to use a cascade of frequency multipliers to bring it up to the operating value. If the FM signal is generated at 200 kHz, the appropriate frequency deviation will be approximately 150 Hz. The corresponding bandwidth to accommodate the first eight sidebands will be 470 Hz.

Figure 4.7 shows a circuit diagram of the preamplifier and its load, the symmetrical clipper. The detailed design of a similar amplifier was presented

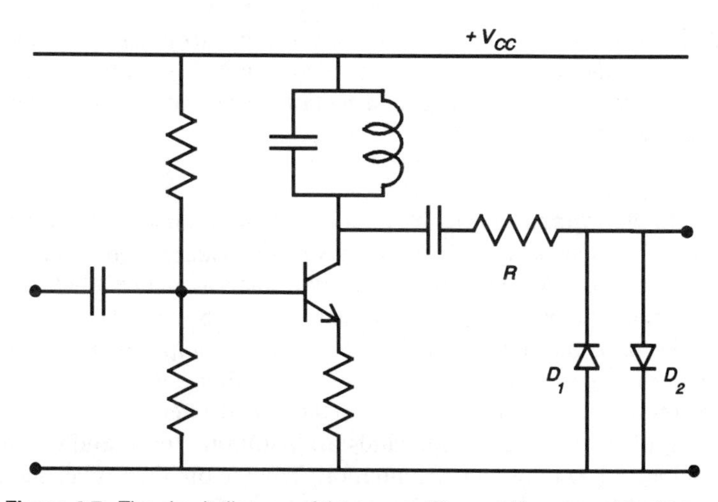

Figure 4.7. The circuit diagram of the preamplifier and the symmetric clipper.

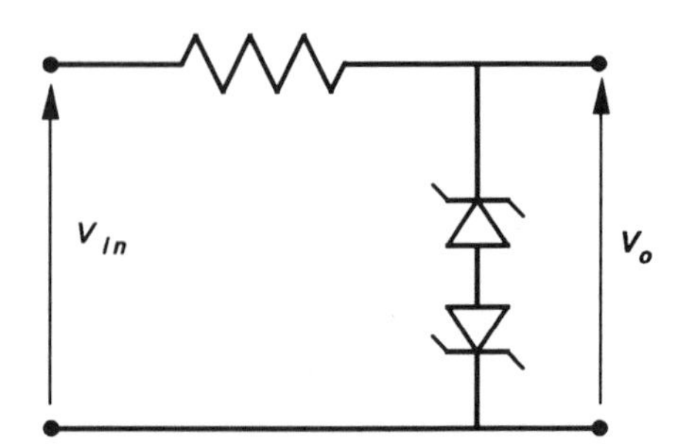

Figure 4.8. The use of two zener diodes connected back to back is an improvement on the symmetric clipper shown in Figure 4.7.

in Section 3.4.2 and will not be repeated here. However, it is worth noting that the diodes can be considered as short circuits and hence the resistance R will be in parallel with the tuned circuit.

4.3.3b The Symmetric Clipper. A simple symmetric clipper is shown in Figure 4.7. The resistance R is chosen to so that the Q factor has a value of approximately 20. The details of the design of a parallel tuned circuit with a specified Q factor was discussed in Section 3.4.2.

The output of the clipper will be approximately 0.7 V when single silicon diodes are used or multiples of 0.7 V depending on how many diodes are connected in series to determine the level of clipping. It is evident that the more severe the clipping is, the more likely it is that the output of the clipper will be a squarewave, but then more gain will be required in the succeeding stages.

A more elegant design when the clipping level is high is to use back-to-back zener diodes, as shown in Figure 4.8. The zener diode provides a much sharper cut-off than an ordinary diode when it is reverse biased. In the forward direction, it is an ordinary diode. With this arrangement, only one of the two zener diodes operates as a zener diode and the other one is conducting current in the forward direction. They switch roles when the applied voltage changes polarity.

4.3.3c The Bandpass Filter. It is reasonable to assume that the output of the clipper is a squarewave. The frequency of the squarewave varies, but it is centered at the carrier frequency. Fourier analysis shows that the most significant harmonic in the squarewave is the third. So with the subcarrier at 200 kHz, the third harmonic will be at 600 kHz. A simple tank circuit tuned to the carrier frequency is all that is necessary to filter out the harmonics. But as before the choice of the Q factor of the tuned circuit must be made to ensure that all the significant sidebands are within the passband of the filter and are subjected to minimum attenuation. The factors that affect the choice of the bandwidth were discussed in Section 4.3.3a.

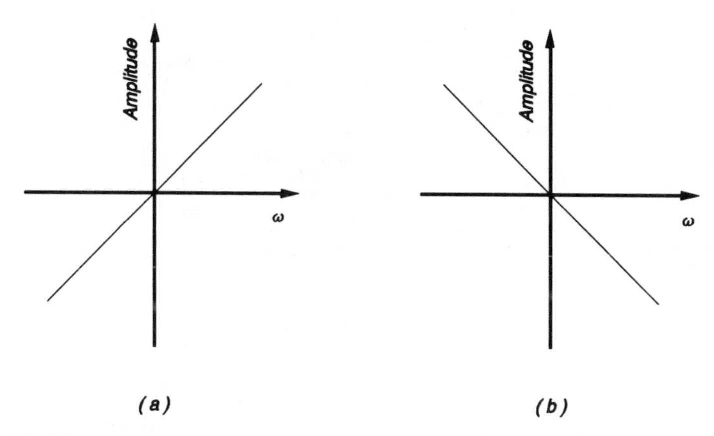

Figure 4.9. The required characteristics of a frequency-to-amplitude converter. (*a*) A high-pass circuit. (*b*) The low-pass version.

4.3.3d The Discriminator. The purpose of the discriminator is to convert the variation of frequency to a variation of amplitude. A frequency-to-amplitude converter followed by an envelope detector is used to recover the message contained in the modulating signal. The transfer characteristics of the two circuits that could be used for the frequency-to-amplitude conversion are shown in Figure 4.9.

The simplest circuit with the characteristics shown in Figure 4.9*a* is a simple *RC* highpass circuit with its corner frequency much higher than the carrier frequency of the FM signal. Figure 4.9*b* shows the response of a simple *RC* lowpass circuit with its corner frequency chosen to be much lower than the carrier frequency. Both circuits should, in principle, convert the FM wave into an AM wave. However, in practice the subcarrier frequency of the system shown in Figure 4.5 will be approximately 200 kHz, whereas the variation will be limited to a maximum of ± 150 Hz. Since both of these circuits have a slope of 6 dB/octave, the variation in amplitude of these circuits will be exceedingly low. The use of a high-gain amplifier following such a circuit will lead to increased noise. A better approach is to find a circuit that has a much greater amplitude-to-frequency slope.

A tuned *RLC* circuit has a rapid change of amplitude with frequency on both sides of the resonance frequency, especially when the Q factor of the circuit is high. The circuit and its response are shown in Figure 4.10. It is evident that this circuit will be prone to produce distortion especially of even harmonics. A variation on the circuit in Figure 4.10 is shown in Figure 4.11. The two tuned circuits made up of L_1–C_1 and L_2–C_2 are tuned to two different frequencies equally spaced from the carrier frequency. The individual and composite characteristics are shown in Figure 4.11. It can be seen that this circuit has a greater dynamic range than the previous circuit and when properly tuned should have no even harmonic distortion products in the output.

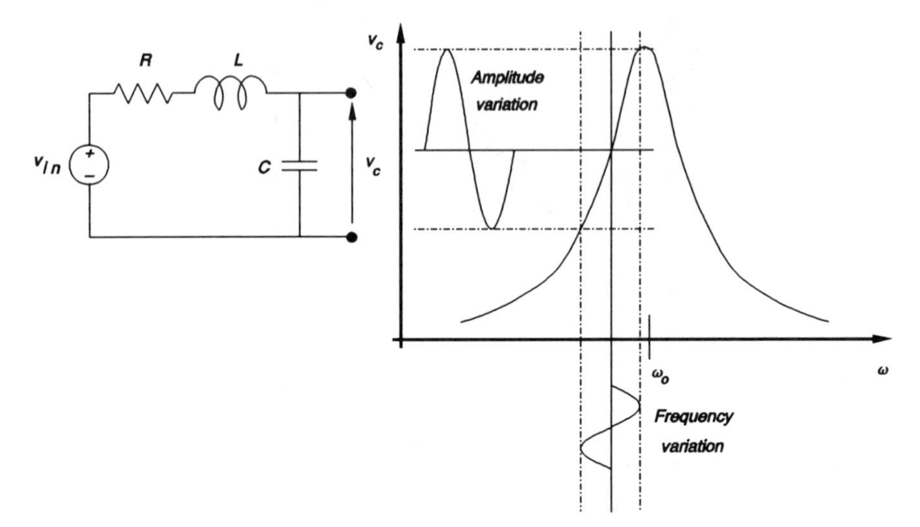

Figure 4.10. A single *LC* tuned circuit used as a frequency-to-amplitude converter. In this case, the skirt with the positive slope (high-pass) has been used. The skirt with the negative slope (low-pass) would be equally effective. Note that because the nonlinearity of both skirts is predominantly a second-order function, the distortion products of the circuit will be mainly even harmonics.

4.3.3e The Envelope Detector. The basic envelope detector was discussed in Section 3.4.6. The importance of the choice of the time constant to avoid distortion of the envelope was discussed at some length. The purpose of this envelope detector is to correct the long-term frequency deviation of the main oscillator while allowing the short-term frequency deviation caused by the modulating signal. Its time constant is therefore chosen on the basis that the lowest frequency present in the modulating signal, $\omega_{s,\min}$ will not cause the frequency correction system to go into operation. This condition is satisfied when the time constant of the detector, τ_{det} is chosen so that

$$\omega_{s,\min}\tau_{\text{det}} \gg 1 \tag{4.3.3}$$

4.4 THE ARMSTRONG SYSTEM

The use of automatic frequency control in the generation of FM signals results in a practical scheme, but one which depends on feedback. For proper operation of the system, all of the circuits must be made to track each other. Furthermore, the voltage-controlled capacitor used in the automatic frequency control circuit has a nonlinearity as well as a temperature characteristic that must be compensated. Evidently, a different scheme with less

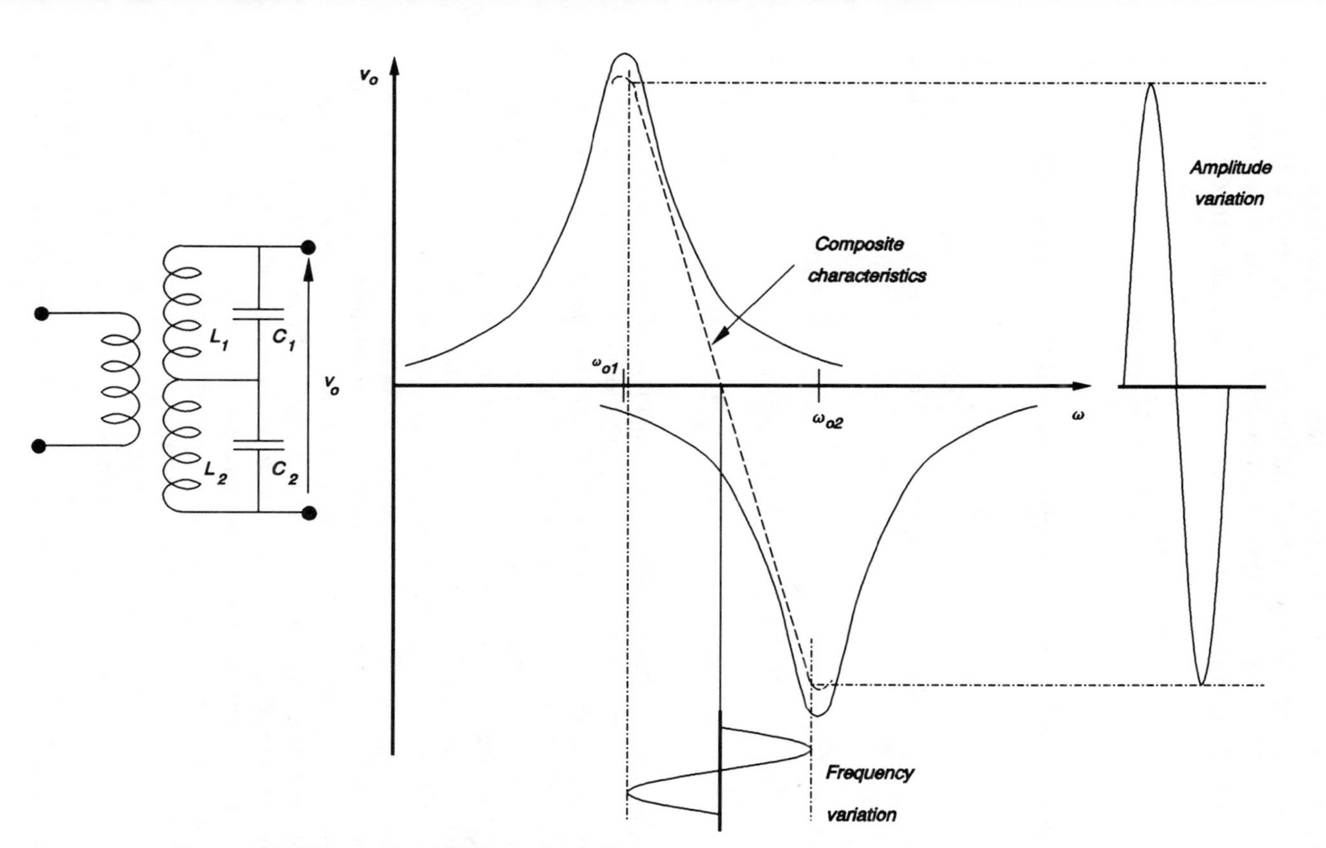

Figure 4.11. The use of a double-tuned circuit provides increased dynamic range and improves the linearity by balancing the nonlinearity of one *LC* circuit with the other. Note that the push–pull arrangement shown here will generate predominantly odd harmonics.

complexity is needed. The Armstrong system, named after its inventor, is one such scheme. The theoretical basis of the Armstrong system is given by Equation (4.2.7). It is clear that in commercial FM systems, the frequency deviation must be small compared to the carrier frequency, that is,

$$\omega_c(t) \gg k\frac{\omega_c}{\omega_s}|\sin \omega_s t| = m_f|\sin \omega_s t| \tag{4.4.1}$$

The following approximations can be applied to Equations (4.2.11)

$$\text{(a)} \qquad \cos(m_f \sin \omega_s t) \approx 1 \tag{4.4.2}$$

$$\text{(b)} \qquad \sin(m_f \sin \omega_s t) \approx (m_f \sin \omega_s t) \tag{4.4.3}$$

to give

$$v_{\text{fm}}(t) \approx A[\sin \omega_c t + (m_f \sin \omega_s t)\cos \omega_c t] \tag{4.4.4}$$

$$= A\{\sin \omega_c t + \tfrac{1}{2}m_f[\sin(\omega_c - \omega_s)t + \sin(\omega_c + \omega_s)t]\} \tag{4.4.5}$$

It should be noted that the above approximation eliminates all the sidebands except the two closest to the carrier. The modulation is then said to be a *narrow band frequency modulation* (NBFM). The advantage it offers is a simple geometric representation of the FM wave as a phasor.

The term $[(m_f \sin \omega_s t)\cos \omega_c t]$ is a *double-sideband suppressed-carrier* (DSB-SC) signal. The addition of the carrier term $(\sin \omega_c t)$ would normally produce an AM signal, but in this case the carrier is 90 degrees out of phase with the DSB-SC signal. The difference between AM and FM is shown in the phasor diagrams in Figure 4.12. Assume a coordinate system in which the phasor $A \sin \omega_c t$ rotates anticlockwise at an angular velocity ω_c. Then, suppose that the rotating phasor is used as a new reference system so it is represented by a horizontal phasor of value A. In terms of this reference system, the term $(\frac{1}{2}A \cdot m_f)\sin(\omega_c + \omega_s)t$ is represented by a phasor of value $\frac{1}{2}A \cdot m_f$ rotating in an anticlockwise direction with an angular velocity of ω_s. Similarly, the term $(\frac{1}{2}A \cdot m_f)\sin(\omega_c - \omega_s)t$ is represented by a phasor of equal value rotating in a clockwise direction with an angular velocity of ω_s. At time $t = 0$, the component of each rotating phasor in the horizontal direction is zero. Therefore the two phasors must be parallel to each other but at right angles to the horizontal phasor A. This situation is depicted in Figure 4.12a. The resultant is a phasor, which apparently has a maximum value

$$A\sqrt{(1 + m_f^2)} \tag{4.4.6}$$

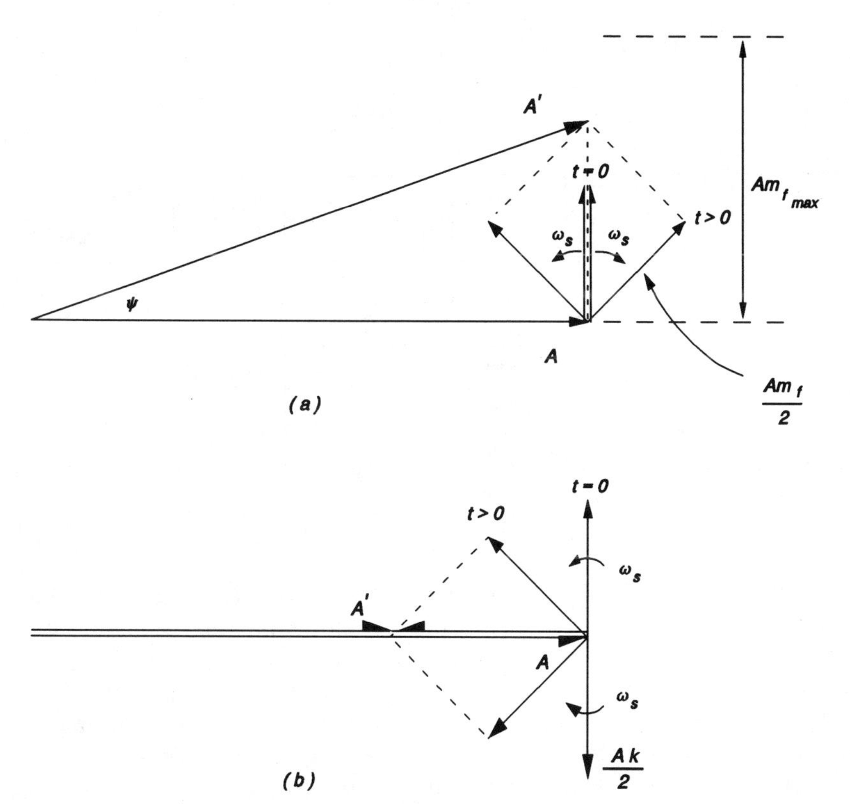

Figure 4.12. (*a*) A phasor diagram for an NBFM showing the two phasors of amplitude $\frac{1}{2}Am_f$ rotating in opposite directions at angular velocity ω_s in quadrature to the phasor A at $t = 0$. (*b*) A phasor diagram for an AM signal showing two counter-rotating phasors of amplitude $\frac{1}{2}kA$ and angular velocity ω_s. Note that the amplitude of the resultant phasor varies from $A(1 - k)$ to $A(1 + k)$.

and a maximum angular displacement of

$$\psi = \tan^{-1}(m_f) \tag{4.4.7}$$

Because m_f is small, ψ is small and

$$\psi \approx m_f \tag{4.4.8}$$

The apparent variation of the amplitude of the FM signal is due to the fact that this analysis has not taken into account the many sidebands present in FM.

For comparative purposes, the AM system is treated similarly. An AM signal is represented by Equation (2.2.7)

$$v_{\text{am}}(t) = A \sin \omega_c t + \frac{kA}{2} \cos(\omega_c - \omega_s)t - \frac{kA}{2} \cos(\omega_c + \omega_s)t \tag{4.4.9}$$

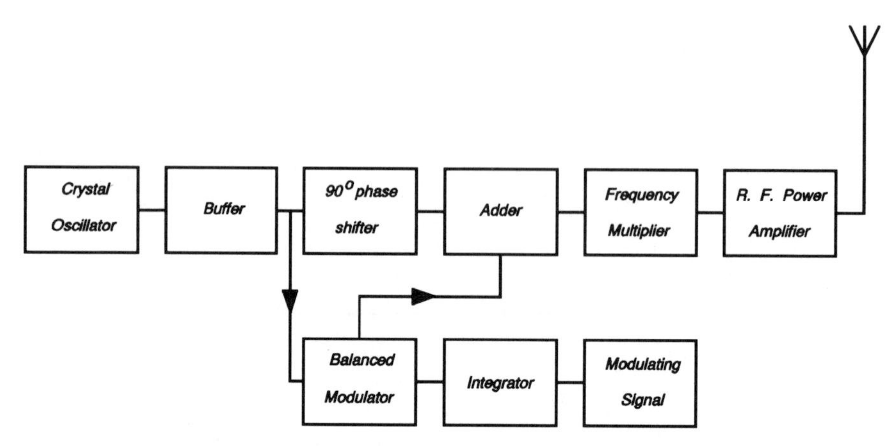

Figure 4.13. Block diagram of the basic Armstrong system FM transmitter. This is an example of an indirect generation of an FM signal.

Again the carrier $A \sin \omega_c t$ is represented by a horizontal phasor of value A and the two sidebands by two counter-rotating phasors of value $kA/2$ rotating at ω_s. At time $t = 0$, the components of both of the sidebands in the horizontal direction are zero, because they are cosine terms and A is a sine term. The phasors representing the two sidebands must be at right angles to the A phasor, as shown in Figure 4.12*b*. Note that one is positive and the other is negative. At some other time $t > 0$, the phasor representing the combination of the sidebands subtracts from or adds to the phasor A. The amplitude of the resultant therefore varies from $A(1 - k)$ to $A(1 + k)$, but at all times it is in the horizontal position.

The conclusion is that in a narrowband ($\Delta\omega_c \ll \omega_c$) FM system, the phase difference between the carrier and the DSB-SC signal is 90 degrees. This is the basis of the Armstrong system. A block diagram of the FM transmitter is shown in Figure 4.13.

4.4.1 Practical Realization

A crystal-controlled oscillator generates the carrier frequency $A \cos \omega_c t$. This signal is fed to two blocks. The first is a 90-degree phase shift circuit, which, as the name suggests, shifts the phase by 90 degrees and therefore its output is $A \sin \omega_c t$. The second block is a balanced modulator. The other input to the balanced modulator is, of course, the modulating signal. But according to Equation (4.2.8), $m_f = \Delta\omega_c/\omega_s$. The audio signal contains a band of frequencies from approximately 20 Hz to 15 kHz and therefore it follows that as ω_s increases, $\Delta\omega$ must increase or the modulation index m_f must decrease. However, in frequency modulation, the frequency of the carrier changes proportionately to the *amplitude* of the modulating signal, not to the *fre-*

quency of the modulating signal. In order to satisfy this condition, the amplitude of the modulating signal has to be reduced by one-half whenever the frequency doubles. This amounts to putting the signal through a low-pass filter with a slope of 6 dB/octave. An integrator with its corner frequency below the lowest modulating frequency is required. The amplitude of the modulating signal frequency band is thereby "predistorted" by the integrator to conform to the above condition. The need for the integrator is evident from Equation (4.2.7).

If the modulating signal is $B \sin \omega_s t$ the output of the integrator will be $(B/\omega_s)\cos \omega_s t$. This signal is fed to the balanced modulator, which produces as an output the product of its two inputs, namely, $(k'AB/\omega_s)\sin \omega_c t \cos \omega_s t$, where k' is a constant. This signal and the output of the 90-degree phase shifter go to the adder, which gives an output

$$v_o(t) = A \sin \omega_c t + \frac{k'AB}{\omega_s} \sin \omega_c \cos \omega_s t \tag{4.4.10}$$

The signal goes through a series of multipliers to bring the frequency up to the operating value. A power amplifier raises the signal to the proper level for radiation by the antenna.

In a practical FM transmitter, the carrier is generated at a lower frequency and the modulation process is carried out. The frequency is then multiplied up to the operating value. If it is assumed that the subcarrier frequency is again 200 kHz—a frequency at which a suitably robust crystal can be found —the required multiplication factor to place it in the middle of the commercial FM frequency band is then approximately 500 ($2^9 = 512$). The maximum frequency deviation for a subcarrier at 200 kHz was calculated in Section 4.3.4 to be 150 Hz.

The ratio of the subcarrier frequency to the maximum frequency deviation may not be the most convenient in a practical system. To produce a high-fidelity signal, that is, maintain a high level of linearity, it is an advantage to make the carrier frequency independent of the frequency deviation. This can be done by splitting the multiplication operation in two and inserting a mixer between them so that the carrier frequency can be changed without affecting the frequency deviation. To help explain the process, an example is given in Figure 4.14.

Suppose the subcarrier frequency is 200 kHz and the maximum deviation is 20 Hz. To get the subcarrier frequency up to the normal operating frequency of approximately 100 MHz, a cascade of frequency multipliers equal to 2^9 (512) will be required. Multiplication of the carrier also multiplies the frequency deviation so that the corresponding frequency deviation will be (512×20 Hz) 10,240 Hz. This is much lower than the limit allowed, which is 75 kHz. On the other hand, to convert the 20-Hz frequency deviation into the required 75 kHz, the multiplication factor is 3750 (approximately $3 \times 2^{10} =$

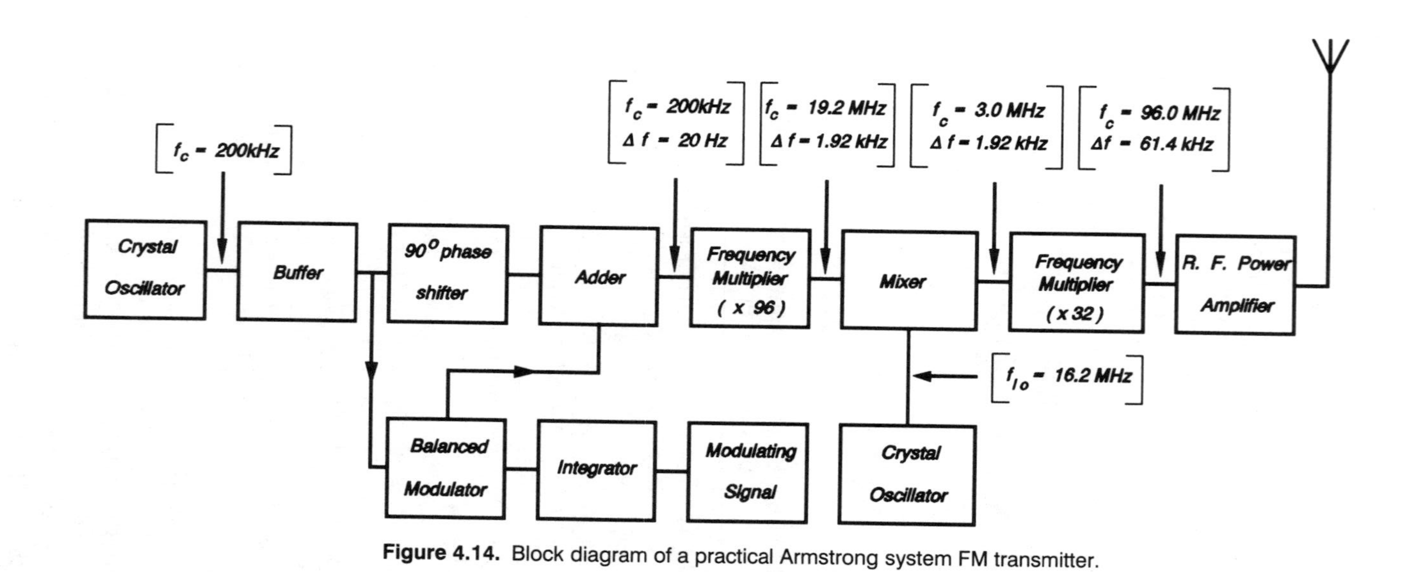

Figure 4.14. Block diagram of a practical Armstrong system FM transmitter.

3072). However, multiplying the subcarrier by this factor will give a carrier frequency of 614.4 MHz, which is outside the commercial FM radio band. To correct this, the multiplication process is split into two stages and a mixer is inserted between them. The first stage of multiplication may be 96 (3×2^5). The carrier frequency is then 19.2 MHz and the corresponding frequency deviation is 1.92 kHz. The signal is now mixed with a "local oscillator" whose output frequency is 16.2 MHz to produce a carrier frequency of 3.0 MHz, but the frequency deviation remains unchanged at 1.92 kHz. The second stage of multiplication ($32 = 2^5$) produces a carrier frequency of 96.0 MHz and a frequency deviation of 61.4 kHz. This is not quite at the allowed limit but close enough for practical purposes. The "local oscillator" signal may be derived from the 200 kHz crystal-controlled oscillator by using a suitable frequency multiplier ($81 = 3^4$).

4.4.2 Component Circuit Design

In the block diagram of the Armstrong system given in Figure 4.14, four new components were added, namely the 90-degree phase shifter, the balanced modulator, the integrator and the adder. The design of these circuit components now follow.

4.4.2a The 90-Degree Phase Shift Circuit. In the circuit shown in Figure 4.15*a*, the resistances R_C and R_E are equal. Since the collector and emitter currents are for all practical purposes equal, it follows that the voltages v_C and v_E must be equal in magnitude. But the phase angle between them is 180 degrees. The voltages v_C and v_E are represented on the phasor diagram shown in Figure 4.15*b*. The voltage, which appears across R and Z is

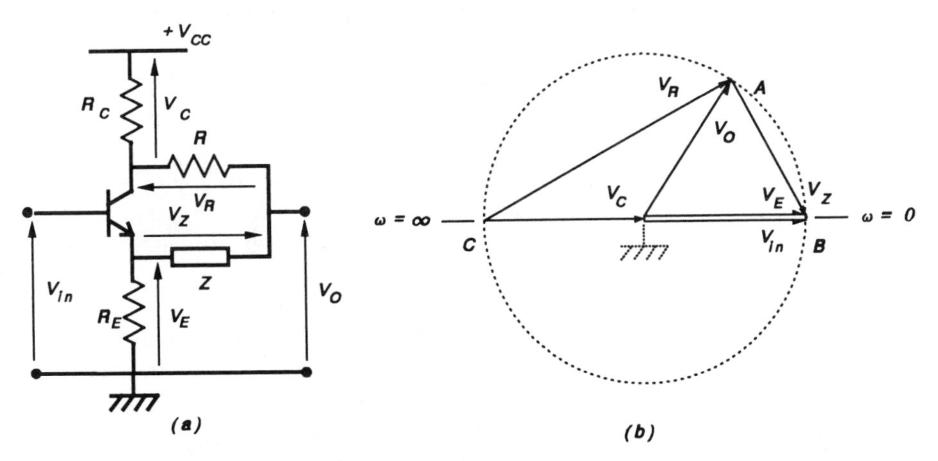

Figure 4.15. (*a*) An active 90-degree phase shift circuit. (*b*) Phasor diagram for the circuit in *a*.

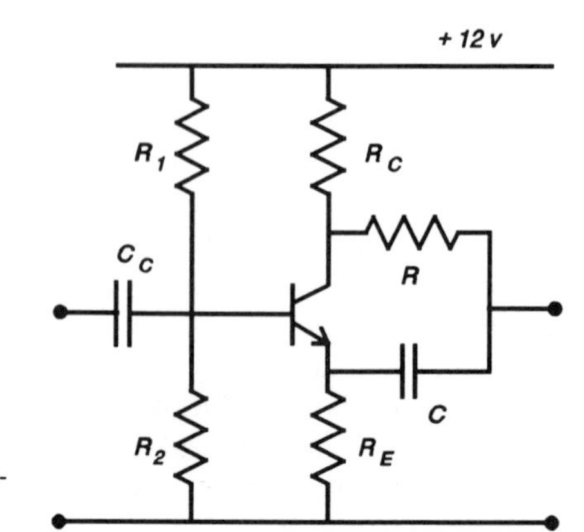

Figure 4.16. Circuit diagram used in the example on the 90-degree phase shift circuit.

$(v_C + v_E)$. If the impedance Z is purely reactive (capacitive or inductive), it follows that the phasors v_R and v_Z will always be at right angles to each other. It follows then that the point A must lie on a circle whose radius is equal to the magnitude of v_C as shown. The output voltage is then v_o. As frequency goes from zero to infinity, point A will trace out a semicircular locus. The output voltage v_o changes its phase angle from zero to 180 degrees, but its magnitude remains constant. It can be seen that with the appropriate choice of R and Z at a given frequency, v_o, can be made to lead or lag v_{in} by 90 degrees. One major disadvantage of this circuit is that its load must be a very high impedance, preferably open circuit, otherwise the magnitude of the output voltage v_o will not remain constant for all frequencies. This circuit is a simple example of a class of networks known as all-pass filters.

Example 4.4.1 90°-phase shift circuit. Design an all-pass circuit with a phase shift of 90 degrees at 200 kHz. The dc supply voltage is 12 V, and it may be assumed that the circuit drives a load of very high impedance. The β of the transistor is 100.

Solution. A suitable circuit for the phase shifter is shown in Figure 4.16 with the impedance Z replaced by a single capacitance C. Since the gain is unity $v_C = v_E$ and the stage will have the maximum dynamic range when the transistor is biased with V_E at $V_{CC}/4$ and V_C at $3V_{cc}/4$, that is, at 3 and 9 V, respectively. The maximum signal voltage that can be applied to the RC series circuit is then 12 volts peak-to-peak. The equivalent circuit is then shown in Figure 4.17. Let the peak current in the RC circuit be approximately equal to 1 mA. Then $R = 4.2\ \text{k}\Omega$ and the reactance of the capacitor must also be $4.2\ \text{k}\Omega$ so that at 200 kHz, $C = 190$ pF.

The dc collector current in the transistor must be chosen so that the 1 mA taken by the RC circuit will have negligible effect on the transistor operation.

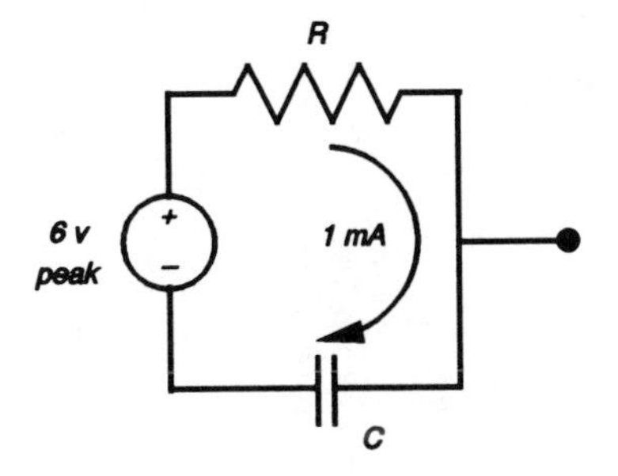

Figure 4.17. An equivalent circuit of the 90-degree phase shift circuit shown in Figure 4.15*a*.

A collector current of 10 mA will be adequate. Under quiescent conditions, the emitter voltage is 3 V. Hence $R_E = 300\ \Omega$ and $R_C = 300\ \Omega$. Since the β of the transistor is 100, the base current is 100 μA. Allowing a current 10 times the base current in the resistive chain (i.e., 1 mA) makes $(R_1 + R_2) =$ 12 kΩ. But the voltage at the base is 3.7 V and therefore $R_2 = 3.7$ kΩ and $R_1 = 8.3$ kΩ. The coupling capacitor has to be chosen so that it is essentially a short circuit at 200 kHz; a 0.1-μF capacitor is adequate.

4.4.2b The 90-Degree Phase Shift Circuit: Operational Amplifier Version. When the frequency of operation is within the bandwidth of an operational amplifier, the circuit shown in Figure 4.18 can be used to realize a 90-degree phase shift. The analysis of the circuit is simplified by the application of superposition. The first step is to separate the inverting input from the noninverting input. With the noninverting input connected to ground, and v_{in} connected to the inverting input, it can be shown the output

$$v_{01} = -\frac{R_2}{R_1} v_{in} \tag{4.4.11}$$

With the inverting input grounded and v_{in} applied to the noninverting input, it can be shown that the output

$$v_{02} = \frac{R_3}{R_3 + \dfrac{1}{j\omega_c}} \left(1 + \frac{R_2}{R_1}\right) v_{in} \tag{4.4.12}$$

The output when v_{in} is connected to both inputs

$$v_o = v_{01} + v_{02} \tag{4.4.13}$$

$$\frac{v_o}{v_{in}} = -\frac{R_2}{R_1} + \frac{j\omega CR_3}{1 + j\omega CR_3}\left(1 + \frac{R_2}{R_1}\right) \tag{4.4.14}$$

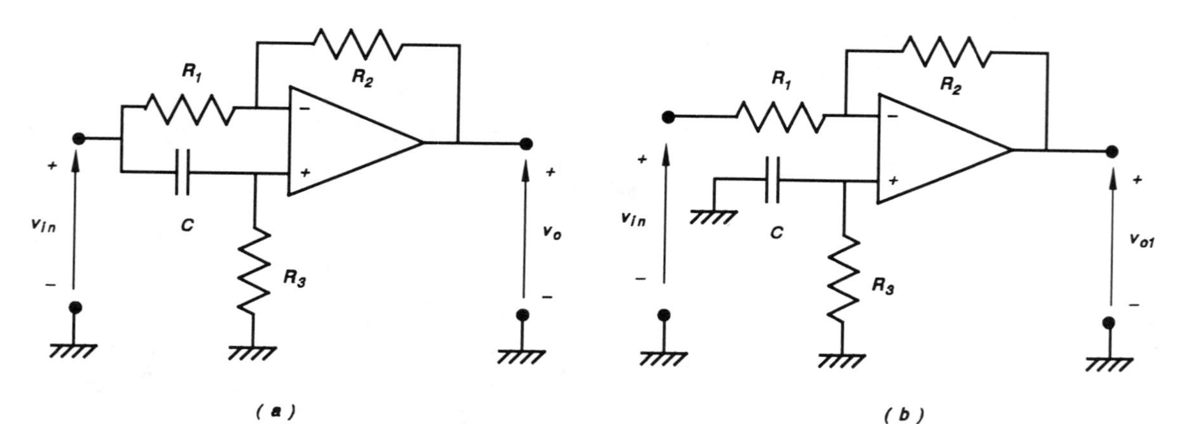

Figure 4.18. (*a*) An operational amplifier version of the 90-degree phase shift circuit. The advantage of this circuit is that its output impedance is low. (*b*) The circuit in (*a*) when superposition is applied.

When $R_1 = R_2$ and $\omega C R_3 = 1$,

$$\frac{v_o}{v_{\text{in}}} = j1 \tag{4.4.15}$$

The gain is unity and the output leads the input by 90 degrees. The main advantage of this approach is that the output impedance of the circuit is low enough to drive the balanced modulator directly.

4.4.2c The Balanced Modulator. A balanced modulator is a circuit that accepts two input signals of different frequencies, a carrier $A \cos \omega_c t$ and a modulating signal $B \cos \omega_s t$, and produces an output proportional to the product of the two inputs

$$v_o(t) = kAB \cos \omega_c t \cos \omega_s t \tag{4.4.16}$$

The balanced modulator is therefore a two-input analog multiplier. It is called a balanced modulator because it produces an AM output with the carrier and modulating signals suppressed. It is also described as a double-sideband suppressed carrier (DSB-SC) modulator.

In Section 2.6.3, the four-quadrant analog multiplier was used to obtain an AM wave. From Equation (2.6.26), the output current is

$$\begin{aligned} I_2 = {} & m_1 V_{\text{dc}} I_{\text{dc}} + m_2 V_{\text{dc}} I_c \cos \omega_c t + m_3 V_s I_{\text{dc}} \cos \omega_s t \\ & + m_4 V_s I_c \cos \omega_c t \cos \omega_s t \end{aligned} \tag{4.4.17}$$

where the ms are constants.

To obtain the required output, the first three terms of Equation (4.4.17) have to be removed. The four-quadrant analog multiplier is realized only in integrated circuit form and one of the basic properties of circuit integration is that the process produces transistors and other components whose characteristics match very closely. Common-mode signals can be removed from the output by providing two identical circuits in parallel and taking the output as the difference between two corresponding points in the two circuits. The basic balanced modulator circuit is shown in Figure 4.19.

It should be noted that the major difference between the circuit of Figure 4.19 and that of Figure 2.33 is that in Figure 4.19, both transistor combinations Q_1–Q_3 and Q_2–Q_4 have a collector load R_L and the output is taken differentially. For specific biasing and detailed information, the reader may refer to the specification of the MC 1595 (four-quadrant multiplier produced by Motorola Semiconductor Products Inc.).

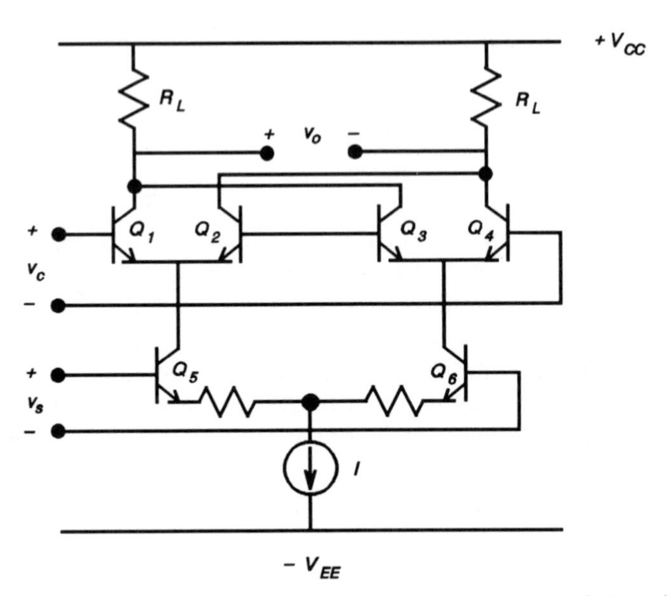

Figure 4.19. The four-quadrant multiplier used as a balanced modulator (DSB-SC). Note the difference between this and Figure 2.33.

4.4.2d The Integrator. The integrator is in fact a simple low-pass filter with a 6-dB/octave slope (20 dB/decade) with its corner frequency chosen to be below the lowest frequency present in the modulating signal. The basic integrator circuit is shown in Figure 4.20a with the time constant $\tau = RC$ chosen so that

$$\omega_{s,\,\min}\tau > 1 \qquad (4.4.18)$$

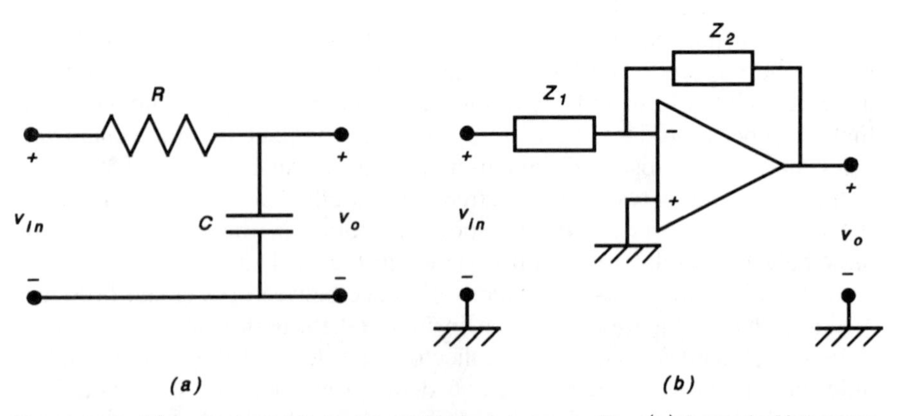

Figure 4.20. (a) A simple RC "integrator," that is, a low-pass filter. (b) A practical integrator can be realized by connecting a suitable feedback circuit to an operational amplifier.

Since the modulating signal contains frequencies spanning almost three decades, the highest frequency present will be attenuated by approximately 60 dB. Moreover the very long time constant required for this application makes the simple *RC* low-pass circuit unattractive.

A more attractive form of the low-pass filter is shown in Figure 4.20*b*. The output voltage of the operational amplifier is

$$v_o = -\frac{Z_2}{Z_1} v_{\text{in}} \tag{4.4.19}$$

When $Z_1 = R_1$ and $Z_2 = 1/(j\omega C)$,

$$v_o = -\frac{1}{j\omega CR_1} v_{\text{in}} \tag{4.4.20}$$

It can be shown that in the time domain

$$v_o(t) = -\frac{1}{(CR_1)} \int_0^t v_{\text{in}}(t)\, dt \tag{4.4.21}$$

The name integrator is therefore appropriate.

For an ideal operational amplifier (open-loop gain = infinity), the gain of the circuit shown in Figure 4.20*b* has to be infinite at dc since Z_2 is an open-circuit at dc. In a practical operational amplifier, the open-loop gain although large (about 10,000) is not precisely defined. To define the dc gain, a resistance R_2 is connected in parallel with C so that the dc gain is equal to $-(R_2/R_1)$. The output voltage is then modified to

$$v_o = -\frac{R_2}{R_1}\frac{1}{1 + j\omega CR_2} v_{\text{in}} \tag{4.4.22}$$

Example 4.4.2 The Integrator. Design an integrator to satisfy the following specifications:

(1) Gain of 0 dB at 1 kHz.
(2) Corner frequency at 10 Hz.
(3) What is the gain at 10 Hz and 15 kHz?

Solution. The required response is as shown in Figure 4.21. The relationship between two frequencies f_1 and f_2 on the basis of decades of frequency is given by

$$f_1 \times 10^n = f_2$$

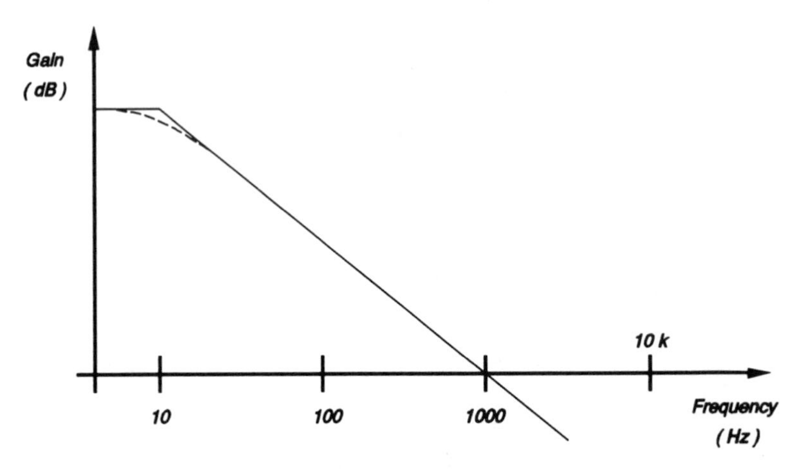

Figure 4.21. The required characteristics of the integrator specified in the example.

where n is the number of frequency decades. When $f_1 = 10$ Hz and $f_2 = 1$ kHz, $n = 2$.

The gain at 10 Hz $= 2 \times 20$ dB $= 40$ dB. The voltage gain of 40 dB corresponds to a voltage ratio of 100. Therefore

$$\frac{R_2}{R_1} = 100$$

Choosing $R_2 = 1$ MΩ makes $R_1 = 10$ kΩ. From Equation (4.4.2) it follows that at the corner frequency $\omega C R_2 = 1$ and C is therefore equal to 0.016 μF.

Using the results for 1 kHz, 15 kHz corresponds to 1.18 decades of frequency. Therefore, the gain at 15 kHz is

$$-20 \times 1.18 = -23.6 \text{ dB}$$

4.4.2e The Adder. The addition of two signals is most easily accomplished by using a resistive T network as shown in Figure 4.22. This circuit has two major disadvantages: The output signal is a fraction of the input signal and there will be a tendency for one signal to pull the other. To increase the isolation between the two signals and hence reduce the possibility of pulling, R_1 and R_2 must be increased relative to R_3. This aggravates the loss of signal power at the output of the circuit.

An adder circuit that can scale as well as amplify the signals to be added is shown in Figure 4.23. Equation (4.4.19) can be extended to give

$$v_o = -\frac{R_f}{R_1}v_1 - \frac{R_f}{R_2}v_2 - \frac{R_f}{R_3}v_3 \qquad (4.4.23)$$

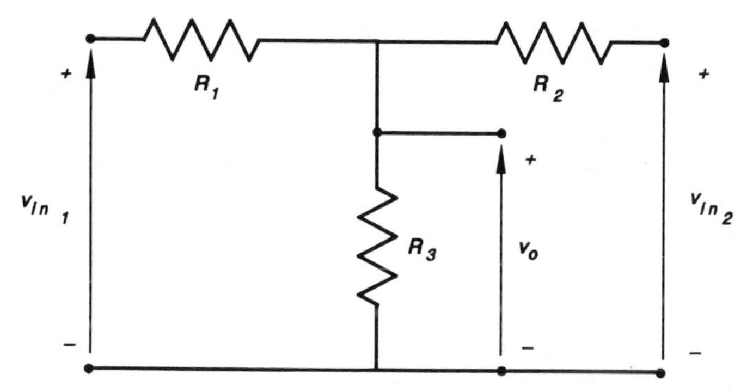

Figure 4.22. A simple resistive adder. The use of such a circuit would lead to severe signal attenuation.

The relative values of R_1, R_2, and R_3 determine the proportion of v_1, v_2, and v_3 in v_o, respectively, whereas the ratio R_f to R_1 determines the voltage gain for the voltage v_1, and so on.

4.5 STEREOPHONIC FM TRANSMISSION

In an effort to create a realistic sound presentation from recorded music, two microphones are used; one to record the sound as it is perceived on the right side and the second on the left side. During playback the right and left signals have to be fed to the right- and left-hand loudspeakers, respectively. The listener sitting in front of the speakers can distinguish the sounds of the different instruments as coming from their proper positions when the music was recorded. When this technique is applied to the cinema and television, a car approaching from one side of the screen can be heard in the proper position presented in the picture and the sound appears to move with the

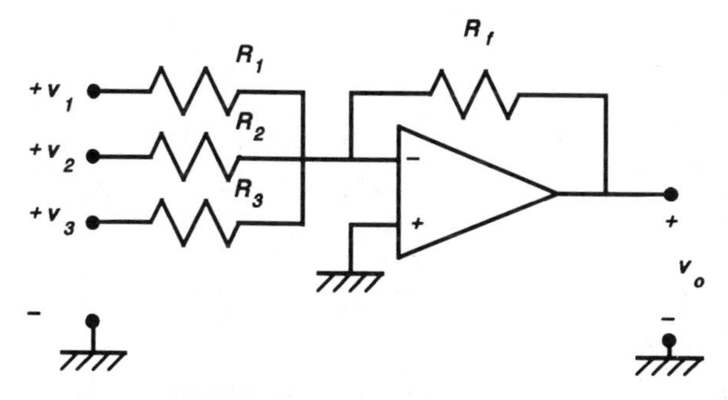

Figure 4.23. An operational amplifier type adder. This circuit enables the circuit designer to choose the gain required as well as the multiplication constant.

picture as the car moves across the screen. In a normal recording process, two separate channels are required for a stereophonic system. So it is expected that a stereophonic FM system will require twice the bandwidth of a monophonic channel. In fact it requires more bandwidth because the system had to be designed in such a way that listeners who own a monophonic FM receiver can tune in the stereophonic transmission and receive the same performance as they would if the original program had been monophonic.

4.5.1 System Design

The system diagram shown in Figure 4.24 is used to preprocess the modulating signal before it goes to frequency modulate a subcarrier using the Armstrong technique discussed in Section 4.4. The left signal $L(t)$ and the right signal $R(t)$ are fed to an adder and a subtracter (the signal is first inverted and then added) to produce $[L(t) + R(t)]$ and $[L(t) - R(t)]$, respectively. The $[L(t) + R(t)]$ signal is fed into one of the inputs of a three-input adder as shown. The $[L(t) - R(t)]$ signal is one of the inputs to a balanced modulator. The other input is a 38 kHz signal originally generated at 19 kHz (pilot frequency) and then doubled. The output of the balanced modulator is then the product

$$[L(t) - R(t)] \cdot A \cos 2\omega_p t \tag{4.5.1}$$

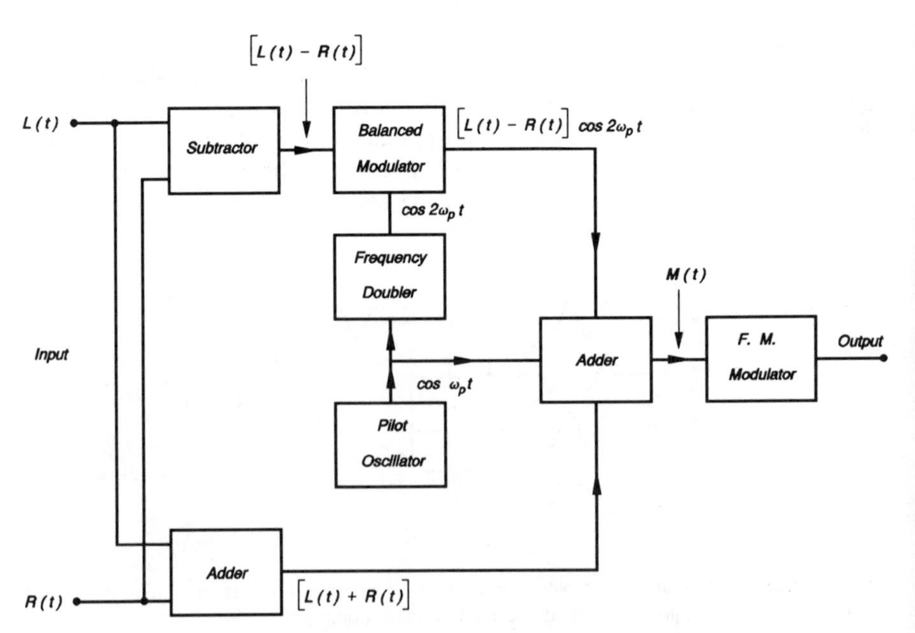

Figure 4.24. A block diagram showing the stages of processing of a stereophonic modulating signal for the Armstrong system FM transmitter.

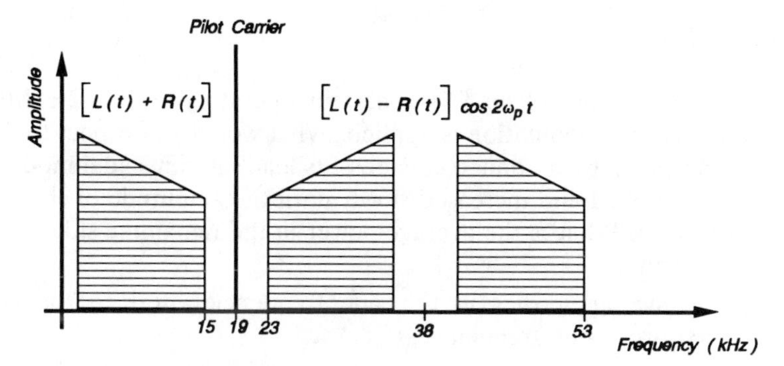

Figure 4.25. The spectrum of the composite modulating signal used in the stereophonic FM transmitter.

where $2\omega_p$ is the angular frequency of the doubled pilot oscillator frequency and A its amplitude. The other two inputs of the three-input adder are the pilot signal and the output of the balanced modulator. The output of the adder is then

$$M(t) = [L(t) + R(t)] + [L(t) - R(t)] \cdot A \cos 2\omega_p t + B \cos \omega_p t \tag{4.5.2}$$

where B is the amplitude of the pilot oscillator. The design of the circuits in all the component blocks shown in Figure 4.24 have been discussed in this and earlier chapters. The baseband spectrum of the stereophonic signal is shown in Figure 4.25. The processing of the stereophonic signal and its separation into right and left signals will be discussed in Chapter 5.

BIBLIOGRAPHY

K. K. Clarke and D. T. Hess, *Communication Circuits: Analysis and Design*, Addison-Wesley, Reading, MA, 1978.

J. J. DeFrance, *Communications Electronics Circuits*, Holt, Rinehart and Winston, Inc., 1966.

R. A. Moline and G. F. Foxhall, "Ion-Implanted Hyperabrupt Junction Voltage Variable Capacitors," *IEEE Trans. Electron Devices*, ED-19 (2), 267–273, 1972.

M. H. Norwood and E. Shatz, "Voltage Variable Capacitor Tuning: A Review," *Proc. IEEE*, 56 (5), 788–798, 1968.

F. G. Stremler, *Introduction to Communication System*, 2nd ed., Addison-Wesley, Reading, MA, 1982.

H. Taub and D. N. Schilling, *Principles of Communication Systems*, McGraw-Hill, New York, 1971.

PROBLEMS

4.1 The power output of an FM generator operating at $100/2\pi$ MHz is 10 W when no modulation is applied. What would you expect to see on a spectrum analyzer connected across its load? A sinusoidal modulating signal is applied and increased slowly until the amplitude of the carrier goes to zero. What is the average power in the remaining side frequencies? Calculate

(1) The average power in the side frequencies next to the carrier (first-order side frequencies).

(2) The average power in the second-order side frequencies.

(3) The average power in the third-order side frequencies.

(4) The average power in the remaining side frequencies.

4.2 A direct FM generator uses a varactor diode for which

$$C = \frac{C_0}{\sqrt{1 + 2V}}$$

where V is the reverse-bias voltage across the p–n junction. Assuming that the oscillator has a parallel LC tuned circuit such as shown in Figure 4.3,

(1) Write an expression for the frequency of operation as a function of V.

(2) Write another expression for the frequency of oscillation, f_0 when the dc voltage applied to the p–n junction is V_0.

(3) Assuming that v is a small time-varying voltage superimposed on V_0, calculate the slope (Hz/volt) of the frequency-voltage characteristics of the circuit about the oscillating frequency f_0 when $f_0 =$ 25 MHz and $V_0 = 5$ V.

4.3 A discriminator (frequency-to-voltage converter) is to be designed to operate at $10/2\pi$ MHz. You are provided with a center-tapped transformer whose secondary self-inductance is L and series resistance r for each half. Choose two capacitors of suitable values, to tune the secondary for minimum second harmonic distortion, given that $L = 0.2$ mH and $r = 40\ \Omega$ (some iteration may be necessary).

4.4 In an indirect FM transmitter, the carrier frequency f_c is 250 kHz with the maximum frequency deviation $\Delta f = 50$ Hz. This signal is to be frequency-multiplied in two stages (with a mixer between them) to bring the carrier frequency up to 100 MHz. Stage one has a multiplication factor of 64. Choose a suitable local oscillator frequency and the multiplication factor of the second multiplier. Calculate the center frequencies and bandwidths as the signal progresses through the frequency multiplier/mixer. Will this transmitter meet the frequency deviation specifications for FM stations?

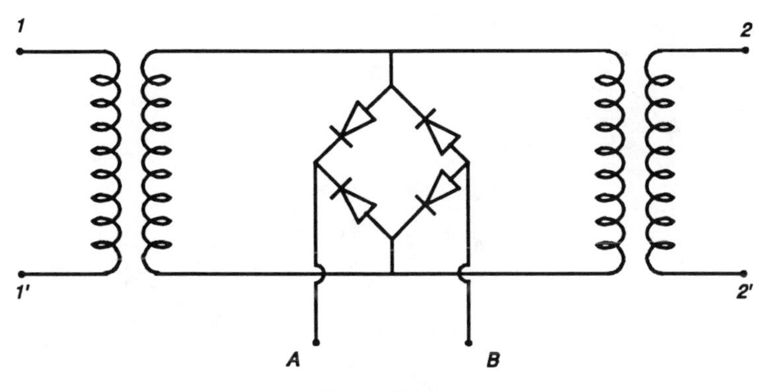

Figure P4.1.

4.5 In the circuit of Figure 4.18, C and R_3 are interchanged. Under what conditions will the circuit still behave like a 90-degree phase shifter?

4.6 If in an FM transmitter operating at 100 MHz, a sinusoidal modulating signal of 15 kHz causes the carrier frequency to deviate by 3 kHz, derive an expression for the modulated wave taking into consideration only the first two pairs of side frequencies. Calculate the amplitudes of the side-frequencies relative to the unmodulated carrier.

4.7 In the circuit shown in Figure P4.1, a sinusoidal signal

$$v_s = 1.5 \sin \omega_s t$$

is applied to terminals 1 and 1′ of the transformer T_1, where $\omega_s =$ 5 krad/sec. A second sinusoidal signal

$$v_c = 10 \sin \omega_c t$$

is applied to the terminals A and B, where $\omega_c = 50$ krad/sec. Assuming that the diodes have ideal characteristics,

(1) Sketch the waveform of the signal which appears across terminals 2 and 2′ of the transformer T_2 in relationship to v_s and v_c.

(2) Derive an expression for the output voltage v_o in terms of v_s and v_c.

(3) Calculate the amplitude of the component of v_o at angular frequency $n\omega_c$ where n is an integer.

(4) Comment on your result from (3).

(5) Suggest an application for the circuit.

You may assume that for a squarewave

$$f(t) = \frac{1}{2} + \frac{2}{\pi} \sum_{n=1}^{\infty} \frac{1}{n} \sin n\omega t$$

5

THE FREQUENCY MODULATED RADIO RECEIVER

5.1 INTRODUCTION

In amplitude modulation, the frequency of the carrier is kept constant while its amplitude is changed in accordance with the amplitude of the modulating signal. In frequency modulation, the amplitude of the carrier is kept constant, and its frequency is changed in accordance with the amplitude of the modulating signal. It is evident that, if a circuit could be found that would convert changes in frequency to changes in amplitude, the techniques used for detecting AM can be used for FM as well.

In Section 4.3.3d, three frequency-to-amplitude conversion circuits were discussed and their performance in terms of linearity and dynamic range were examined. It therefore follows that the FM receiver must have the same basic form as the AM receiver. The structure of the FM receiver is shown in Figure 5.1.

The superheterodyne technique is used in FM for the same reasons it is used in AM; it translates all in-coming frequencies to a fixed intermediate frequency at which the filtering process can be carried out effectively. The antenna is responsible for capturing part of the electromagnetic energy propagated by the transmitter. The basic rules of antenna design apply, but, because in commercial FM radio the frequency of the electromagnetic energy is between 88 and 108 MHz, it is practical to have antennas whose physical dimensions are within tolerable limits.

The radio frequency amplifier raises the power level to a point where it can be used in a mixer or frequency changer to change the center frequency to a lower frequency—the intermediate frequency. The mixer in conjunction with the local oscillator translates the in-coming radio frequency to an intermediate frequency of 10.7 MHz. There is nothing special about an intermediate frequency of 10.7 MHz except that it is a relatively low fre-

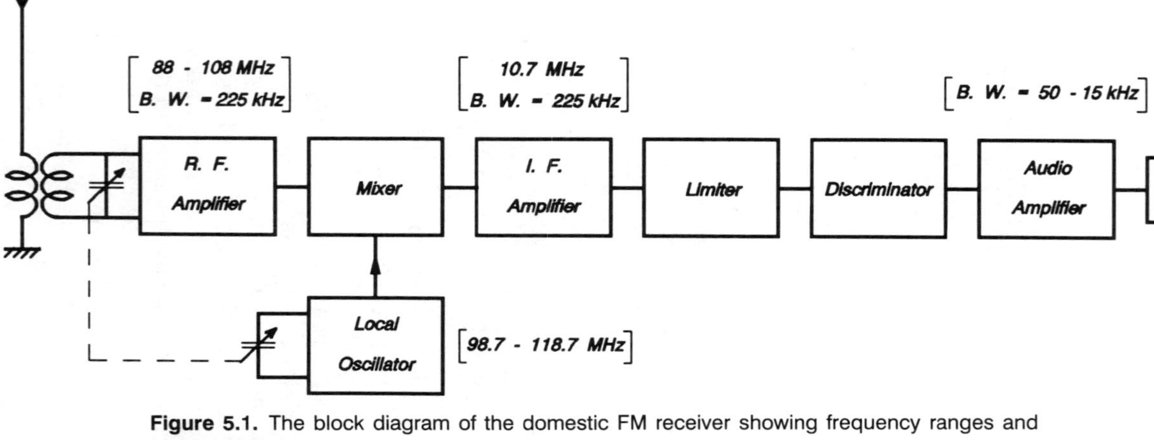

Figure 5.1. The block diagram of the domestic FM receiver showing frequency ranges and bandwidths.

quency at which the required values of Ls and Cs are large enough to reduce the effects of circuit strays. It is at this fixed frequency that filtering to remove the unwanted products of the mixing process and other interfering signals and noise takes place. The filtered signal then proceeds to the amplitude limiter. The need for the limiter becomes evident when one recalls that the FM signal is usually converted into an AM signal in the discriminator before it is detected. This means that any variation in the amplitude of the FM signal will be superimposed on the proper signal from the discriminator and hence will cause distortion. The amplitude limiter very severely clips the signal to a constant amplitude and also filters out the unwanted harmonics that are produced by the limiter. The signal then proceeds to the frequency discriminator (frequency-to-amplitude convertor) and onto the envelope detector. The audio amplifier raises the output of the envelope detector to a level suitable for driving a loudspeaker.

Although the structures of AM and FM receivers are similar, there are very important differences, which require different design and construction approaches. These are the following:

1. The higher carrier frequencies (88–108 MHz) used in FM requires small values of both L and C in the tuned the circuits used. This means that stray inductances and capacitances will constitute a larger percentage of the designed value and hence have a much greater effect on all tuned circuits. Although measures can be taken to incorporate the effects of fixed circuit strays in the design, there are other changes in element values due to such factors as temperature and vibration that can cause sufficient drift to necessitate retuning of the receiver during a program. The local oscillator is most vulnerable to stray elements, especially because it has to operate at a frequency 10.7 MHz above the carrier frequency. To ensure the stability of the oscillator, high-circuit Q factors, negative temperature-coefficient capacitors and automatic frequency control (AFC) are used. Some radio frequency amplifiers for FM front ends use distributed parameter circuit components such as coaxial and transmission lines.

2. With the intermediate frequency set at 10.7 MHz, the band from which image interference can originate is from 109.4 to 129.4 MHz. This frequency band is reserved for aeronautical radionavigation systems. It follows that one FM station cannot cause image interference for another, but the aeronautical radionavigation systems can. As in AM, image interference can be reduced by differentially amplifying the desired signal relative to the image signal. A high Q factor tuned radio frequency amplifier is used for this purpose.

3. The use of a high Q factor tuned amplifier in the radio frequency stage requires a very stable local oscillator frequency that will accurately track the incoming radio frequency and produce a minimum variation from the selected intermediate frequency. The local oscillator by itself is not capable of

this, but, used in conjunction with the AFC and other stabilizing measures, it can perform satisfactorily.

4. The ideal intermediate frequency filter should have a bandpass characteristic that is flat with infinitely steep sides. The flat top is required to avoid any frequency-dependent amplitude variation; the steep sides are required to eliminate interference from the unwanted products of the mixing process and from adjacent channels. Two techniques, both involving a number of successive stages of tuned circuits, are used to obtain an approximation of the ideal filter characteristics. In the first, all the tuned circuits have the same resonant frequency—this is known as *synchronous tuning*. In the second the resonant frequencies are placed at different points in the passband—this is known as *stagger tuning*.

The above qualitative as well as quantitative changes from the AM system discussed in Chapter 3 make a separate consideration necessary.

5.2 COMPONENT DESIGN

5.2.1 Antenna

An important point to remember is that an antenna is a reciprocal device; that is, it can be used both for transmitting signals, as well as for receiving them. An antenna structure that produces a good ground wave radiation pattern will have the same response when used in the receiving mode.

Commercial FM receivers commonly use two types of antennas: the vertical whip antenna commonly used for car radios and the dipole and folded dipole antennas used with other types of portable and nonportable FM radios. Assuming that the vertical whip antenna approximates a vertical grounded antenna, a half-wavelength antenna for the middle of the FM band will be approximately 1.5 m long. Such an antenna can be conveniently mounted on a vehicle. The field patterns of vertical grounded antennas were given in Figure 2.54. It can be seen that when the height h is less than $\lambda/2$, the response is limited to the ground wave. Commercial FM stations are designed to operate within a local area and therefore have antennas that ensure that most of the radiated power goes into the ground wave. An FM receiver with an antenna whose height is equal to or less than $\lambda/2$ will have a good response. The dipole and its variation, the folded dipole, are commonly used with nonportable FM receivers. They can be used in conjunction with directors and/or reflectors to increase the gain of the antenna. Such an arrangement is called a Yagi-Uda antenna.

Antenna design is beyond the scope of this text. However the interested reader is encouraged to refer to any standard text on antennas.

5.2.2 The Radio Frequency Amplifier

The purpose of the radio frequency amplifier is to boost the power of the incoming signal relative to all the other signals picked up by the antenna to a level that can be used in the frequency changer. A second function of the radio frequency amplifier is to act as a matched load to the antenna, so that the antenna signal is not reflected at the interface leading to the loss of efficiency.

The bandwidth of an FM signal in commercial radio was calculated to be approximately 240 kHz in Section 4.3.3a. With a carrier frequency of approximately 100 MHz, the required Q factor is therefore approximately 400. Such a high Q factor cannot normally be achieved in a simple tunable radio frequency amplifier and the practical approach is to use a lower Q circuit and to correct for this in the intermediate frequency stage that follows. An alternative technique uses a number of cascaded stages separated by buffer amplifiers. For the superheterodyne system to work, the local oscillator frequency must be set equal to the radio frequency plus the intermediate frequency for the commercial FM band, the standard intermediate frequency is 10.7 MHz. The frequency of the local oscillator must be variable from 98.7 to 118.7 MHz. This is normally not a problem since the ratio of the high frequency to the low frequency is only 1.2 : 1. The more important point is that the centre frequency of the radio frequency amplifier and the frequency of the local oscillator must maintain the difference of 10.7 MHz throughout the FM frequency band. In the AM system, the frequencies were low and some drift could be tolerated without serious deterioration of the signal. In the FM system, the frequencies are much higher and small percentage changes in one or both radio frequency and local oscillator frequency can cause large changes in the intermediate frequency. To overcome this problem, a system for automatic frequency control (AFC) of the local oscillator is used. The basic operation is similar to that described in Section 4.3.2.

5.2.3 Local Oscillator

The local oscillator can take any of the usual oscillator forms with a bipolar transistor as the active element. It must produce enough power to drive the mixer. The values of the inductors and capacitors have to be chosen to minimize the effects of circuit strays. As mentioned earlier, the local oscillator incorporates an AFC circuit for stable tracking with the radio frequency amplifier.

5.2.4 Frequency Changer

The basic frequency changer was discussed in Section 3.4.3. For this application, the dual-gate FET mixer has the advantage of low leakage of the local oscillator signal to the antenna via the radio frequency amplifier. Such a leakage and its subsequent radiation can cause interference with other communication and radionavigation equipment.

5.2.5 Intermediate Frequency Stage

The intermediate frequency for commercial FM radio is 10.7 MHz. The required bandwidth of the filter is approximately 240 kHz centered at 10.7 MHz giving a Q factor of approximately 45. It is usual to realize the filter in two or more stagger-tuned stages with suitable buffer amplifiers between them.

5.2.6 Amplitude Limiter

The radio frequency amplifier, the mixer, and the intermediate frequency amplifier should have in theory a flat amplitude response in their pass bands. In practice, this is not so. The result is that the signal emerging from the intermediate frequency amplifier has some variation of amplitude with respect to frequency. This is a form of AM and it must be removed if distortion is to be avoided.

The amplitude limiter was discussed in Sections 4.3.3a and 4.3.3b. It is worth noting that sometimes the amplitude limiter is preceded by an automatic gain control. This reduces the severity of the clipping action and hence the required signal power and the number of harmonics produced.

5.2.7 Frequency Discriminator

The purpose of the frequency discriminator is to convert relatively small changes of frequency (in a very high-frequency signal) to relatively large changes in amplitude with respect to time. The signal can then be demodulated using a simple envelope detector as discussed in Section 3.4.6. Two basic frequency discriminators discussed in Section 4.3.3d, serve to illustrate the concepts. In practice, a number of more sophisticated discriminators are used. Some of these will now be discussed.

5.2.7a Foster-Seeley Discriminator. The Foster-Seeley discriminator [1, 2] is similar to the balanced slope discriminator shown in Figure 4.11. The major difference is that it has two tuned circuits instead of three and both are tuned to the same frequency. This is a major advantage when the receiver is being aligned. A secondary advantage is that it has a larger linear range of operation than the slope discriminator. The basic circuit of the Foster-Seeley discriminator is shown in Figure 5.2.

Connected to the collector of the transistor, L_1 and C_1 are tuned to resonate at the frequency f_0 (the intermediate frequency of the receiver). The inductance L_1 is mutually coupled to a center-tapped inductor L_2. The

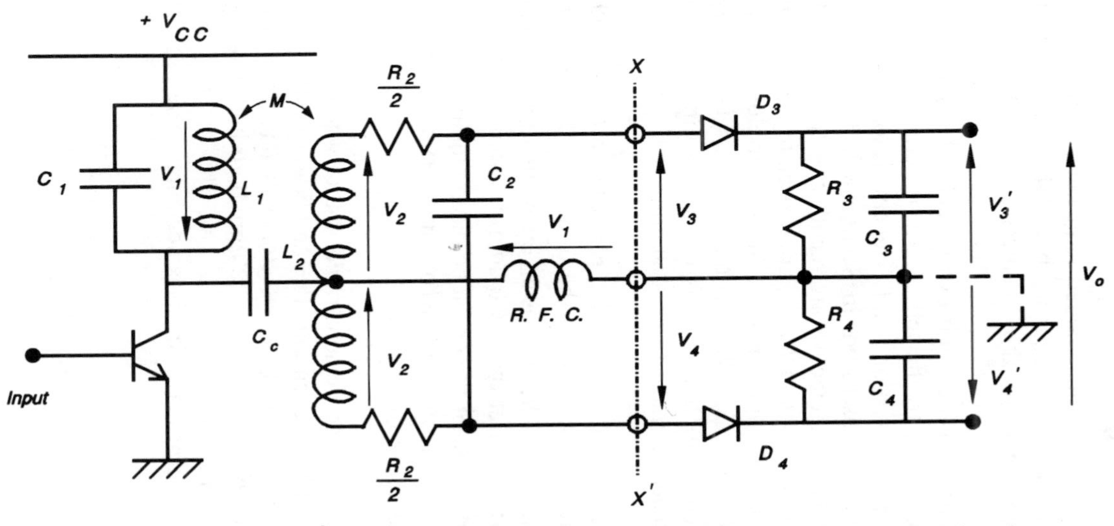

Figure 5.2. The Foster-Seeley discriminator. The line XX' divides the circuit into high (radio) and low (audio) frequency.

center-tapped inductor is connected by a coupling capacitor C_c to the collector of the transistor. The inductor L_2 and C_2 are tuned to resonate at frequency f_0. Two identical circuits consisting of a diode in series with a parallel combination of a resistance and a capacitance (D_3–R_3–C_3 and D_4–R_4–C_4) are connected across L_2 to form a symmetric circuit. A radio frequency choke (RFC, a high-valued inductance that can be considered to be an open circuit at high frequency but a short circuit at low frequency) connects the center-tap to the "neutral" node of the D_3–R_3–C_3 and D_4–R_4–C_4 circuits.

The circuit can be divided into two parts by the line connecting X—X'. The parts of the circuit to the left of XX' operate at a high-frequency f_0 with a relatively small deviation $\pm\Delta f$. The circuit to the right of XX' are two envelope detectors, as discussed, in Section 3.2. The high-frequency input voltages to the envelope detectors are rectified by the diodes D_3 and D_4 and the time constants R_3–C_3 and R_4–C_4 are chosen to smooth out the half-wave pulses but follow any slow changes in the envelope (amplitude) of the half-wave pulses. These slow changes represent low (audio) frequency.

Before proceeding to the analysis of the circuit, it is necessary in the interest of simplicity, to make three assumptions:

(1) The impedance of the coupling capacitance C_c is small enough to be considered a short circuit at the frequency of operation.
(2) The impedance of the RFC is an open circuit at the high-frequency f_0 but a short circuit at the low (audio) frequency.
(3) The neutral node of the envelope discriminators can be considered to be grounded, because the secondary circuit, including the envelope discriminators, is symmetric.

Now, all that needs to be demonstrated is that when a signal of frequency $f_0 \pm \Delta f$ is applied to the circuit, the amplitude of the voltage appearing across the inputs of the envelope detectors will vary proportionally to $\pm\Delta f$.

The tuned circuits L_1–C_1 and L_2–C_2 are both high Q factor circuits, but their mutual coupling coefficient M is low. This means that the secondary load coupled into the primary circuit is negligible. The primary current

$$I_1 \approx \frac{V_1}{j\omega L_1} \tag{5.2.1}$$

The voltage induced in the secondary

$$2V_2 = \pm j\omega M I_1 \tag{5.2.2}$$

where the $\pm$ sign depends on the relative directions of the primary and secondary windings. Assuming the positive sign and substituting for I_1

$$2V_2 = \frac{MV_1}{L_1} \tag{5.2.3}$$

Since the secondary circuit is tuned to resonance, the secondary current

$$I_2 = \frac{MV_1}{R_2 L_1} \tag{5.2.4}$$

where R_2 is the series resistance of the secondary circuit.

The voltage across the capacitor C_2 is

$$2V_2 = \frac{I_2}{j\omega_o C_2} \tag{5.2.5}$$

Substituting for I_2,

$$2V_2 = \frac{MV_1}{j\omega_o C_2 R_2 L_1} \tag{5.2.6}$$

The secondary voltage applied to one envelope discriminator is given by

$$V_2 = \frac{M}{j2\omega_o C_2 R_2 L_1} V_1 \tag{5.2.7}$$

It is now clear that at resonance, the primary voltage V_1 is at right angles to the secondary voltage V_2. The phasor diagram of the voltage applied to the inputs of the envelope discriminators is as shown in Figure 5.3(a). This can be modified by reversing the direction of one of the phasors representing V_2 as shown in Figure 5.3b.

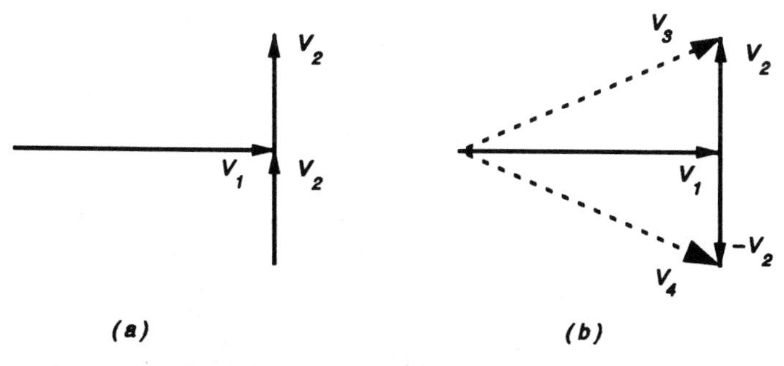

Figure 5.3. (a) The basic phasor diagram of the discriminator. (b) The phasor diagram when the FM carrier is unmodulated.

The discriminator input voltage phasors V_3 and V_4 are equal in magnitude, and, because the outputs of the envelope discriminators are proportional to the magnitude of the applied voltages, the output is zero when it is taken differentially. This means that the Foster-Seely discriminator has 0 V output at the resonant frequency.

The impedance of the secondary tuned circuit, at any given frequency ω, is

$$Z_2 = R_2 + j\left(\omega L_2 - \frac{1}{\omega C_2}\right) \tag{5.2.8}$$

However at resonance, $C_2 = 1/(\omega_0^2 L_2)$. Eliminating C_2 from Equation (5.2.8) gives

$$Z_2 = R_2 + j\left(\omega L_2 - \frac{\omega_0^2 L_2}{\omega}\right) \tag{5.2.9}$$

If we now define the Q factor at resonance as $Q_0 = \omega_0 L_2/R_2$ then Equation (5.2.9) can be written as

$$Z_2 = R_2 + j\omega_0 L_2\left(\frac{\omega}{\omega_0} - \frac{\omega_0}{\omega}\right) \tag{5.2.10}$$

Now consider relatively small changes of frequency about the resonant frequency ω_0 and define fractional detuning δ as

$$\delta = \frac{\omega - \omega_0}{\omega_0} \tag{5.2.11}$$

then

$$\frac{\omega}{\omega_0} = 1 + \delta \tag{5.2.12}$$

hence,

$$\frac{\omega}{\omega_0} - \frac{\omega_0}{\omega} = 1 + \delta - \frac{1}{1+\delta} = \delta\left(\frac{2+\delta}{1+\delta}\right) \tag{5.2.13}$$

For a high Q circuit at a frequency near the resonance, the fractional detuning, δ, is much smaller than 1; therefore,

$$\frac{\omega}{\omega_0} - \frac{\omega_0}{\omega} \approx 2\delta \tag{5.2.14}$$

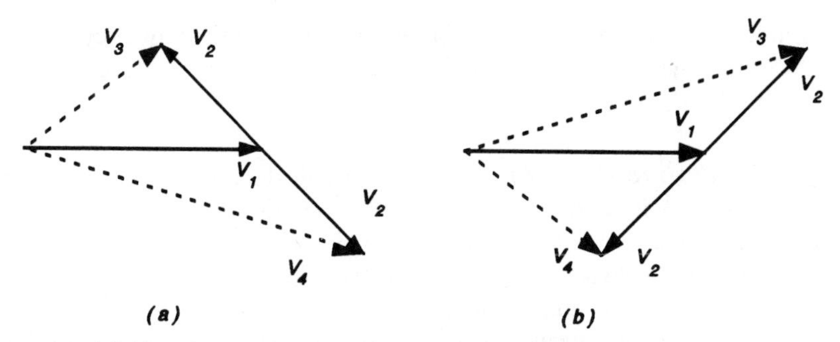

Figure 5.4. (*a*) The phasor diagram when the signal frequency is lower than the carrier frequency. (*b*) The phasor diagram when the signal frequency is higher than the carrier frequency.

Substituting into Equation (5.2.10),

$$Z_2 = R_2(1 + j2Q\delta) \tag{5.2.15}$$

Replacing R_2 in Equation (5.2.7) by Z_2 gives

$$V_2 = \frac{M}{j2\omega_0 C_2 L_1 R_2(1 + j2Q_0\delta)} V_1 \tag{5.2.16}$$

It is evident that V_1 and V_2 are no longer at right-angles to each other; the angle between them depends on the magnitude and sign of δ. When δ is positive, the angle between V_1 and V_2 is less than 90 degrees, and, when δ is negative, the angle is greater than 90 degrees or vice versa. The phasor diagrams for positive and negative values of δ are shown in Figure 5.4*a* and *b*, respectively. It is clear from these that the magnitudes of the input voltages V_3 and V_4 are unequal when δ has any value other than zero.

The amplitude-frequency characteristics of the Foster-Seeley discriminator are shown in Figure 5.5. A variation on the Foster-Seely discriminator

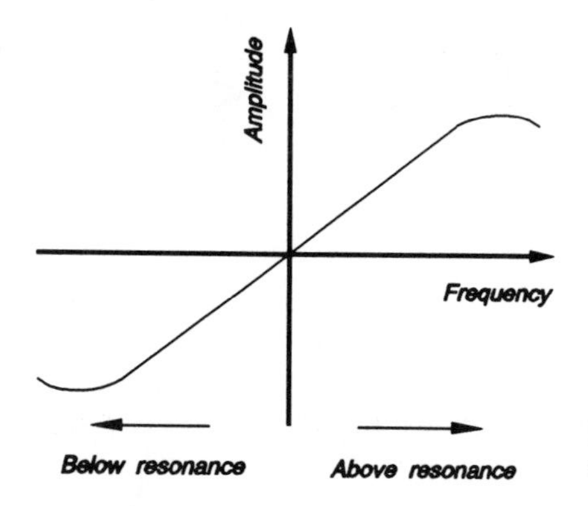

Figure 5.5. The amplitude-frequency characteristics of the Foster-Seeley discriminator.

that combines the functions of the amplitude limiter and frequency discriminator is called the ratio detector [3]. Its performance, however, leaves much to be desired.

5.2.7b Quadrature Detector. The basic circuit diagram of the quadrature detector is shown in Figure 5.6. All biasing circuit components have been omitted for the clarity of its operation. The circuit consists of a tuned amplifier Q_1, with a very high Q factor collector load. The input to the circuit is the output from the amplitude limiter, which is a frequency-modulated squarewave (i.e., a squarewave of fixed frequency with relatively small deviations in its zero crossings). Due to the high Q factor of the tuned circuit, the output from the amplifier is a sinusoid at a fixed frequency. The same squarewave is fed to the base of Q_4, which is a constant current source for the differential pair Q_2–Q_3. Therefore, current flows in the differential pair only when Q_4 is switched on. The sinusoid applied to Q_2 determines what proportion of the constant current in Q_4 flows through Q_2 as opposed to Q_3. It can be seen from Figure 5.7*a* that when the input signal is unmodulated, that is, when the phase difference between the sinusoid and the squarewave is fixed, the circuit can be adjusted so that Q_2 and Q_3 conduct equal currents giving a constant voltage across C_3. The time constant R_3C_3 holds the base of Q_5 at a dc value and the output remains constant.

When the input to the circuit is modulated, the sinusoid driving Q_2 is no longer coincident with the squarewave and Q_3 now conducts for a period proportional to the "phase shift" between the two signals. This can be seen

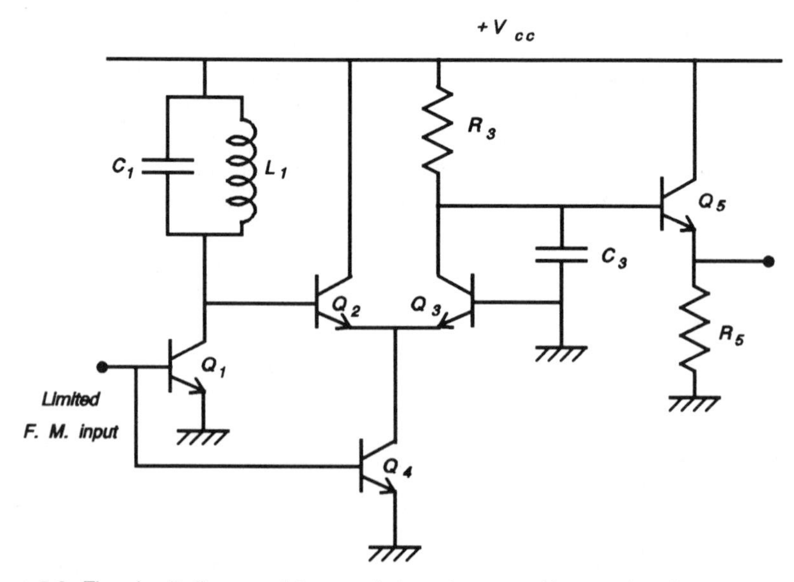

Figure 5.6. The circuit diagram of the quadrature detector with an emitter-follower to give a low impedance output.

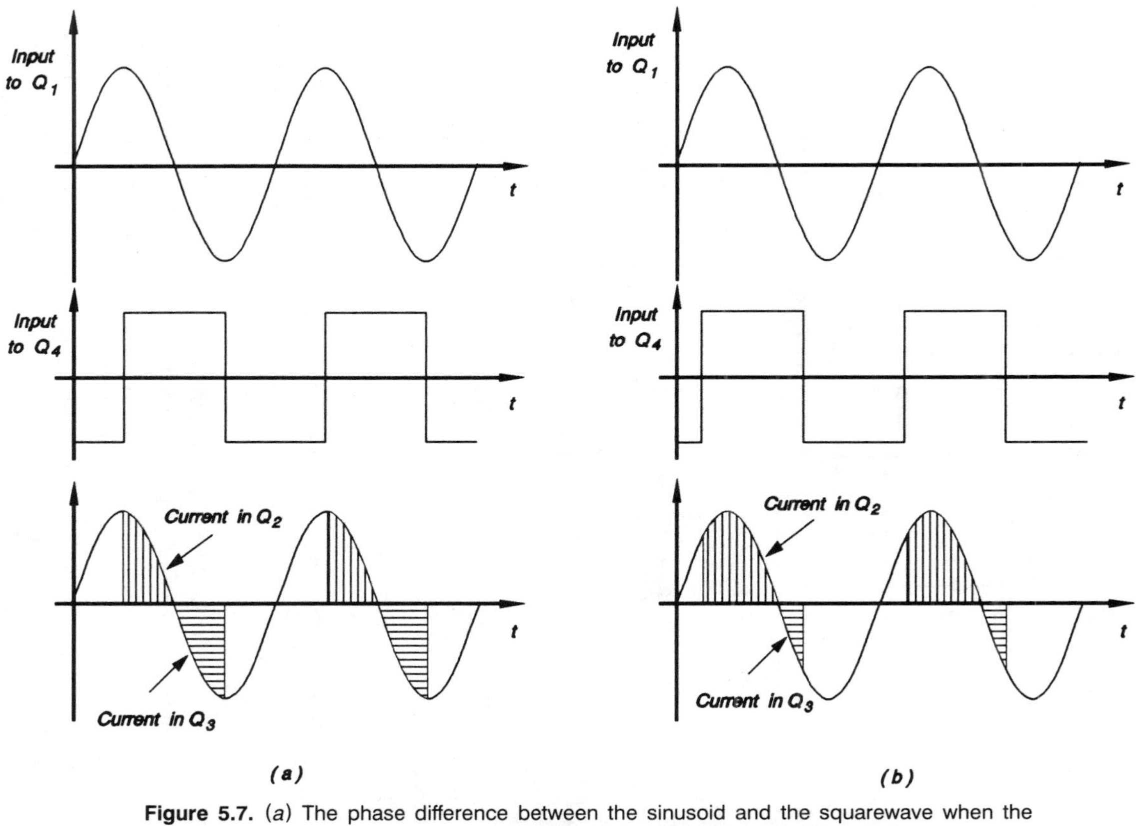

Figure 5.7. (*a*) The phase difference between the sinusoid and the squarewave when the carrier is unmodulated is such that currents of equal magnitude flow through Q_2 and Q_3 giving a dc output. (*b*) When the carrier is modulated the relative phase between the sinusoid and the square shifts and the currents in Q_2 and Q_3 are no longer equal; the dc changes its value — the changing dc is the audio frequency signal.

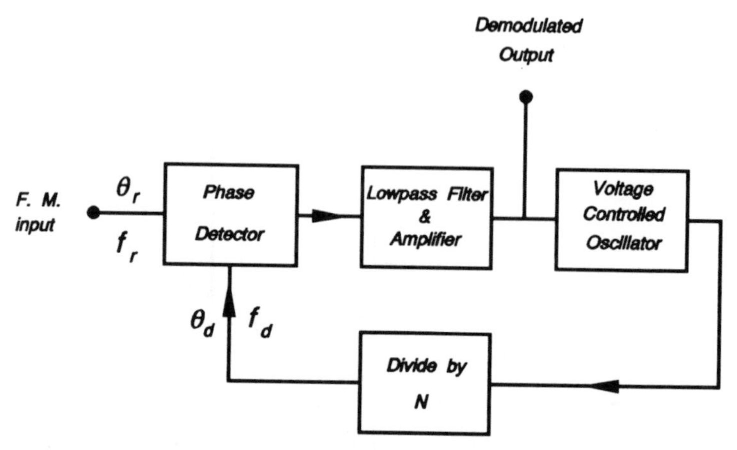

Figure 5.8. The block diagram of the phase-locked loop FM detector. Note that the relatively slow-varying dc required to keep the loop in lock is the audio frequency signal.

in Figure 5.7*b*. The voltage across C_3 is a slowly varying direct current and the output is in fact the audio frequency that was used to frequency modulate the radio frequency.

The following data should be noted.

(1) The amplifier Q_5 provides a low output impedance for the circuit.
(2) The structure of the circuit is suitable for realization in integrated circuit form.
(3) The time constant R_3C_3 is chosen to follow the changes in the amplitude of the audio frequency signal.
(4) When the two signals are 90 degrees out of phase (in quadrature), Q_2 and Q_3 conduct equally and this condition may be used as a datum.

5.2.7c Phase-Locked Loop FM Detector. The phase-locked loop [4] FM detector is the most complex of FM detectors, but it has the advantage that it can be realized in integrated circuit form where complexity is not necessarily a disadvantage. The basic system is as shown in Figure 5.8.

It consists of a phase detector that generates an output signal proportional to the difference between the phases of its two input signals (error signal). The output signal is amplified, low-pass filtered, and used to control a voltage-controlled oscillator (VCO) which usually operates at a higher frequency than the input signal. The output of the oscillator is divided by a suitable factor N to bring it to the same frequency as the input signal. This is the second input to the phase detector.

The error signal fed to the VCO causes it to change frequency, so that f_d moves closer to f_r. When the two frequencies are close to each other, the system *locks*; that is, the two frequencies become equal and their phase difference is zero. The control voltage from the low-pass filter is then dc. When the incoming signal changes its frequency and hence phase, the control voltage changes its value to keep the system locked. The excursions of the control voltage is, in fact, the required demodulated output.

The phase-locked loop FM detector has the advantage of having no *LC* tuned circuits. In its integrated-circuit form, it requires a number of resistors and capacitors for its proper operation. This information is usually provided by the manufacturer.

5.3 STEREOPHONIC FREQUENCY-MODULATED RECEPTION

The baseband frequency spectrum of the stereophonic FM signal was described in Section 4.5.2 and shown in Figure 4.25. The use of the signal to frequency modulate a suitable carrier remains essentially the same as in the case of monophonic transmission, except for the fact that the stereophonic spectrum has a larger bandwidth—more than three times larger. At the receiver, the signal is demodulated and the baseband information is recovered. Figure 5.9 shows a scheme for separating the left $L(t)$ and right $R(t)$ information and passing them on to their respective loudspeakers.

The first bandpass filter has a passband from 23 to 53 kHz and it is used to separate the double-sideband suppressed-carrier (DSB-SC) signal which contains the left-minus-right signal $[L(t) - R(t)]$. The second bandpass filter is a narrowband filter centered on 19 kHz, and it separates the pilot carrier. The third filter is a low-pass filter with a cut-off frequency of 15 kHz, and it separates the left-plus-right $[L(t) + R(t)]$ (i.e., the monophonic) signal from the other two.

The DSB-SC signal cannot be demodulated unless the carrier is reinstated. Because the DSB-SC signal was obtained in the transmitter from the pilot oscillator followed by a frequency doubler, the process is repeated in the receiver to obtain the carrier at 38 kHz. The carrier is then fed into the synchronous demodulator to produce a baseband signal containing the $L(t) - R(t)$ information, and, after an amplification of two, it is combined with the $L(t) + R(t)$ signal in an adder and subtracter, respectively, to produce $L(t)$ and $R(t)$.

5.3.1 Synchronous Demodulation

Synchronous demodulation of a DSB-SC signal can be achieved simply by multiplying the DSB-SC signal $[L(t) - R(t)]\cos 2\omega_p t$ by the synchronized

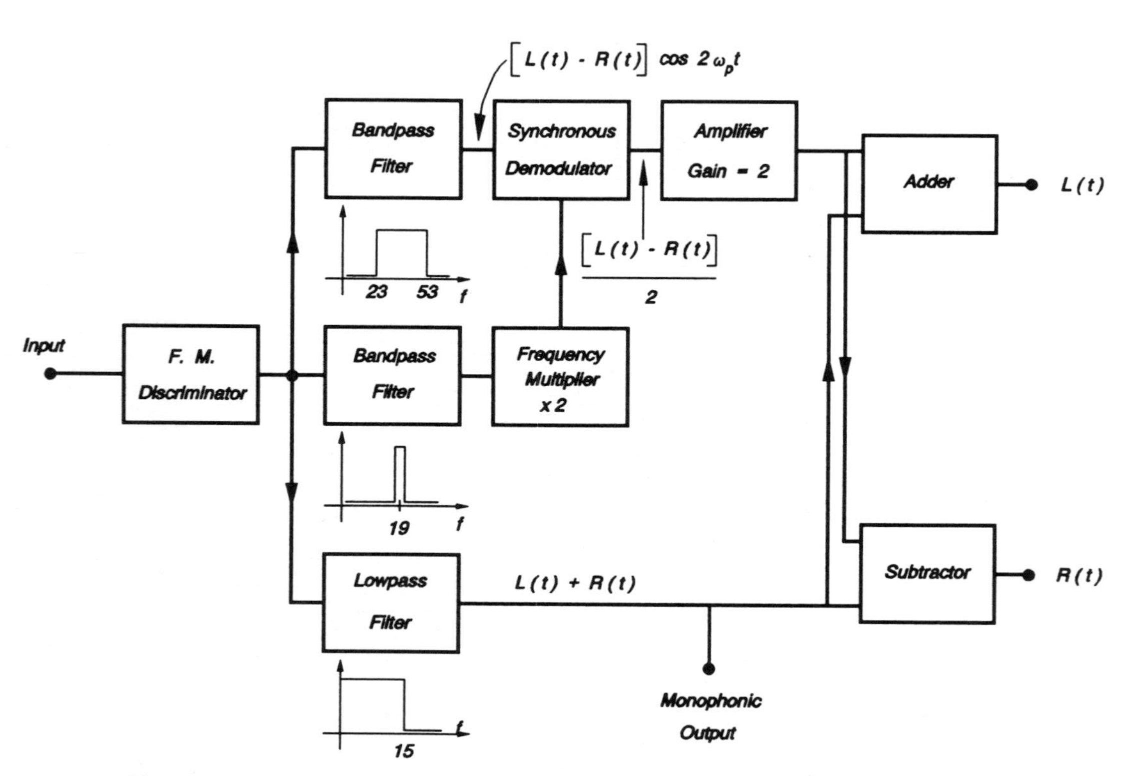

Figure 5.9. The block diagram of the stereophonic FM receiver showing the ideal characteristics of the various filters required and the access point for listeners with monophonic receivers.

signal $\cos 2\omega_p t$. The result is

$$v(t) = [L(t) - R(t)]\cos^2 2\omega_p t \tag{5.3.1}$$

$$= \tfrac{1}{2}[L(t) - R(t)](1 + \cos 4\omega_p t) \tag{5.3.2}$$

The required signal $\frac{1}{2}[L(t) - R(t)]$ is easily separated from the signal at $4\omega_p$.

5.3.2 Stereophonic Receiver Circuits

Filter design is beyond the scope of this book. A list of references on filter design can be found at the end of Chapter 3. The design of frequency multipliers was discussed in Section 2.5. The design of the synchronous demodulator (two-input analog multiplier) was discussed in Section 4.4.3c. The adder/subtracter was the subject of Section 4.4.3e.

REFERENCES

1. D. E. Foster and S. W. Seeley, "Automatic Tuning Simplified Circuits and Design Practice," *Proc. IRE*, 25, 289, 1937.
2. S. W. Seeley, *Radio Electronics*, McGraw-Hill, New York, 1956.
3. S. W. Seeley and J. Avins, "The Ratio Detector," *RCA Rev.*, 8, 201, 1947.
4. R. E. Best, *Phase-Locked Loops*, McGraw-Hill, New York, 1984.
5. J. J. DeFrance, *Communication Electronics Circuits*, 2nd ed. Rinehart Press, San Francisco, 1972.
6. H. Stark and F. B. Tuteur, *Modern Electrical Communications*, Prentice-Hall, Englewood Cliffs, NJ, 1979.

PROBLEMS

5.1 A bipolar transistor operating at a frequency of 100 MHz has the following Y parameters: $y_{11} = 9.84 + j9.1$, $y_{12} = 0.01 - j0.68$, $y_{21} = 60.7 - j95.8$, and $y_{22} = 0.60 + j1.8$ measured in mSiemens in the common emitter configuration with 0.6 V peak-to-peak appearing across the base emitter junction. Using this device, design an oscillator to supply maximum power to a suitable conductive load. What modifications would you make to the oscillator to obtain a local oscillator for an FM receiver operating in the range from 88 to 108 MHz?

5.2 Describe with the aid of a block diagram the superheterodyne technique for the reception of commercial FM signals. You have an AM/FM receiver and you wish to make changes to the configuration of the receiver by introducing a second mixer to translate the intermediate frequency from 10.7 MHz to the standard intermediate frequency of 455 kHz used in the AM part of the receiver. Redesign the block diagram, giving all the necessary details to achieve your wish. What are the advantages and disadvantages of your new system?

5.3 Using the characteristics of the FET given in Figure P3.2, design an amplifier tuned to 100 MHz with a voltage gain of 15. The dc supply voltage is 15 V and you may allow 2 mA dc to flow through the drain. The -3-dB bandwidth should be no larger than 250 kHz when the amplifier is connected to a second stage whose input impedance is 250 Ω resistive.

5.4 An FM receiver is tuned to a transmitter operating at 100 MHz. The modulation is a sinusoidal signal of frequency 15 kHz. The intermediate frequency is 10.7 MHz and the Q factor of the intermediate frequency amplifier is 100. Estimate the maximum allowable drift in the local oscillator frequency assuming that the $J_3(m_f)$ terms must lie within the -3-dB bandwidth of the intermediate frequency amplifier.

5.5 Give two reasons why a double-sideband suppressed-carrier (DSB-SC) modulation system is not commonly used in communication systems. Name one communication system in which it is used giving the relevant details. An in-coming signal in a communication receiver is given by

$$\nu(t) = \tfrac{1}{2}\beta[\cos(\omega_c - \omega_m)t - \cos(\omega_c + \omega_m)t]$$

where ω_c is the carrier angular frequency and ω_m is the modulating signal angular frequency. What do you need to demodulate the signal and how would you carry out the demodulation? Support your answer with suitable diagrams and derived formulas. Can you think of a second demodulation technique for the signal?

5.6 The dual-gate FET whose characteristics are given in Figure P5.1 is to be used in the design of a frequency changer in a commercial FM receiver whose intermediate frequency is 10.7 MHz. The output of the radio frequency amplifier is applied to gate 1 and it is 0.4 V peak-to-peak at a frequency of 100 MHz. The signal from the local oscillator applied to gate 2 is a squarewave whose voltage varies from +4 V to a negative value sufficient to cut off the drain current. Design a suitable mixer with the following specifications:

(1) Direct current power supply voltage, 12 V

(2) Average drain current, 4 mA

(3) -3-dB bandwidth, 240 kHz

Choose all other circuit values and in every case justify your choice. Calculate the ratio of the intermediate frequency power output to the radio frequency power input.

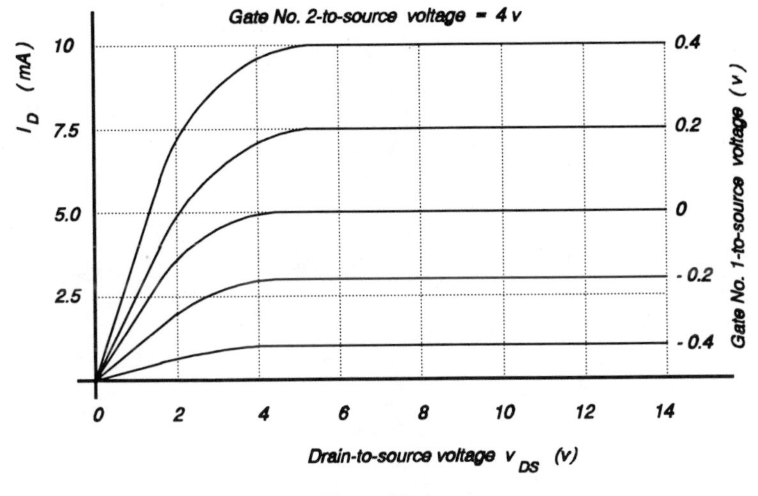

Figure P5.1.

5.7 The purpose of the discriminator in an FM receiver is to convert variation in frequency to variation in amplitude. Describe the operation of the Foster-Seeley discriminator and demonstrate its frequency-to-amplitude conversion ability. What measures should be taken prior to the signal entering the discriminator to reduce distortion and why?

PART 2

TELEVISION

6

THE TELEVISION TRANSMITTER

6.1 INTRODUCTION

The transmission of video images depends on a scanning device that can break up the image into a grid and measure the brightness of each element of the grid. This information can be sent serially or in parallel to a distant point and used to reproduce the image. It is evident that the smaller the size of the grid element, the better the definition of the image.

One of the simplest devices that can measure the brightness of light is the phototube. It consists of a cathode coated with a material that gives off electrons when light shines on it and an anode that can collect the emitted electrons when a suitable voltage is applied to it. The cathode and anode are enclosed in an evacuated glass envelope. The number of electrons emitted by the cathode is proportional to the intensity of the light impinging on it. Assuming complete collection of the electrons, the current in the resistor R shown in Figure 6.1 will be proportional to the light intensity and so will the voltage across R.

A primitive video signal can be generated by using a 3×3 matrix made up of phototubes as shown in Figure 6.2. For simplicity we assume that the tree is black and its background is white. A suitable lens focuses the image of the tree onto the matrix of phototubes. It is clear that the voltage output from phototubes $(1, 1)$, $(1, 3)$, $(3, 1)$ and $(3, 3)$ will be high; all others will be low. The voltages so obtained can be transmitted and used to control the brightness of a corresponding 3×3 matrix of lights at a distant point giving a vague idea of what the tree looks like! The picture detail can be improved by increasing the number of elements in the matrix so that each element corresponds to the smallest area possible. The assumption of a black tree on a white background is no longer necessary since with increasing detail different shades of grey can be accommodated.

The information may be sent along individual wires linking the phototube to the light matrix (parallel transmission) but this would be very expensive and impractical for any system other than the simple one described here. A

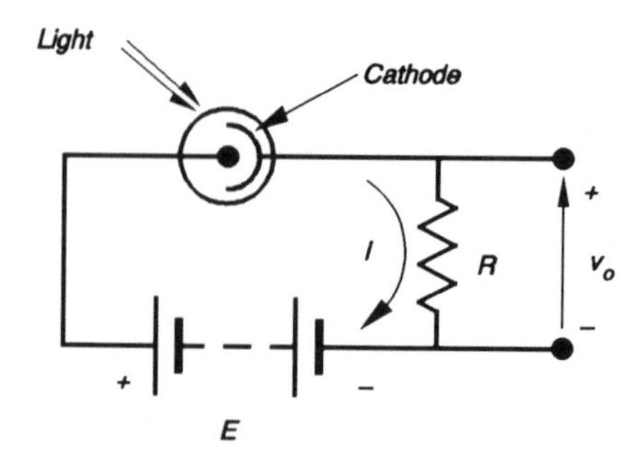

Figure 6.1. The phototube with its added circuitry to convert light intensity to voltage.

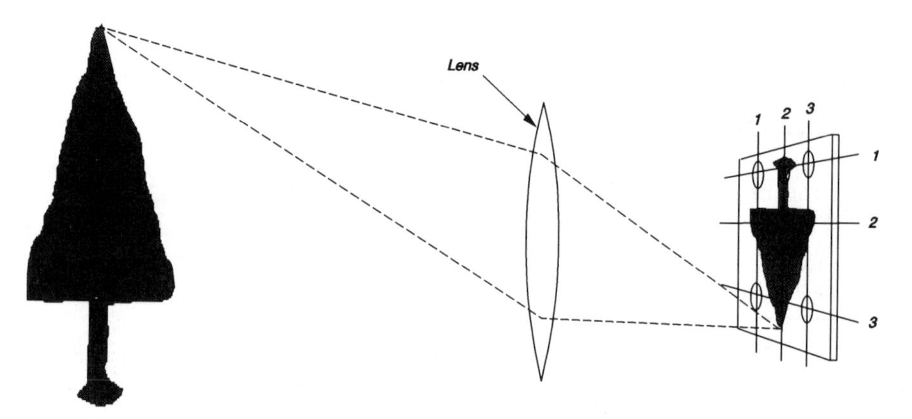

Figure 6.2. Generation of a primitive video signal using a 3 × 3 matrix of phototubes.

better system would one in which the voltage from each phototube is scanned in some given order and the voltage and position of each phototube are sent on a single wire to the receiving end for reconstruction (serial transmission). The price to be paid for reducing the number of wires is the increased complexity introduced by the scanner and a system for coding and decoding the voltage and position information at the transmitter and receiver.

6.2 SYSTEM DESIGN

Figure 6.3 shows the basic components of a television transmitter. A system of lenses focuses the image onto a camera tube, which collects and codes the information about the brightness and position of each element of the matrix forming the picture by scanning the matrix. A pulse generator supplies pulses to the camera to control the scanning process. The output from the camera goes to a video amplifier for amplification and for the addition of extra pulses to be used at the receiver for decoding purposes.

A microphone picks up the sound associated with the picture and after amplification the signal is fed to the audio terminal of a frequency modula-

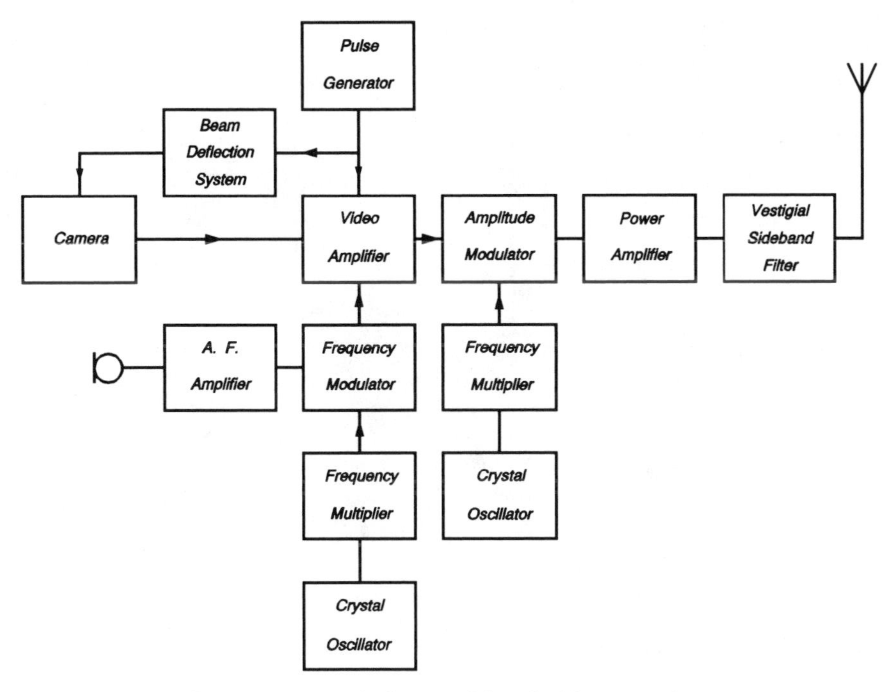

Figure 6.3. A block diagram of the television transmitter.

tor. The carrier signal supplied to the modulator is a 4.5 MHz signal, generated by a crystal-controlled oscillator at a lower frequency and multiplied by an appropriate factor. The FM signal carrying the audio information is added to the video signal. The output of the video amplifier consisting of the video signal, receiver control pulses, and the FM signal is fed to the amplitude modulator. The carrier of the amplitude modulator is supplied by a second crystal oscillator and associated multiplier, which produce a signal of frequency within the band 54 to 88 MHz (VHF). The radio frequency power amplifier boosts the power to the legally determined value, and the vestigial sideband filter removes most of the lower sideband signal before it goes to the antenna for radiation.

6.3 COMPONENT DESIGN

6.3.1 Camera Tube

6.3.1a Iconoscope. The first practical video camera tube invented by the American scientist Vladimir Zworykin [4, 5] is best viewed as a progression from the primitive phototube arrangement discussed earlier. In Zworykin's iconoscope, he replaced the matrix of phototubes with what he called the *mosaic*. This was made up of a very large number of droplets of photosensi-

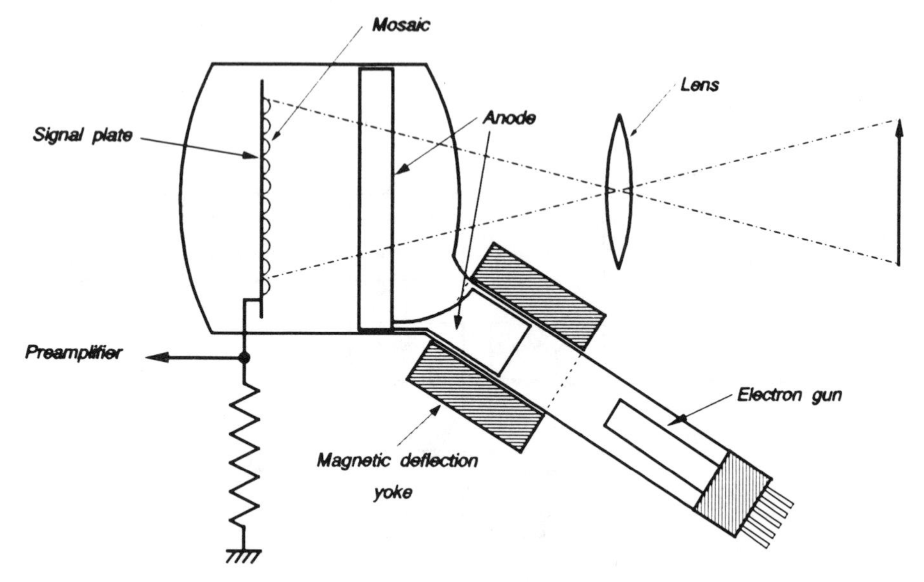

Figure 6.4. A cross-sectional view of the iconoscope. The two parts of the anode are cylinders formed on the inner wall of the tube and neck. Reprinted with permission from H. Pender and K. McIlwain (eds.), *Electrical Engineers Handbook*, 4th ed., Wiley, 1967, pp. 15–21.

tive material on one side of a sheet of mica. The other side of the mica sheet was covered with a very thin layer of graphite—the signal plate. A cross section of the iconoscope is shown in Figure 6.4.

The mosaic and the signal plate constitute a large number of tiny capacitors, which share one common plate. When the image is projected onto the mosaic, each individual droplet of photosensitive material emits electrons proportional to the light intensity falling on it. These electrons are collected by an anode placed close to the mosaic. The mosaic is now a picture painted with electric charge. The transformation of the charge into a voltage output is carried out by the electron gun and the associated circuitry.

The electron gun produces a very narrow beam of electrons focused on the mosaic. The arrival of these new electrons have the effect of discharging the tiny capacitor on which they fall. Since the number of electrons required to discharge the capacitor is proportional to the charge induced by the light intensity, the electron gun current is a function of the charge present and hence of the light intensity. The electron gun current is then proportional to the voltage that appears across the resistor R.

To transform the picture into a video signal, it is necessary make the beam of electrons sweep across the mosaic in a series of orderly lines. The electron gun has two deflection systems for scanning the picture. The first moves the electron beam at a constant speed along a horizontal straight line and returns it very quickly to the start ready for the next sweep (horizontal trace). The second system controls the vertical position of the beam and ensures that each line is swept before returning the beam to the top of the picture ready for the next frame (vertical trace).

Two major disadvantages of the iconoscope are the high light intensity required to obtain acceptable quality images and the production of secondary electrons from the photosensitive material during the scanning process. The secondary electrons cause noise (false information) in the video signal. In due course, the iconoscope was replaced by the image orthicon, another invention of Zworykin.

6.3.1b Image Orthicon. In the image orthicon [6] the two functions of the mosaic in the iconoscope, which are (a) the production of the image in terms of charge and (b) the target for the electron beam scan, are separated. Figure 6.5 shows a cross section through the image orthicon.

The camera lens focuses the image onto the photocathode. Every small segment of the photocathode emits electrons proportional to the amount of light falling on it. The electrons are attracted to the target by the positive voltage applied to it. The focus coil creates a magnetic field, which ensures that the electrons travel to the target in straight parallel paths. The target is a very thin sheet of glass that has a high conductivity through its thickness but low conductivity across points on its surface. The electrons striking the target dislodge secondary electrons from the photocathode side and are immediately captured by the very fine wire mesh, called the *target screen*, which has a relatively low positive voltage. This leaves a positive charge on the photocathode side of the target. Due to the high conductivity through the thickness of the target an identical pattern of positive charge is produced on its other side.

The electron gun positioned at the other end of the evacuated tube produces a narrow stream of electrons, which are accelerated toward the target by the high positive voltage on the accelerator anode. The *accelerator anode* is a graphite coating on the inside of the neck of the tube. If the electrons were allowed to strike the target without any control of their speed, it is clear that they would produce secondary electrons. To control the speed of the electrons on arrival at the target, a second ring of graphite coating, called the *decelerator grid*, is provided. By adjusting the voltages on the accelerator anode and the decelerator grid, the speed of the electron stream at the target can be controlled to ensure that no secondary electrons are emitted. The electron stream striking an elemental area of the target uses some of the electrons to neutralize the positive charge left there by the image. The remaining electrons are attracted in a backward direction by the accelerating anode, but they take a different path from before. By placing an electron collector in the appropriate position in the neck of the tube, the returning electrons can be collected and amplified by an electron multiplier. The output of the electron multiplier is converted into a voltage inversely

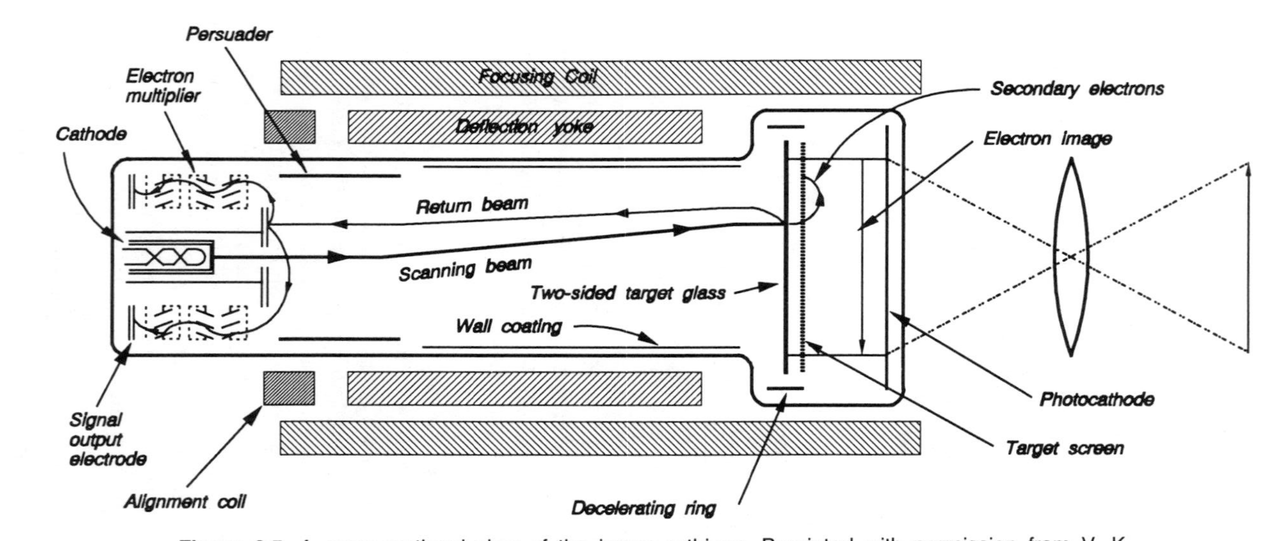

Figure 6.5. A cross-sectional view of the image orthicon. Reprinted with permission from V. K. Zworykin and G. Morton, *Television*, 2nd ed., RCA, 1954, p. 367.

proportional to the amount of light impinging on the elemental area of the photocathode. By using a scanning system in conjunction with the electron gun (deflection coil), the complete image can be produced at the output in the form of a varying voltage, which is a function of the brightness of all the elements of the image projected on the photocathode.

6.3.1c Vidicon. The vidicon is not as sensitive as the image orthicon, but it is generally smaller in size and therefore well adapted to field applications such as surveillance, where high resolution is not critical. Its principle of operation is different from both the image orthicon and the earlier iconoscope as it relies on a change of photoconductivity as a function of light intensity. An electron-scanning system is used to extract the video information.

6.3.2 Scanning System

In Section 6.3.1, the role of the electron beam scanning system in the production of the video signal was discussed. A series of pulses are generated that control the initiation of the horizontal sweep of the electron beam across the target (horizontal trace) from left to the right. A blanking pulse is used to cut off the beam while it is returned to the left-hand side ready for the next sweep (horizontal retrace). During this period, the vertical trace circuit moves the beam down just the right distance for the second line to be swept. When the whole frame has been scanned, a reset pulse returns the beam to the top left-hand corner ready to repeat the process. The beam is blanked during the reset. All the control pulses are added to the video signal and used at the receiver to synchronize the receiver to the transmitter. All the pulses are derived from a 60-Hz power supply or from a 31.5-kHz crystal-controlled oscillator and divided down to the appropriate frequency.

Black-and-white television in North America uses 525 horizontal line scans to cover the image. If the lines were scanned sequentially, the persistence of the phosphor on the first line would have faded before the last line would have been scanned at the receiver. This would lead to an annoying flicker on the picture tube, especially at high levels of brightness. The scan is therefore interlaced; that is, all odd numbered lines are scanned first from 1 to 525 and then the beam is returned to line 2 and all even numbered lines are then scanned. So each frame has $525/2$ lines and they are scanned 60 times per second. This means that there are 262.5×60 lines scanned per second. The horizontal scan therefore operates at 15.75 kHz. It must be pointed out that not all the 525 lines are used for picture production; some are used for vertical retrace and other controls as well as for equalization and synchronization.

The period for the horizontal scan is 63 μsec and is divided into 56 μsec for the trace and 7 μsec for the retrace. The beam deflection system is

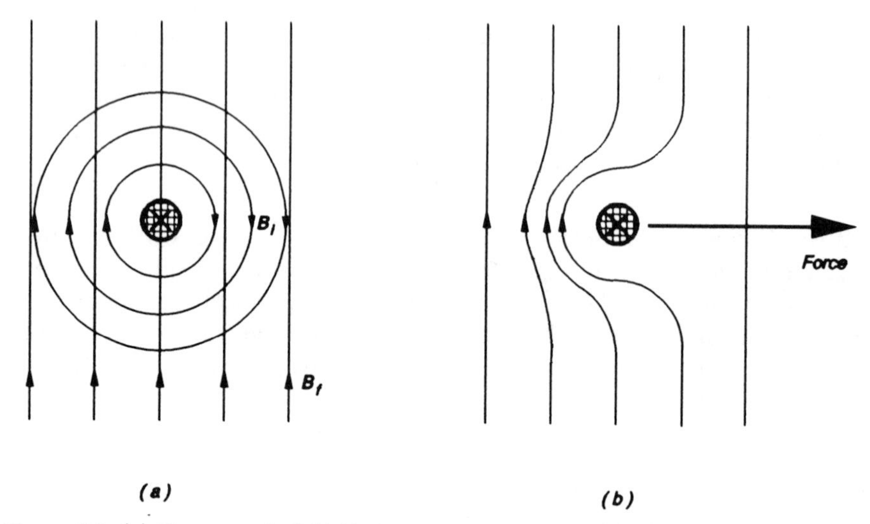

Figure 6.6. (*a*) The magnetic field B_i due to a current i flowing perpendicular to the page superimposed on a uniform magnetic field B_f with no interaction. (*b*) The combined effect of the two magnetic fields generates a force at right angles to the conductor and magnetic field.

therefore required to move the beam from the left of the picture to the right in 56 μsec and return it in 7 μsec.

The mechanism for deflecting the electron beam relies on the fact that a conductor experiences a force mutually orthogonal to the field and the direction of the current when it is placed in a magnetic field. Motion will occur if the conductor is free to move. In the camera tube, the electron gun and its associated components produce a stream of electrons. Each of these electrons carries a charge, and a moving charge is a current. The electron beam, therefore, experiences a force on it, and it will be deflected from it original path because it is free to move. Figure 6.6*b* illustrates the situation.

Consider the current in the conductor (the electron beam) to be flowing into the plane of the page. The right-hand rule gives the direction of the magnetic flux B_i associated with the current in the conductor as concentric circles in a clockwise direction. Assuming the conductor is placed in a magnetic flux B_f, which is in the vertical upward direction, the two fluxes will interact to produce the distorted field shown in Figure 6.6*b*. The flux on the left of the conductor is strengthened, whereas that on the right is weakened; the conductor therefore moves to the right. When trying to determine the direction of electron deflection, it is necessary to remember that electron flow is opposite to current flow.

The deflection of the electron beam, therefore, depends on designing a circuit to supply the appropriate current to the horizontal deflection coil so that it produces a magnetic field that is a linear ramp with respect to time.

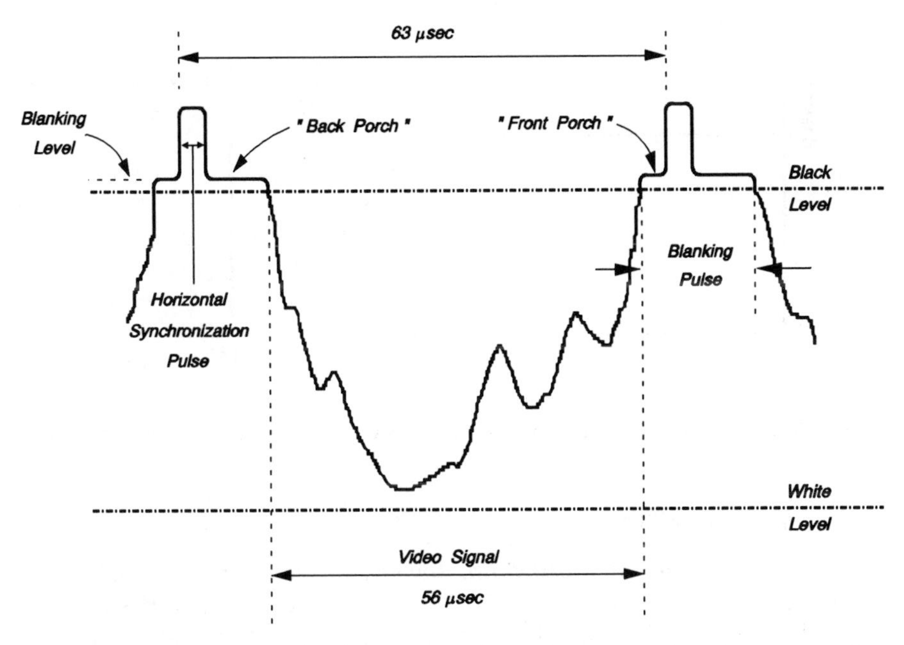

Figure 6.7. A typical composite video signal with blanking and horizontal synchronization pulses.

The vertical deflection system also has to be supplied with a current that produces a linear ramp magnetic field, and its operation has to be synchronized to the horizontal trace so that it repeats the process after 262.5 cycles.

The video signal, the control pulses, and the output of the FM modulator are combined in the video amplifier. A typical composite video signal is shown in Figure 6.7. Note the following:

(1) Only horizontal synchronizing pulses are shown (~ 5 μsec); vertical synchronization pulses (~ 190 μsec) are distinguished from the horizontal ones by the difference in their width.
(2) The blanking pulse starts before and ends after the horizontal synchronization pulse.
(3) The video signal contains the radio frequency carrier of 4.5 MHz frequency modulated by the audio signal. The spectrum of the composite signal at the output of the video amplifier is shown in Figure 6.8.

6.3.3 Audio Frequency and FM Circuits

Various audio frequency amplifiers were discussed in Section 2.7. The design of the crystal oscillator was discussed in Section 2.4.7. and the frequency multiplier in Section 2.5. Design details for the FM modulator can be found in Section 4.4.

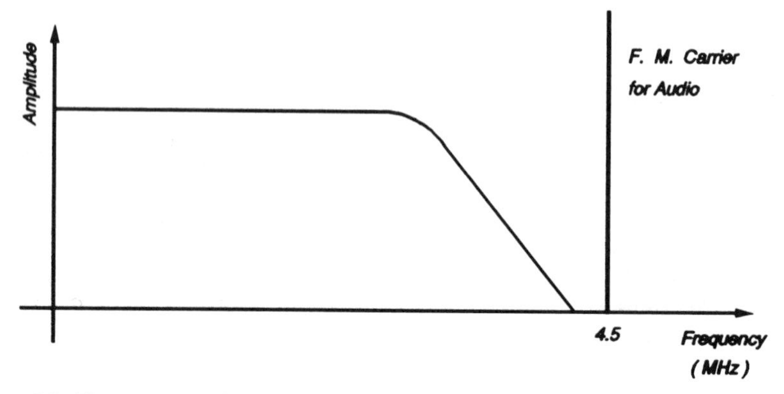

Figure 6.8. The spectrum of the composite video showing the FM carrier for the audio signal.

6.3.4 Video Amplifier

6.3.4a Calculation of Bandwidth. Ideally, a video signal consists of frequencies from zero (dc) to some high frequency. The dc response is required when large areas of black, white, or other intermediate shades have to be transmitted. The limit of the high-frequency response is determined by the resolution required for a single vertical black line on a white background or vice versa. The minimum visible horizontal line has a height (width) equal to the width of the horizontal trace. To keep the same resolution for the horizontal as the vertical line, the video signal must go from white to black in

$$56/525\ \mu\text{sec} = 0.11\ \mu\text{sec}$$

The lowest frequency that can change from peak positive to peak negative is a sinusoid as depicted in Figure 6.9. Note that the period of the sinusoid is 2×0.11 μsec, or 0.22 μsec.

It has been determined from psychological experiments that a picture that has a ratio of horizontal-to-vertical dimension less than 4 : 3 is not considered to be pleasing to look at. This ratio has been adopted as standard for all video equipment. It is referred to as the *aspect ratio*. Taking the aspect ratio

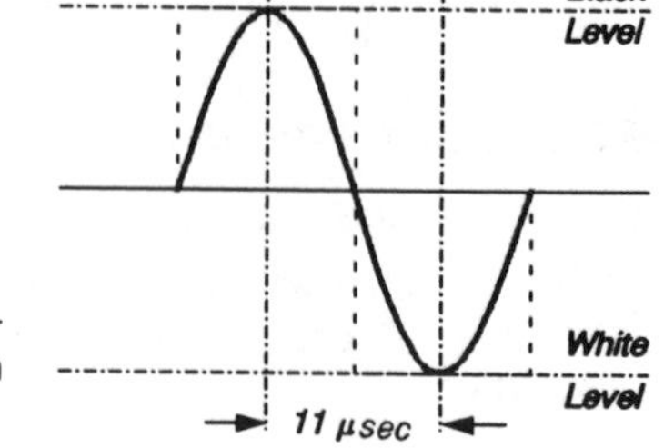

Figure 6.9. The sinewave shown represents the highest frequency required to follow the transition from black to white in 11 μsec.

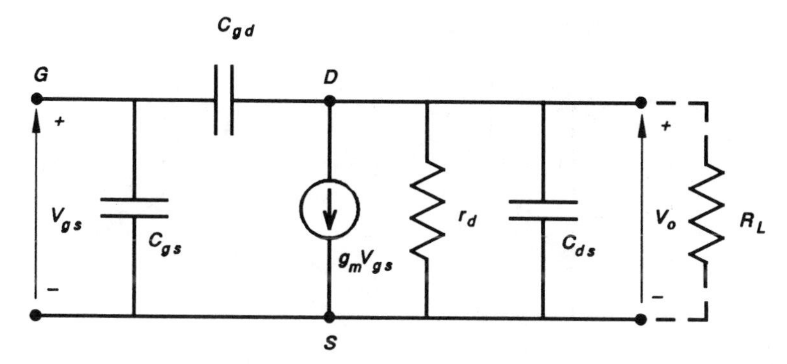

Figure 6.10. The equivalent circuit of common-source FET.

into account, one cycle of the sinusoid shown in Figure 6.9 must take $0.22 \times 3/4$ μsec, that is, the period $T = 0.16$ μsec. The lowest frequency that must be present in the video signal for it to be able to reproduce the vertical black line on a white background with the same resolution as a horizontal black line on a white background is given by

$$f = 1/T = 6.25 \text{ MHz} \tag{6.3.1}$$

This means that all the video circuits in the system must have bandwidths greater than 6.25 MHz. Note that the bandwidths must be greater than 6.25 MHz because of the effect of cascading on bandwidth.

6.3.4b Video Amplifier Design. Because the video signal contains frequencies from zero to some high frequency, a video amplifier must be designed to have a flat response from direct current to the high frequency in question. Ideally, all circuits in the video chain must be directly coupled. In practice, an amplifier with a bandwidth from approximately 30 Hz to approximately 4.0 MHz is considered to be a video amplifier.

Consider an FET video amplifier [1] in which the FET is modeled in the common-source mode as shown in Figure 6.10. Since $r_d \gg R_L$ and $|X_{Cds}| \gg r_d$, both r_d and C_{ds} can be neglected as shown in Figure 6.11. From Figure 6.11,

$$I_1 = j\omega C_{gs} V_{gs} \tag{6.3.2}$$

$$I_2 = j\omega C_{gd}(V_{gs} - V_o) \tag{6.3.3}$$

$$I_3 = I_1 + I_2 = j\omega V_{gs}\left[C_{gs} + \left(1 - \frac{V_o}{V_{gs}}\right)C_{gd}\right] \tag{6.3.4}$$

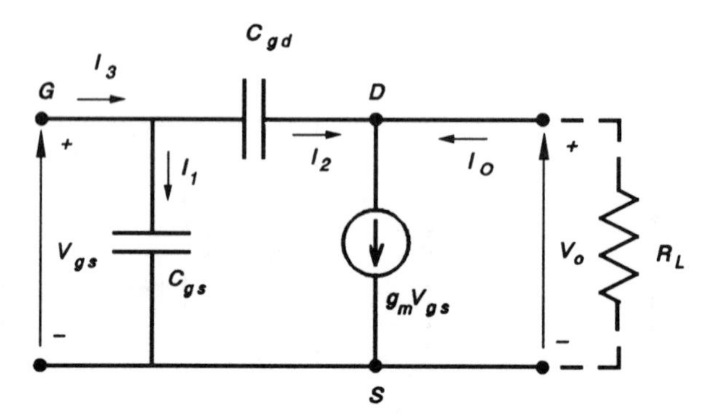

Figure 6.11. A simplified equivalent circuit of the FET with r_d and C_{ds} excluded.

Now $I_1 \gg I_2$, therefore,

$$V_o = -g_m V_{gs} R_L \tag{6.3.5}$$

Substituting into Equation (6.1.4),

$$I_3 = j\omega V_{gs}\left[C_{gs} + (1 + g_m R_L) C_{gd}\right] \tag{6.3.6}$$

$$\frac{V_{gs}}{I_3} = \frac{1}{j\omega\left[C_{gs} + (1 + g_m R_L) C_{gd}\right]} \tag{6.3.7}$$

The FET can now be represented by Figure 6.12, in which

$$C_g = \left[C_{gs} + (1 + g_m R_L) C_{gd}\right] \tag{6.3.8}$$

Consider the two-stage video amplifier shown in Figure 6.13. The resistor R_1 is used to bias the gate of Q_1, and it is usually very large (MΩs). The time constant $C_g R_1$ is therefore very large compared to $C_g R_D$. To a first approximation, the time constant $C_g R_D$ determines the high-frequency response of

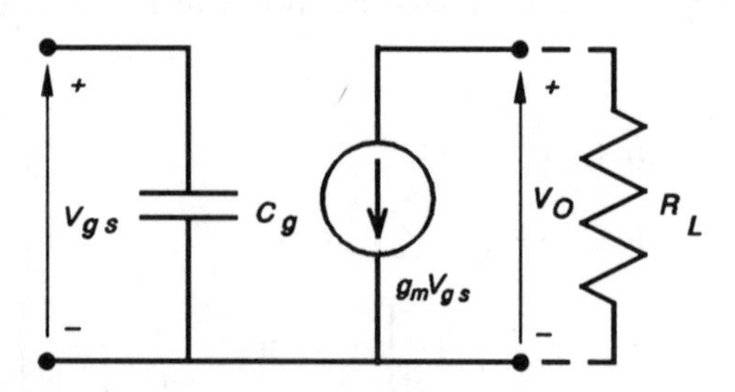

Figure 6.12. A modified form of the FET shown in Figure 6.11.

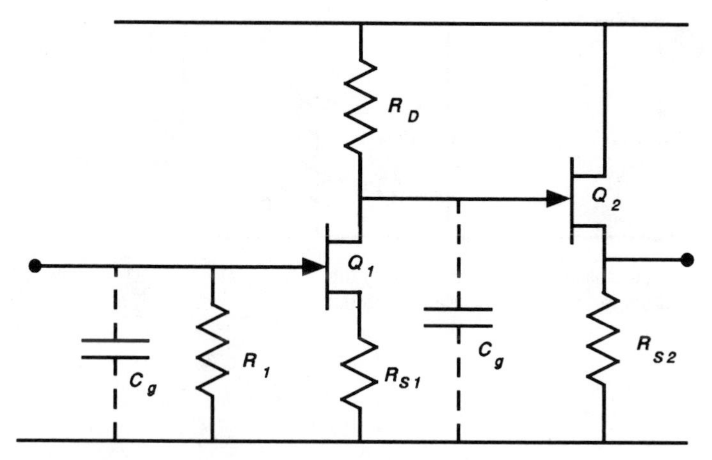

Figure 6.13. A two-stage video FET amplifier showing C_g for both devices.

the amplifier. The −3-dB frequency

$$\omega_2 \approx \frac{1}{R_D C_g} \tag{6.3.9}$$

The bandwidth ω_2 can be increased by connecting an inductance in series with the drain resistance R_D as shown in Figure 6.14. The equivalent circuit is shown in Figure 6.15.

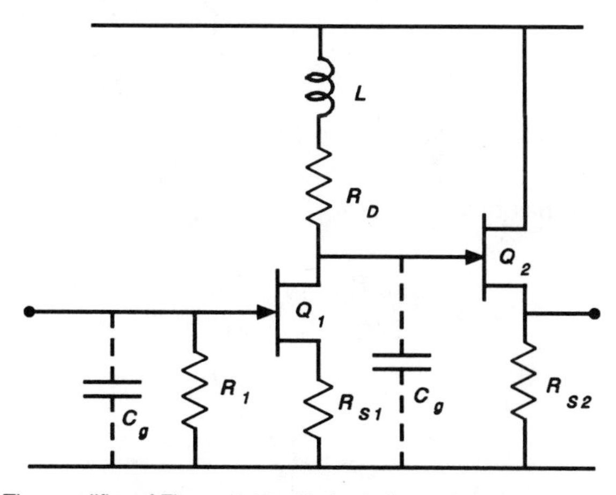

Figure 6.14. The amplifier of Figure 6.13 with the inductor L connected in series with R_D.

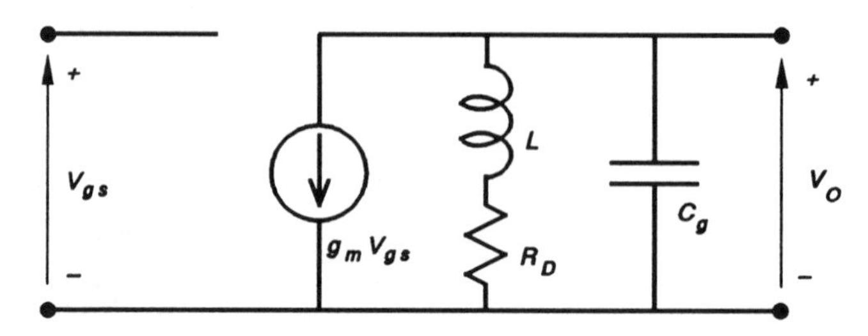

Figure 6.15. A simplified equivalent circuit of the first stage of the FET amplifier with the C_g of the second stage connected across the output.

The load

$$Z_L = \frac{1}{(R_D + sL)^{-1} + sC_g} \tag{6.3.10}$$

where

$$s = \sigma + j\omega \tag{6.3.11}$$

and

$$V_o = -g_m V_{gs} Z_L \tag{6.3.12}$$

The gain

$$V_o/V_{gs} = -g_m Z_L \tag{6.3.13}$$

$$A = \frac{V_o}{V_{gs}} = -\frac{g_m}{C_g} \frac{s + (R_D/L)}{s^2 + s(R_D/L) + \omega_o^2} \tag{6.3.14}$$

where

$$\omega_o^2 = \frac{1}{LC_g} \tag{6.3.15}$$

The gain function has two poles at

$$s_1, s_2 = -\frac{R_D}{2L} \pm \sqrt{\frac{R_D^2}{4L^2} - \omega_o^2} \tag{6.3.16}$$

When

$$\omega_o^2 = \frac{R_D^2}{4L^2} \tag{6.3.17}$$

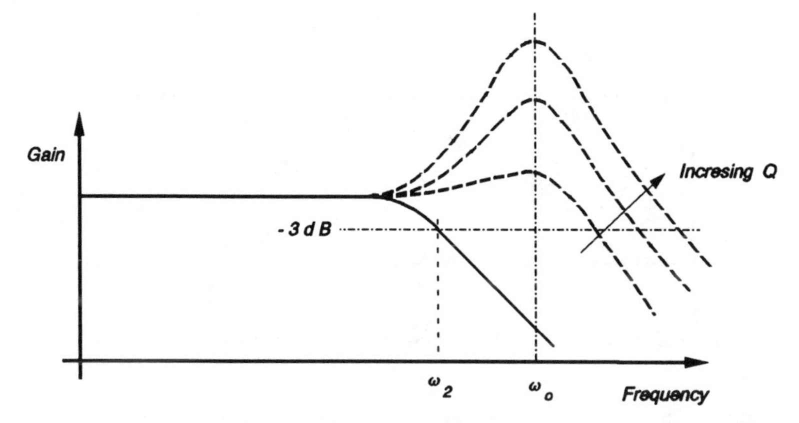

Figure 6.16. The frequency response of the amplifier after the addition of the inductor. Note the resonant effect that increases the gain at high frequency, thereby increasing the bandwidth.

the gain function poles are conjugate on the s plane and these increase the gain at the resonant frequency ω_o.

The critical value of L is

$$L_{\text{crit}} = \frac{R_D^2 C_g}{4} \tag{6.3.18}$$

The response is then as shown in Figure 6.16.

The shape of the gain response can be controlled by varying the Q factor of the circuit. Let

$$q = \frac{\omega_2 L}{R_D} = \frac{L}{R_D^2 C_g} = \frac{L}{4L_{\text{crit}}} \tag{6.3.19}$$

Substituting q and $s = j\omega$ into Equation (6.1.14) gives

$$|A| = \bar{A}\sqrt{\frac{1 + q^2(\omega/\omega_2)^2}{1 + (1 - 2q)(\omega/\omega_2)^2 + q^2(\omega/\omega_2)^4}} \tag{6.3.20}$$

The gain response can, in general, be represented by the ratio of two polynomials in ω^2

$$H(\omega) = \frac{1 + a_1\omega^2 + a_2\omega^4 + \ldots}{1 + b_1\omega^2 + b_2\omega^4 + \ldots} \tag{6.3.21}$$

which gives a flat gain–frequency response when $a_1 = b_1$, $a_2 = b_2$, and so on.

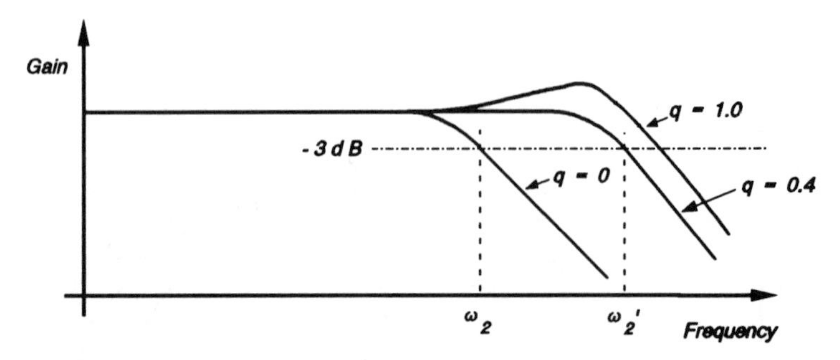

Figure 6.17. The frequency response of the amplifier for various values of q.

Equating the coefficients of $(\omega/\omega_2)^2$ in Equation (6.3.21) gives

$$\begin{aligned} q &= \pm\sqrt{2} - 1 \\ &= 0.414 \text{ or } -2.414 \end{aligned} \tag{6.3.22}$$

The negative value has no physical meaning, and it is neglected. Substituting into Equation (6.3.19),

$$q = 0.414 = \frac{L}{4L_{\text{crit}}} = \frac{L}{R_D^2 C_g} \tag{6.3.23}$$

$$L = 1.656\ L_{\text{crit}} = 0.414\ R_D^2 C_g \tag{6.3.24}$$

The frequency response for three values of q are shown in Figure 6.17. The new −3-dB frequency

$$\omega_2' = 1.72\ \omega_2 \tag{6.3.25}$$

Example 6.3.1 Video Amplifier. A video amplifier uses FETs and has two stages as shown in Figure 6.18. The FETs have the following parameters:

$$g_m = 0.004\ \text{S}, \qquad C_{gs} = 1.0\ \text{pF}, \qquad C_{gd} = 1.5\ \text{pF}$$

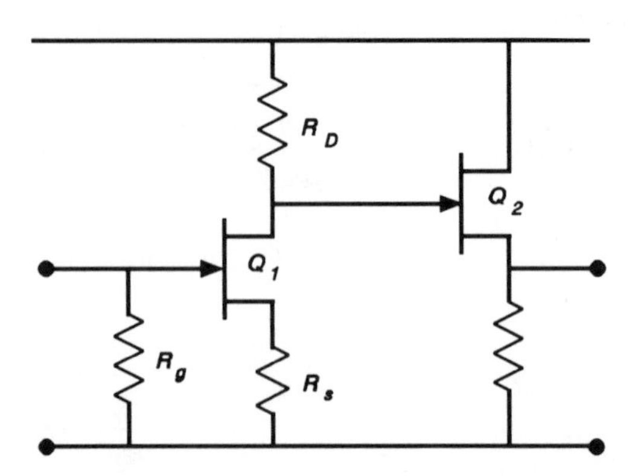

Figure 6.18. The video amplifier used in the example.

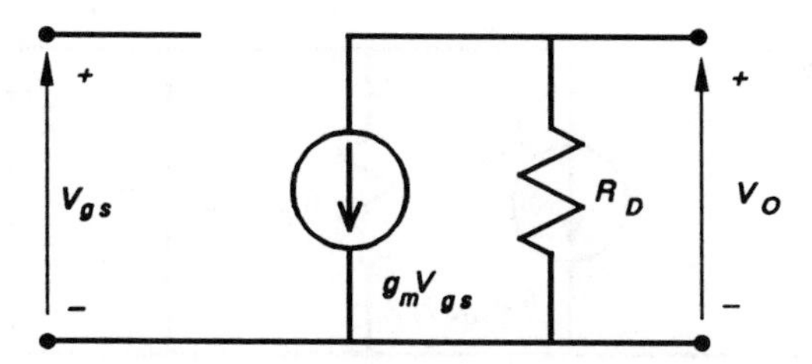

Figure 6.19. The equivalent circuit of stage 1 of the amplifier when all capacitances have been neglected.

Assuming that only the Miller-effect capacitance affects the high-frequency response, calculate the −3-dB frequency of the amplifier when the gain is 15. Calculate the value of an inductance that, when connected in series with R_D, will extend the −3-dB cut-off frequency as high as possible while keeping the gain response flat. Calculate the new cut-off frequency.

Solution. Neglecting all capacitance effects, the equivalent circuit of the amplifier is shown in Figure 6.19, where

$$V_o = -g_m V_{gs} R_D$$

and the voltage gain is

$$V_o/V_{gs} = -g_m R_D = 15$$

$$R_D = 15/0.004 = 3{,}750\ \Omega$$

But the voltage gain $\simeq R_D/R_s = 15$, so that

$$R_s = 250\ \Omega$$

and R_g can be made arbitrarily large, because no current flows in it. The Miller-effect capacitance of Q_1 will have negligible effect on the frequency response especially if the internal resistance of the driving source is small. The Miller-effect capacitance of Q_2 will appear across R_D as shown in Figure 6.20, where

$$C_g = C_{gs} + (1 + g_m R_D) C_{gd}$$

$$= 1.0 + (1 + 15) \times 1.5 = 25\ \text{pF}$$

The −3-dB frequency ω_2 is determined by

$$\omega_2 \tau = 1$$

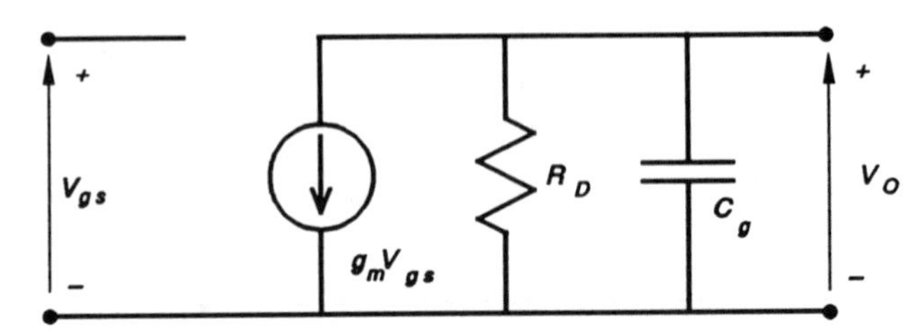

Figure 6.20. The equivalent circuit of stage 1 of the amplifier with the C_g of stage 2 included.

where $\tau = R_D C_g$ and

$$f_2 = \frac{1}{2\pi R_D C_g}$$

$$f_2 = 1.67 \text{ MHz}$$

When the inductance L is connected in series with R_D and L has the critical value,

$$L = 4qL_{\text{crit}} = qR_D^2 C_g$$

When $q = 0.414$,

$$L = 145.5 \ \mu\text{H}$$

The new -3-dB cut-off frequency is

$$f_2' = 1.72 \times f_2 = 2.87 \text{ MHz}$$

6.3.5 Radio Frequency Circuits

The composite video signal from the output of the video amplifier is used to amplitude modulate the radio frequency carrier obtained at a lower frequency from a crystal oscillator and multiplied by a suitable factor by a frequency multiplier. Amplitude modulators are discussed in Section 2.6, frequency multipliers in Section 2.5, and crystal-controlled oscillators in Sections 2.3 and 2.4.

6.3.6 Vestigial Sideband Filter

From Figure 6.8, it can be seen that the composite signal occupies a bandwidth of approximately 0 to 4.5 MHz. If both sidebands were transmitted, it would require a bandwidth of just over 9.0 MHz. Not only would that consume scarce spectra, but it is also unnecessary, because the information in the upper and lower sidebands are the same. Single sideband (SSB) transmission could be used, but the filter needed to remove one of the sidebands is fairly sophisticated and the required demodulation equipment is complex and

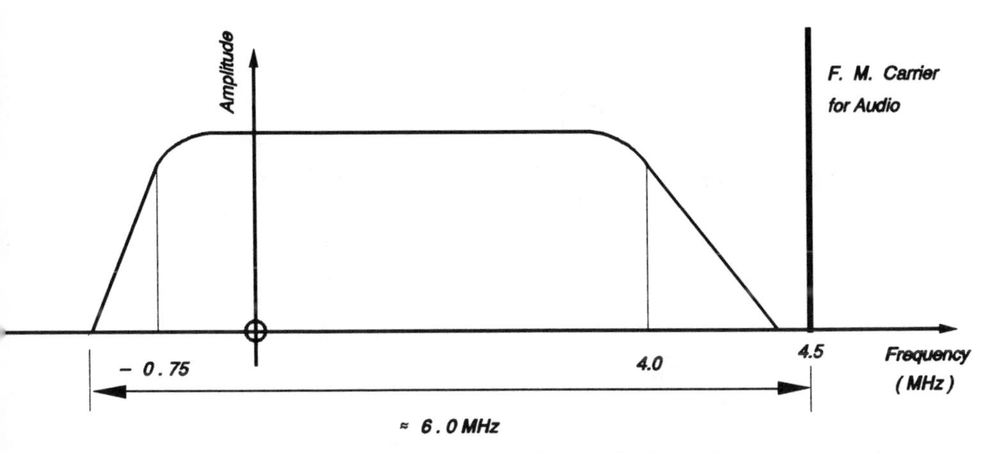

Figure 6.21. The spectrum of the video signal after amplitude modulation and vestigial sideband filtering. Note that zero frequency is, in fact, the AM carrier frequency.

difficult to maintain. A compromise is to remove most of the lower sideband using a bandpass filter. As shown in Section 7.2.4, a simple envelope detector is adequate for the demodulation of such an AM signal if the index of modulation is low. The frequency spectrum at the output of the vestigial sideband filter is shown in Figure 6.21.

6.3.7 Antenna

Television broadcast frequencies are either in the very-high-frequency (VHF) band, which is from 30 to 300 MHz, or in the ultra-high-frequency (UHF) band, which is from 300 to 3000 MHz. At these frequencies, antennas have highly directional properties. To get the circular radiation pattern in the horizontal plane necessary for broadcasting television signals, several arrangements of antenna arrays can be used. The most popular is the *turnstile array*: The basic principle is that when two or more radiating elements such as dipoles are placed in close proximity to each other, they interact to produce a radiation pattern that is the vector addition of the individual elements. By varying the relative physical positions of the elements and the phase angle of the signal, it is possible to use the interactive properties to create a radiation pattern that is approximately circular in the horizontal plane.

6.4 COLOR TELEVISION

The transmission of video signals in color is a subject that could take up several volumes. However, because color television is so common today, a simplified explanation of how it works is now offered.

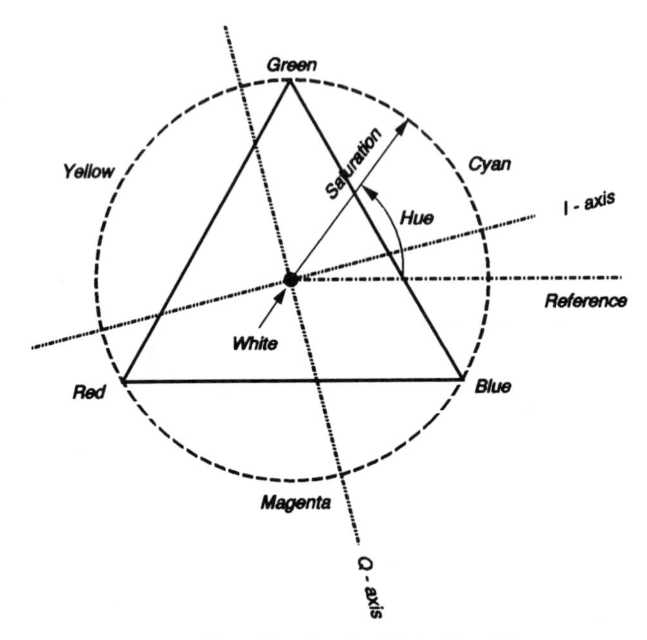

Figure 6.22. The color triangle with the *I* and *Q* axes.

The first step is to discuss some of the properties of color and the results of mixing them. There are three primary colors red, blue, and green and using appropriate proportions of these, all other colors perceived by the human eye can be obtained. A simple framework makes it easy to understand the properties of color the color triangle [2, 3], which is shown in Figure 6.22.

The primary colors occupy the apices of the triangle. A mixture of equal proportions of red and blue produces the color magenta, that of red and green produces yellow, and finally blue and green produce cyan. These fit into the scheme as shown. The center of the triangle represents white, because equal proportions of red, blue, and green produce white. Two pieces of information are required to code a video signal in color. The first is brightness; the better technical term is *luminance*, which is a measure of the energy in the light. Luminance is the only information required for monochromatic television. The second is *chrominance* or simply color. Chrominance is best represented on the diagram by a polar coordinate scheme whose origin is at the center (white) of the triangle. When one moves along a straight line from the origin towards the red apex one sees at first white light followed by a slowly increasing presence of red (pink). At the apex (pure red light) the color red is said to have reached *saturation*. The distance from the center of the triangle (the magnitude of the complex number) is then a measure of the saturation of the particular color. The angle of the complex number, with respect to some arbitrary reference line, is called the

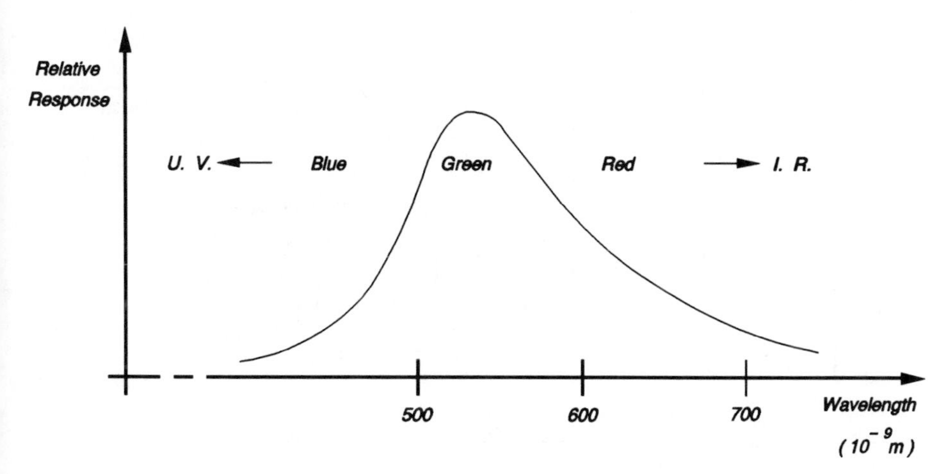

Figure 6.23. The relative sensitivity of the human eye to the wavelength of light.

hue of the light. The combination of chrominance and luminance will produce a cylindrical coordinate system in which luminance will be on the axis of the cylinder. In this scheme, the coding of color must have three coordinates, luminance, saturation, and hue.

The above scheme does not take into account the sensitivity of the human eye across the color spectrum, and it needs to be modified since the human eye is the final arbiter (detector) of the video signals transmitted. The response of the human eye is shown in Figure 6.23.

A mixture of equal proportions of the primary colors is therefore perceived by the human eye as a greenish-yellow color. In fact to obtain white light the mixture has to contain 59% green, 30% red, and only 11% blue.

The simplest system for coding video information for television transmission is simply to scan the picture for the three primary colors and process each one as if it were monochromatic. Some multiplex system could then be worked out to accommodate all three signals. Such a scheme was tried in the United States in late 1940s and early 1950s but was found to be incompatible with the black-and-white system then in existence.

In 1953, a scheme proposed by the National Television System Committee (NTSC) was accepted by the Federal Communications Commission (FCC), and it has since been adopted as a standard by television broadcasting authorities in many countries. The NTSC scheme separates the signal into chrominance and luminance (the chrominance information is further separated indirectly into I and Q components, which correspond approximately to hue and saturation). This will be explained later. The actual system acquires the luminance information by taking the weighted sum of the three primary colors to obtain white light as perceived by the human eye as

$$L_w = 0.59L_g + 0.30L_r + 0.11L_b \qquad (6.3.26)$$

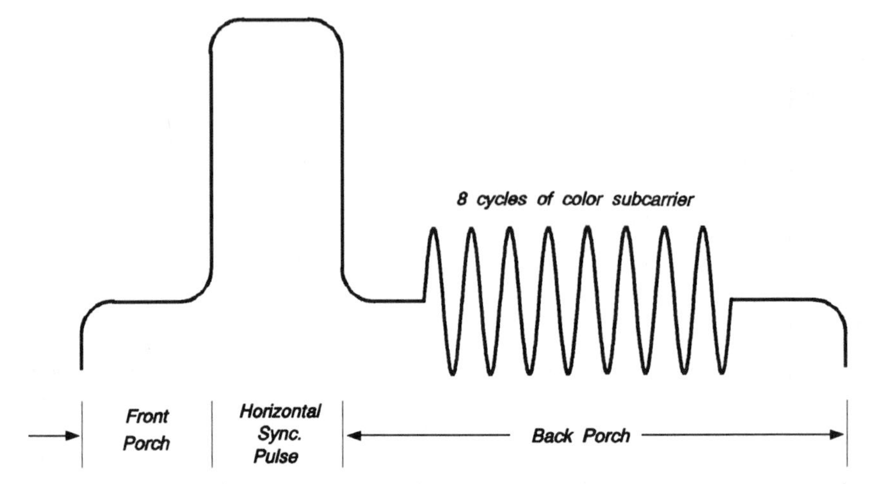

Figure 6.24. A modified back porch of the blanking pulse used to transmit the color barrier.

L_w then contains all the information for the black-and-white picture transmission. For color pictures, the three signals L_w, $(L_r - L_w)$ and $(L_b - L_w)$ are sent. The third possible signal $(L_g - L_w)$ is not used because it is the smallest of the three difference signals and therefore more likely to be corrupted by noise. The two difference signals $(L_r - L_w)$ and $(L_b - L_w)$ are quadrature multiplexed, a form of DSB-SC modulation, with the color subcarrier of frequency ω_o. The two difference signals have the form:

$$(L_b - L_w)\cos \omega_o t + (L_r - L_w)\sin \omega_o t \tag{6.3.27}$$

The signal

$$L_w + (L_b - L_w)\cos \omega_o t + (L_r - L_w)\sin \omega_o t \tag{6.3.28}$$

is used to amplitude modulate a suitable carrier in the usual vestigial sideband scheme.

At the receiving end, the two difference signals are demultiplexed by multiplying them by the output of a local oscillator, which is synchronized to the color subcarrier of frequency ω_o in the transmitter. The color subcarrier signal is sent in "bursts" during the trailing portion of the horizontal blanking pulse called the *back porch*, as shown in Figure 6.24.

The demodulation system is shown in Figure 6.25. The recovered signals L_w, $(L_b - L_w)$, and $(L_r - L_w)$ are used to reconstruct the missing difference signal $(L_g - L_w)$ as follows:

$$(L_g - L_w) = -\frac{0.30}{0.59}(L_r - L_w) - \frac{0.11}{0.59}(L_b - L_w) \tag{6.3.29}$$

The four signals are used to control the appropriate grids on the picture tube to reproduce the original scene.

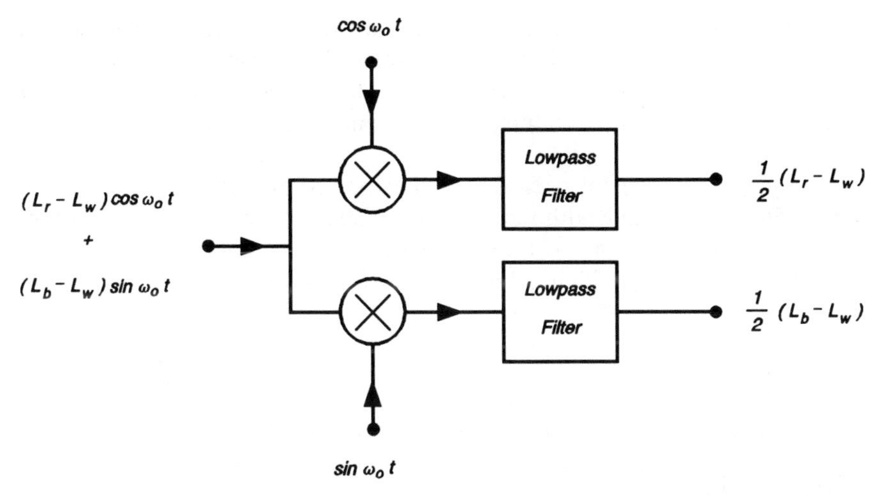

Figure 6.25. A scheme for demultiplexing the color information.

In the above discussion, the bandwidth requirements of the three signals L_w, $(L_b - L_w)$, and $(L_r - L_w)$ were assumed to be the same. This is in fact not true. The human eye has a higher resolution in black-and-white than it does in color. It is adequate to present the uniform, large areas of the picture (low frequency) in color; the fine details (high frequency) can be presented in black-and-white. The luminance information L_w must have the whole bandwidth available. The difference signal containing the chrominance information requires much less bandwidth.

Further advantage can be taken of the greater sensitivity of the human eye to green-yellow light by deemphasizing the green-yellow component by using

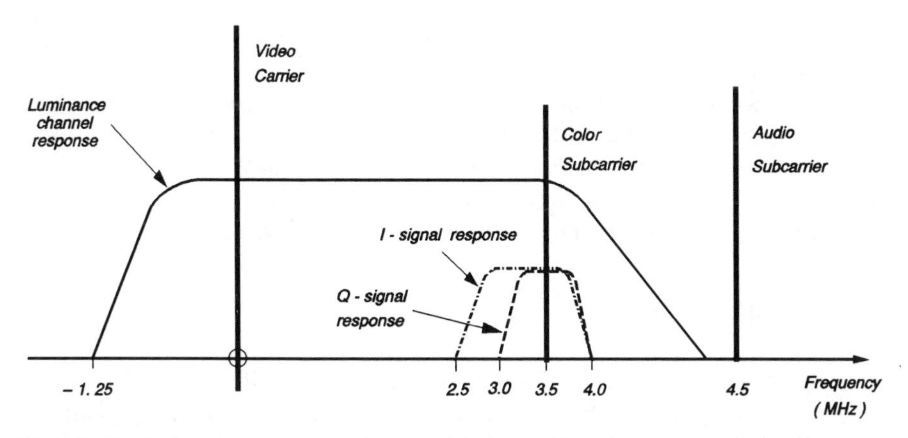

Figure 6.26. The full spectrum of the color TV signal. Note that 0 frequency is the AM (video) carrier frequency.

a smaller bandwidth for transmitting that part of the spectrum. This is called the Q signal. The blues and the reds to which the human eye is less sensitive are given a relatively larger bandwidth. This is called the I signal. The comparative spectra for the luminance, Q and I signal are shown in Figure 6.26.

Because the Q signal occupies a total bandwidth of 1 MHz and it fits within the overall bandwidth limits of the signal, both sidebands are transmitted. The upper sideband of the I signal lies outside the overall bandwidth and part of the upper sideband is removed with a vestigial sideband filter.

BIBLIOGRAPHY

D. G. Fink, *Television Engineering Handbook*, McGraw-Hill, New York, 1957.

D. G. Fink and D. Christiansen, *Electronics Engineers' Handbook*, 2nd ed., McGraw-Hill, New York, 1982.

A. Shure, *Basic Television*, J. F. Rider, New York, 1958.

W. A. Stover (Ed.), *Circuit Design for Audio, AM/FM and TV*, McGraw-Hill, New York, 1967.

F. G. Stremler, *Introduction to Communication System*, 3rd ed., Addison-Wesley, Reading, MA, 1990.

REFERENCES

1. J. D. Ryder, *Electronic Fundamentals and Applications: Integrated and Discrete Systems*, 5th ed., Prentice-Hall Inc., New Jersey, 1976.
2. J. H. Lambert, *Beschreibungen einer Farbenpyramide*, Berlin, 1772; and "Momoire sur la Partie Photometrique de l'Art de Peinture," Preussiche Akademie der Wissenschaften, Berlin; *Histoire de l'Acad. Royal des Science et Belles Lettres*, 24, 80–108, 1768.
3. J. C. Maxwell, "On the Theory of Three Primary Colours," in W. D. Niven (Ed.), *The Scientific Papers of James Clerk Maxwell*, Cambridge University Press, Cambridge, 1890.
4. V. K. Zworykin, G. A. Morton, and L. E. Flory, "Theory and Performance of the Iconoscope," *Proc. IRE*, 25(8), 1071–1092, 1937.
5. V. K. Zworykin and G. Morton, *Television*, 2nd ed., John Wiley, New York, 1954.
6. A. Rose, P. K. Weimer, and H. B. Law, "The Image Orthicon—a Sensitive Television Pickup Tube," *Proc. IRE*, 34(7), 424–432, 1946.

PROBLEMS

6.1 What is a video signal? Explain why a video amplifier has to be used in conjunction with a television camera tube. Use a block diagram to illustrate the configuration and function of each block of a typical television transmitter. Indicate the type of signal you expect at the output of each block in terms of waveform, frequency, and bandwidth.

6.2 Explain with the help of a suitable diagram the operation of the iconoscope. What conditions have to be met for it to operate optimally?

6.3 Explain with the help of a suitable diagram the operation of the image orthicon. Discuss its performance relative to the iconoscope.

6.4 A television system has the following features:

(1) 625 lines per frame with interlacing,

(2) 25 frames per second,

(3) 5% of the trace period is used for the retrace,

(4) the aspect ratio is 4 : 3

Assuming equal resolution on the vertical and horizontal axes of the picture tube, calculate the bandwidth required to handle the signal.

6.5 The voltage applied to the anode of a picture tube accelerates an electron to a velocity of 10×10^6 m/sec along the axis of the tube (x axis). The electron then passes between two parallel (deflection) plates of length 2.5 cm (in the x direction). The distance between the plates in the y direction is 2.0 cm and a dc voltage of 270 V is maintained between them. Assuming that the electrostatic field is uniform and confined to the space between the plates, calculate the following:

(1) The velocity of the electron in the y direction as it emerges from the electrostatic field.

(2) The displacement of the electron in the y direction when it hits a target placed 5 cm away from the end of the deflection plates.

You may assume the following

(a) The force (in Newtons) on the electron due to the electrostatic field is given by

$$F = eE$$

where e is the charge (in Coulombs) on the electron and E is the electric field intensity (in volts per meter).

(b) The mass of the electron is 9×10^{-31} kg and the charge is 1.6×10^{-19} Coulombs.

6.6 The electrostatic field described in Problem 6.5 is replaced by a uniform magnetic field of width 2.5 cm in the x direction. The magnetic field is oriented in the z direction. The flux density of the field is 1.44×10^{-3} Teslas [Wb/m^2]. Assuming no fringing effects, calculate the following:

(1) the velocity of the electron as it emerges from the magnetic field,

(2) the displacement of the electron when it hits the target placed 5 cm from the edge of the magnetic field.

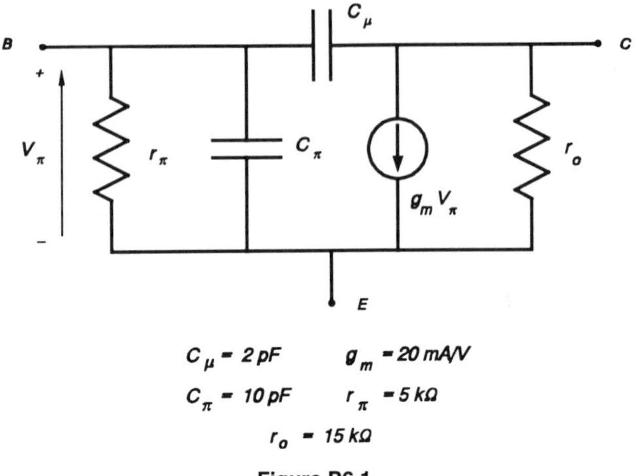

C_μ = 2 pF g_m = 20 mA/V

C_π = 10 pF r_π = 5 kΩ

r_o = 15 kΩ

Figure P6.1.

You may assume the following:

(a) The force F [newton] on a conductor of length l (in meters) carrying a current i (in amperes) in a magnetic field of flux density B [Tesla = Wb/m^2] is given by

$$F = Bli$$

(b) An electron carrying a charge e (in Coulombs) at a velocity of v (in meters per second) is, in fact, a current, $i = e$ (amperes = Coulombs per second) flowing in a conductor of length l (meters).

6.7 A video amplifier with a voltage gain of 10 is to be designed in two stages, such that the first stage provides the required voltage gain and the second provides a current gain with a low output impedance. Design such a circuit and estimate its bandwidth using the hybrid π model given in Figure P6.1. Calculate the value of an inductor that, when connected in series with the collector load of the voltage-gain stage, will optimize the bandwidth of the amplifier.

7

THE TELEVISION RECEIVER

7.1 INTRODUCTION

In Chapter 6, the coding of video signals in a form suitable for transmission over a telecommunication channel was discussed. In this chapter, the techniques for decoding the signals and their presentation on a cathode ray tube will be examined.

The television receiver is almost identical to the radio AM receiver in that it uses the superheterodyne principle. There are a few differences in the details of the signal processing due to the greater complexity of the system. Figure 7.1 shows a block diagram of a typical television receiver.

The antenna picks up the electromagnetic radiation from the transmitter and feeds this to the radio frequency amplifier. After amplification and filtering to attenuate other incoming signals from other transmitters, the signal goes to the mixer, where it is mixed with the output from the local oscillator. As before, the local oscillator and the radio frequency amplifier are tuned to track each other with a constant frequency difference equal to the intermediate frequency. The intermediate frequency for the television receiver is usually 45.75 MHz. The signal is subjected to further filtering before it proceeds to the video demodulator for the recovery of the baseband information in the signal. The next stage is to separate the composite video signal into its three components: namely the video proper, the FM sound subcarrier, and its sidebands, and the vertical and horizontal control pulses. The video signal is amplified to the level required to drive the picture tube by the video amplifier and the vertical and horizontal control pulses are suitably conditioned and used in the deflection systems of the receiver to synchronize it to the transmitter—a condition that must be met for proper reproduction of the images sent. The FM signal is amplified, amplitude limited, and detected, and after some amplification it is used to drive the loudspeaker.

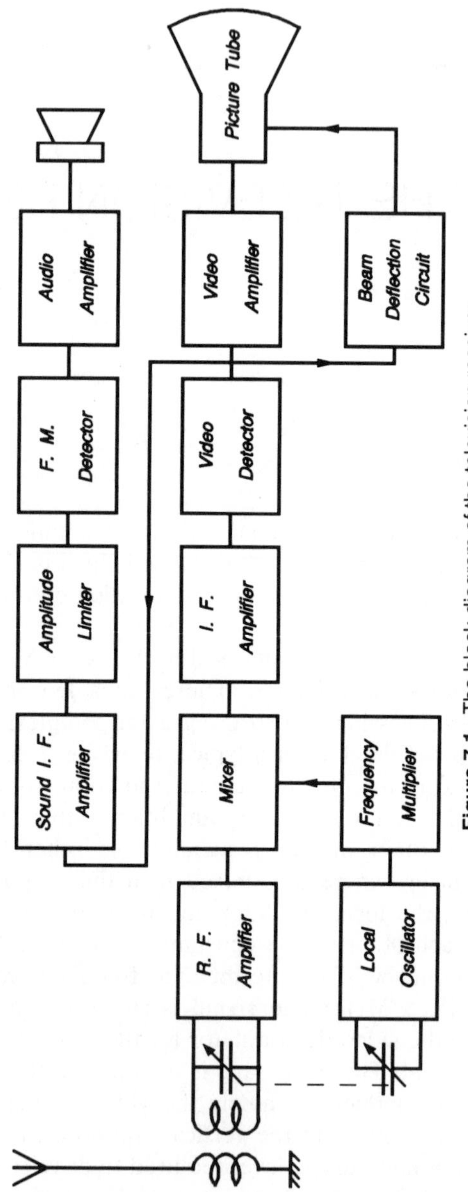

Figure 7.1. The block diagram of the television receiver.

7.2 COMPONENT DESIGN

7.2.1 Antenna

Antenna design is outside the scope of this book. However, a brief qualitative discussion can be found in Section 2.9. Further discussion of antennas for commercial FM reception is presented in Section 5.2.1. Frequencies for commercial FM (88 to 108 MHz) occupy the spectrum between channel 6 and 7 of the VHF television frequencies (54 to 88 MHz and 174 to 216 MHz, respectively). Except for slight differences in the physical dimensions, the antennas tend to take the same form. These are frequency ranges in which a half-wavelength dipole antenna has reasonable physical dimensions (0.7 to 2.7 m).

7.2.2 Superheterodyne Section

The radio frequency amplifier is tunable over the VHF frequency range. This is accomplished with a variable capacitor that is mechanically ganged to the variable capacitor, which tunes the local oscillator. The objective is to generate a local oscillator frequency equal to the radio frequency amplifier center frequency plus the intermediate frequency. In this case, the radio frequency range is from approximately 57 MHz (channel 2) to approximately 85 MHz (channel 6) and from approximately 177 MHz (channel 7) to approximately 213 MHz (channel 13) for VHF television. The bandwidth is nominally 6 MHz and the radio frequency gain is between 20 and 50 times. Because the video intermediate frequency is normally 45.75 MHz, the local oscillator has to be tunable from 102.75 to 258.75 MHz. From Figure 6.21, it can be seen that there are two carriers in the composite video signal: the video carrier and the voice carrier. The voice carrier is 4.5 MHz above the video carrier; after mixing the video intermediate frequency is at 45.75 MHz and the voice intermediate frequency is at 45.75 − 4.50 or 41.25 MHz. In general parlance, the radio frequency amplifier, the local oscillator, and the mixer are called the *television front end* or simply the *television tuner*. These components are generally housed in a separate shielded container in an attempt to control the effects of stray electromagnetic fields and stray capacitive and inductive elements.

The principles to be followed in the design of the circuits in the television front end are the same as discussed earlier in connection with radio. The only things that have changed are the frequency of operation and the bandwidth requirements. Greater attention must be paid to the physical layout of the practical circuits. Radio frequency amplifiers were discussed in Section 2.8. Oscillator design can be found in Section 2.4 and mixer design in Section 3.4.3.

7.2.3 Intermediate Frequency Amplifier

Like all superheterodyne receiver systems, the detailed selection of the desirable and the rejection of undesirable frequencies take place at the intermediate frequency stage. At the same time, some parts of the spectrum may be emphasized to equalize the quality of the low-frequency video (large uniform areas) and the high-frequency video (areas with fine details). The exact frequency response of the video intermediate frequency amplifier is of no importance at this point except to point out that it is designed to compensate for the frequency response of the vestigial sideband filter in the transmitter. To understand the techniques used to achieve this objective requires a good understanding of the theory of filter design, which is beyond the scope of this book. A list of books on filter design is provided in the bibliography of Chapter 3.

The output of the intermediate frequency amplifier must be of the order of several volts to drive the video detector that follows. Intermediate frequency amplifiers were discussed in Section 3.4.4.

7.2.4 Video Detector

It will be recalled that the video signal is amplitude modulated and in theory it requires a simple envelope detector to demodulate it. However, the situation is complicated somewhat by the fact that the input signal to the detector is a vestigial sideband signal.

7.2.4a Demodulation of Vestigial Sideband Signals. When a carrier of frequency ω_c is amplitude modulated by a signal of frequency ω_m, the result is

$$\phi_1(t) = A[1 + m\cos\omega_m t]\cos\omega_c t \tag{7.2.1}$$

$$= A\cos\omega_c t + \frac{mA}{2}\left[\cos(\omega_c + \omega_m)t + \cos(\omega_c - \omega_m)t\right] \tag{7.2.2}$$

when one of the sidebands is removed. It could be either of them, but in this case it is assumed that it is the lower sideband. Also complete removal is assumed. For simplicity, we have:

$$\phi_2(t) = A\cos\omega_c t + \frac{mA}{2}\cos(\omega_c + \omega_m)t \tag{7.2.3}$$

$$= A\cos\omega_c t + \frac{mA}{2}\cos\omega_c t\cos\omega_m t - \frac{mA}{2}\sin\omega_c t\sin\omega_m t \tag{7.2.4}$$

$$= A\left[1 + \frac{m}{2}\cos\omega_m t\right]\cos\omega_c t - \frac{mA}{2}\sin\omega_c t\sin\omega_m t \tag{7.2.5}$$

The amplitude of the carrier signal is

$$|\phi_2(t)| = \sqrt{A^2\left[1 + \frac{m}{2}\cos\omega_m t\right]^2 + \left[\frac{mA}{2}\sin\omega_m t\right]^2} \tag{7.2.6}$$

$$= \sqrt{A^2\left[1 + \frac{m^2}{4}\right] + A^2 m\cos\omega_m t} \tag{7.2.7}$$

When the depth of the modulation is low, $m < 1$ and $m^2/4 \ll 1$

$$|\phi_2(t)| \approx A[1 - m\cos\omega_m t]^{1/2} \tag{7.2.8}$$

Using the binomial expansion, we get

$$|\phi_2(t)| \approx A\left[1 + \frac{m}{2}\cos\omega_m t\right] \tag{7.2.9}$$

This contains the modulating frequency ω_m as well as its higher harmonics whose amplitudes diminish very rapidly.

The conclusion is that a vestigial sideband signal can be demodulated using a simple envelope detector, so long as the modulation index is much less than unity. It is worth noting that the amplitude of the output signal is one-half of what it would have been if both sidebands had been present.

Envelope detectors were discussed in Section 3.2. An example is given in Section 3.4.6. Figure 7.2 shows the input and output waveforms of a typical television envelope detector.

7.2.5 The Video Amplifier

The design of video amplifiers was discussed in Section 6.3.4. In the television receiver, the load of the video amplifier is the grid of the cathode ray tube (usually called the picture tube). This requires voltages between approximately 50 V and 100 V, and, in theory, no current flows in the grid circuit. However, the grid represents a capacitive load and a capacitance requires the movement of charge (current) to change the voltage across it. The output stage of the video amplifier must be capable of providing the necessary current and hence power. Another way of saying the same thing is that the output stage of the video amplifier must have a low output resistance so that the grid capacitance can be charged much faster than the fastest change in voltage present in the video signal.

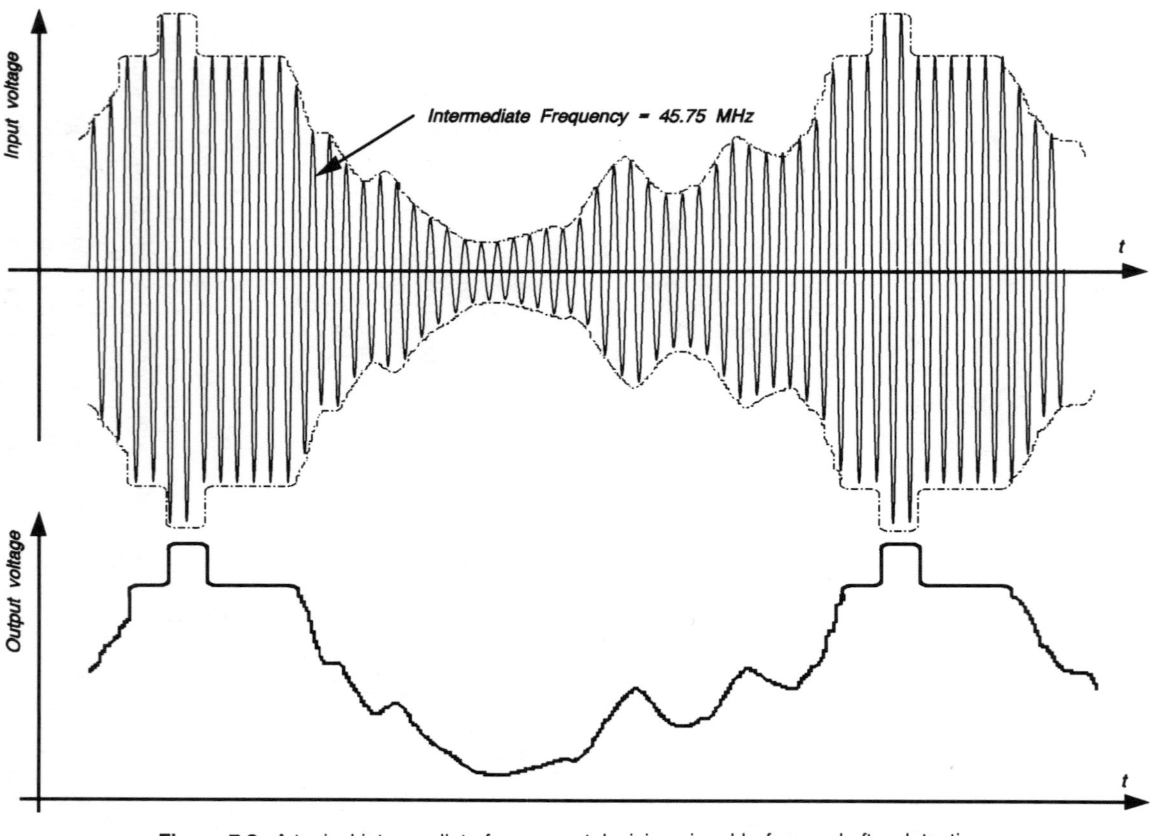

Figure 7.2. A typical intermediate frequency television signal before and after detection.

7.2.6 The Audio Channel

From the output of the video detector, a bandpass amplifier selects and boosts the FM signal centered at 4.5 MHz. The limiter and the FM detector are exactly the same as described in Sections 4.3.3 (a and b) and 4.3.3d, respectively. The audio frequency amplifier was described in Section 2.7. The loudspeaker was discussed in Section 3.4.8.

7.2.7 Electron Beam Control Subsystem

In the television transmitter, the pulse generator output was used to control the vertical and horizontal sweeps of the electron beam, which was used for scanning the mosaic in the camera tube. The same pulses were added to the video signal together with the 4.5 MHz FM voice carrier to make up the composite video. Figure 7.2 shows two such pulses (horizontal sync-pulses only shown). The horizontal synchronization pulses are used to control the initiation of the horizontal sweep of the electron beam in the picture tube so that synchronism with the horizontal sweep of the electron beam in the camera tube is maintained. Similarly, the vertical synchronization pulses are used to keep the camera and picture tubes in step in the vertical direction. It is very important to keep the camera and picture tubes in synchronism in both directions, otherwise no meaningful image appears on the picture tube. The first step is to channel the timing information in the sync-pulses into a separate circuit for further processing. Figure 7.3 is a block diagram of the electron beam control subsystem.

The composite video signal is fed into the sync-pulse separator, which takes out both vertical and horizontal sync-pulses. The output is used to drive the two separate branches of the system. The vertical branch has the vertical sync-separator, which is designed to produce an output only when a vertical pulse is present at the input. The two timing signals undergo essentially identical process steps, namely, synchronization to the oscillator, generation of the sweep signal, and amplification to obtain enough power to drive the deflection coils. It is important to remember that the vertical sync-pulses have a frequency of 60 Hz, whereas the horizontal runs at 15.75 kHz.

The sync-pulses have the following timing characteristics:

Vertical		Horizontal	
Field period:	16.683 msec	Line period:	63.556 μsec
Blanking period:	1.335 msec	Blanking period:	10.5 − 11.4 μsec
Scan period:	15.348 msec	Scan period:	52.156 − 53.056 μsec

7.2.7a Sync-Pulse Separator. The basic sync-pulse separator is a simple transistor invertor, such as shown in Figure 7.4. The ratio of R_1 to R_2 is chosen so that the transistor remains cut off until the applied voltage exceeds

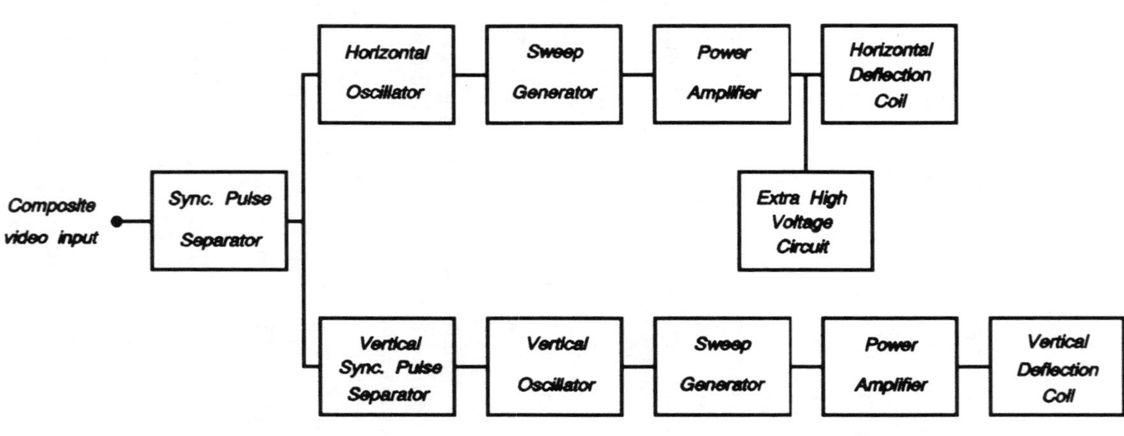

Figure 7.3. The block diagram of the electron beam scan control system.

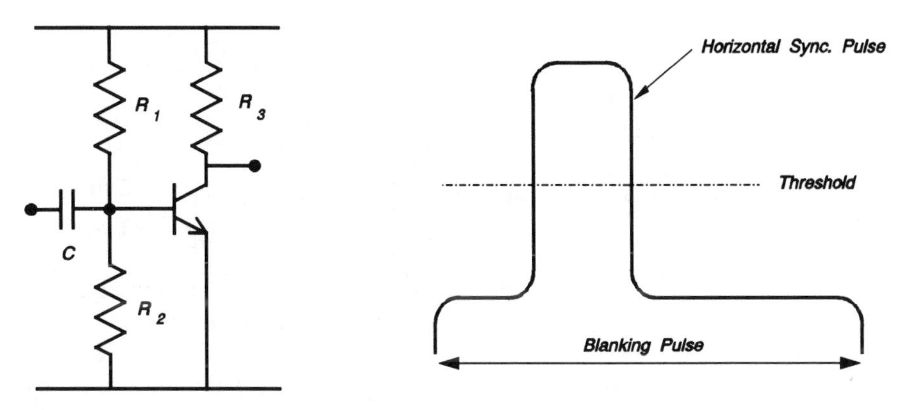

Figure 7.4. The circuit used for separating the horizontal sync-pulse. The threshold is set by the relative values of R_1 and R_2.

the blanking level (black level of the video signal). The transistor then conducts, and, with an appropriate value for R_3, it goes into saturation. The output of the circuit is then a series of rectangular pulses coincident with the sync-pulses but inverted. These pulses are used directly to control the horizontal deflection oscillator.

7.2.7b Vertical Sync-Separator. It must be recalled that the horizontal sync-pulses are 5 μsec long, whereas the vertical are 190 μsec. They are easily separated by using a simple low-pass *RC* filter with a suitable time constant. Figure 7.5 shows a typical vertical sync-separator circuit.

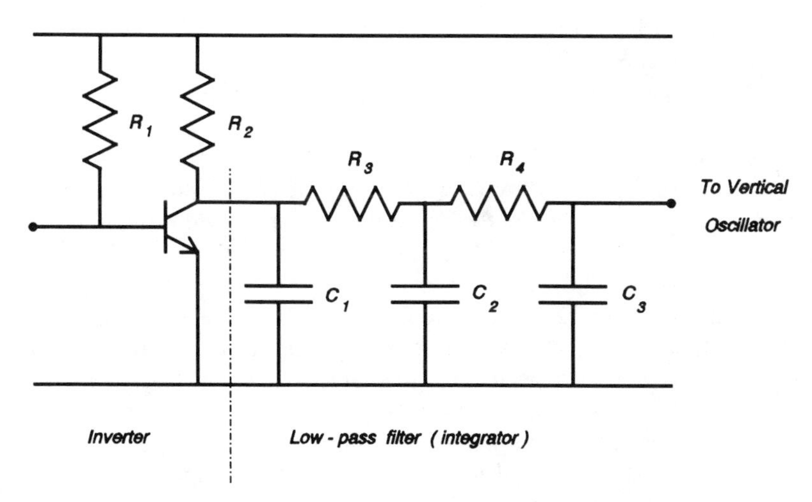

Figure 7.5. The vertical sync-pulse separator. The low-pass filter ensures that it does not react to the horizontal sync-pulse.

The RC low-pass filter shown in Figure 7.5 has three sections, which should give it a higher rate of amplitude change with frequency. The difference between the two frequencies to be filtered (60 Hz and 15.75 kHz) make the filter design simple.

7.2.7c Vertical Deflection Oscillator. The vertical deflection oscillator is an astable multivibrator, which is synchronized to the vertical sync-pulses. Figure 7.6*a* shows the circuit diagram of the astable multivibrator.

When the dc power is first switched on, current is supplied to the bases of both transistors, and they will both tend to conduct. In general, it can be assumed that one of the two transistors (say Q_1) will conduct a little bit better than the other. The voltage at the collector of Q_1 will therefore drop a little faster than that of Q_2. Since it is not possible to change the voltage across the capacitor C_2 instantaneously, the voltage at the base of Q_2 will be forced downward. This will have the effect of reducing the forward bias on the base-emitter junction of Q_2. The collector current of Q_2 will be reduced and this will make the collector voltage of Q_2 go up. Again, the voltage across C_1 cannot change instantly, so the collector voltage of Q_2 will tend to force the base of Q_1 upward. This will cause the base-emitter junction of Q_1 to become more forward biased than it was. More collector current will flow, and the collector voltage of Q_1 will drop even more. We are now back to where we started and the downward change, which started the chain of events, is now magnified. This is evidently a regenerative (positive feedback) process that ends with Q_1 in full conduction and Q_2 cut off. With the appropriate choice of R_{c1} and R_{b1}, Q_1 goes into saturation, and essentially its collector voltage is zero. This phase of the operation is called the regenerative phase and because regenerative phenomena are basically unstable, they tend to happen very fast.

The next phase of the operation, called the relaxation phase, starts as C_2 begins to charge up through R_{b2}. The voltage at the node B_2 will start to rise exponentially and eventually, it will reach a value sufficient to bias the base-emitter junction of Q_2 in the forward direction. A new regenerative phase starts. Q_2 will start to conduct, its collector voltage will drop. The voltage drop will be passed by C_1 to the base of Q_1, which will now conduct less; its collector will come out of saturation (that is go positive) and will be passed by C_2 to the base of Q_2. Q_2 will conduct more heavily and its collector voltage will drop even faster than before. Again, the change that set the chain of events in motion has been magnified. Q_2 will eventually go into full conduction and Q_1 will be cut off. The appropriate choice of the values of R_{b2} and R_{b1} will ensure that Q_2 goes into saturation driving the base of Q_1 to a negative value equal to the dc supply voltage. C_1 starts to charge up exponentially headed for the positive dc supply voltage, but when it gets to the value required to forward bias Q_1, Q_1 starts to conduct again the next regenerative phase repeats itself followed by the next relaxation phase *ad infinitum*.

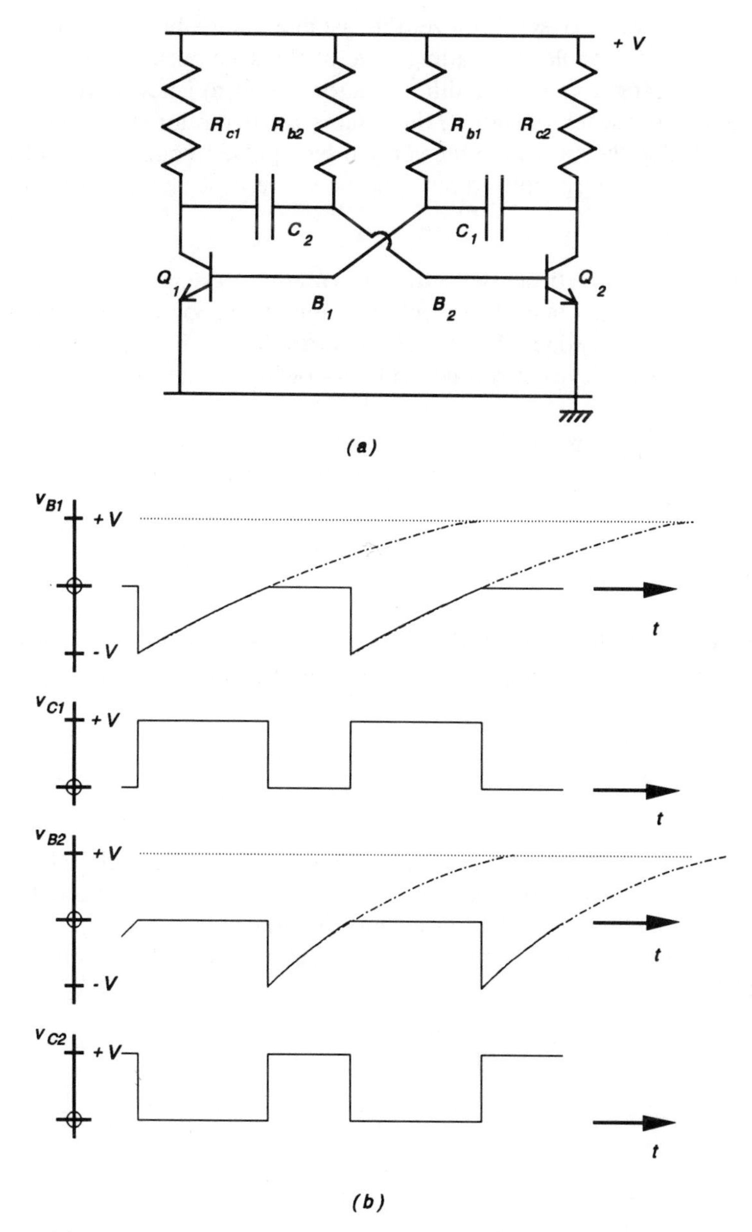

Figure 7.6. (*a*) The circuit diagram of the astable multivibrator. (*b*). The voltage waveforms for the bases and collectors.

The voltage waveforms at the collector and bases are shown in Figure 7.6*b*.

A simple way to synchronize the astable multivibrator to the vertical sync-pulse is to couple the leading edge of the sync-pulse to the base of one of the two transistors using a differentiator circuit to get a sharp clock pulse. To get a stable synchronization, the astable multivibrator should operate at a frequency slightly lower than the vertical sync-pulse frequency when free-running. The design principles discussed above are best illustrated by an example.

Example 7.2.1 Vertical Deflection Oscillator Design. Design an astable multivibrator with a mark-to-space ratio equal to 15.35 : 1.34 and synchronize it to the leading edge of a 60 Hz squarewave derived from the vertical sync-pulse of a television receiver. The following are given:

(1) Supply voltage, 12 V dc
(2) Two NPN silicon transistors, $\beta = 100$ and $V_{be} = 0.7$ V
(3) Load current, 10 mA

Solution. Assuming that

(a) The collector loads of the transistors are the loads of the oscillator
(b) The collector voltage is zero when the transistor is in saturation

The collector resistors

$$R_{c1} = R_{c2} = 12/10 \text{ mA} = 1.2 \text{ k}\Omega$$

The time constant must be chosen, so that one complete cycle takes slightly longer than 1/60 sec, for example, 17.0 msec. This means that one transistor will be in conduction for 15.6 msec and the other for 1.36 msec.

From Figure 7.6*b*, it can be seen that the equation of the voltage on the base of Q_1 is

$$v_{b1} = 2V(1 - e^{-t/\tau}) - V \tag{7.2.10}$$

The time taken by the base voltage of Q_1 to reach 0.7 V is $t = 15.6$ msec. The time constant

$$\tau_1 = 20.71 \text{ msec}$$

Because the transistor has a $\beta = 100$, the base current required to cause saturation in each transistor is 10 mA/100 = 100 μA. The base resistor

$$R_{b1} = R_{b2} = (12 - 0.7)/100\ \mu\text{A} = 113 \text{ k}\Omega$$

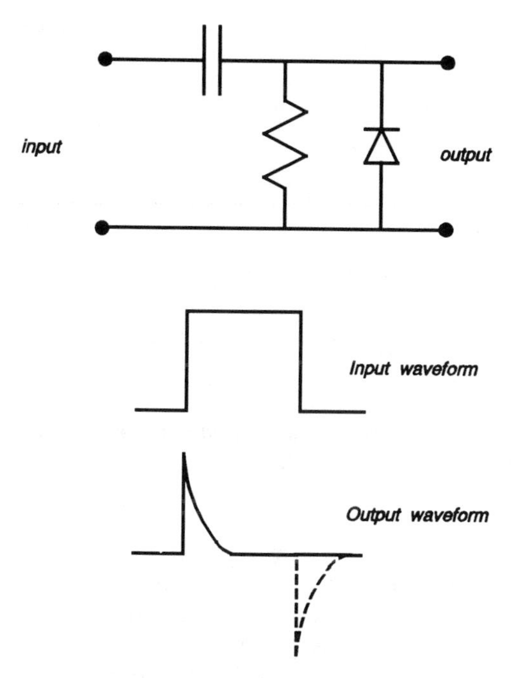

Figure 7.7. The differentiation circuit with its input and output.

The time constant, $\tau_1 = C_1 R_{b1}$; therefore,

$$C_1 = 0.183\ \mu\text{F}$$

Similarly, $\tau_2 = 1.79$ msec $= C_2 R_{b2}$, where

$$C_2 = 0.016\ \mu\text{F}$$

To produce the clock pulse coincident with the leading edge of the squarewave, the differentiation circuit shown in Figure 7.7 is used. The condition for differentiation is that the time constant CR is much smaller than T, the period of the input voltage. The diode clips off the unwanted negative-going spike.

The base voltage of Q_1 is now an exponential curve with the positive clock pulse superimposed on it, as shown in Figure 7.8. It is clear from the diagram that when the clock pulse occurs at position a, the oscillator will not synchronize to the clock pulse. However, when the clock pulse occurs in position b, the voltage at the base of Q_1 will exceed 0.7 V, Q_1 will go into the regenerative mode and the oscillator will be synchronized.

7.2.7d Vertical Sweep Current Generator. A very simple sweep current generator with a fairly linear output current with respect to time is shown in Figure 7.9. The circuit also acts as a class A amplifier and drives the yoke coil of the vertical deflection system.

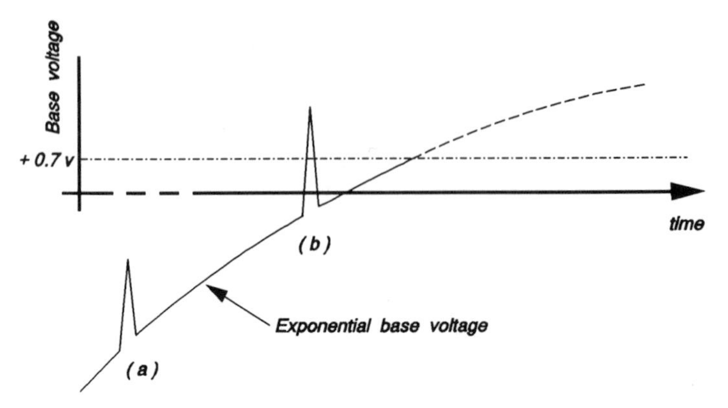

Figure 7.8. The base voltage of the transistor used in the astable multivibrator showing the synchronizing pulses.

The transistor is biased for class A operation. The collector is connected to the dc power supply by a large inductance with negligible resistance; that is, it is an open circuit at the frequency of operation but a short circuit to direct current. When the squarewave from the multivibrator is applied to the base, the collector current of the transistor will be a squarewave. During current transitions, a voltage will be developed across L_c; otherwise the yoke inductor L_y has the dc power supply connected directly across it. Because the dc voltage is a constant, di/dt in L_y is a constant. The capacitor C serves as a block to dc flowing to ground. Because of the low frequency of

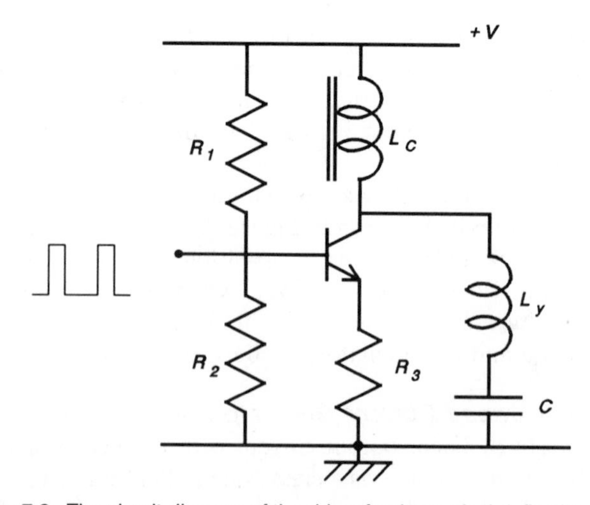

Figure 7.9. The circuit diagram of the driver for the vertical deflection yoke.

operation, this circuit does not consume much power; neither is there a large reactive power circulating in the circuit.

7.2.7e Horizontal Deflection System. From Figure 7.3, it can be seen that the horizontal deflection system has the same modules as the vertical. The description of the modules and their design will not be repeated. In practice, the vertical and horizontal deflection systems are different and the differences arise from the following:

(1) The vertical system operates at 60 Hz, whereas the horizontal runs at 15.75 kHz.
(2) The ratio of the deflection in the horizontal and vertical directions is 4 : 3 (aspect ratio).
(3) The horizontal deflection system is very sensitive to phase modulation, and hence the horizontal oscillator has to have an automatic phase-control loop to keep phase modulation as low as possible.
(4) High power (~ 50 W) is required to run the horizontal deflection coil if the energy stored in it is dissipated at the end of each cycle. By using extra circuitry, it is possible to return most of the energy to the dc source by resonating the inductance of the yoke with a suitable capacitor to cause ringing at approximately 60 kHz.
(5) The high dc voltage (~ 10,000 V) required to run the picture tube can be obtained by using a transformer coupling (auto-transformer) to step up the voltage followed by a rectifier (tube-type). A simplified circuit diagram of the final stage of the horizontal deflection system is shown in Figure 7.10.

The horizontal oscillator is synchronized to the output of the sync separator. The oscillator then drives the base of Q_2 with a string of rectangular pulses at approximately 15.75 kHz. When Q_2 is on, it draws current through the primary of the transformer T_1. This causes a current to flow in the secondary, biasing Q_1 on and putting it into saturation. The dc supply voltage V is therefore connected directly across the primary terminal P of the autotransformer and ground. Q_1 and the primary winding form an emitter follower. The horizontal deflection yoke L_x, which is connected to the autotransformer at node N, has a constant voltage, higher than V, connected across it. The current that flows in L_x is the linear ramp required to produce a linear horizontal deflection. The capacitor C_c blocks dc from flowing in L_x, and it is a short-circuit at the frequency of operation. At the end of the linear current ramp, Q_2 is cut off, a voltage is induced in the secondary of T_1, and this is in the appropriate direction to shut off Q_1 abruptly. Normally, the sudden termination of current flow in the autotransformer and L_x will generate a huge voltage spike in the attempt to dissipate the energy stored in

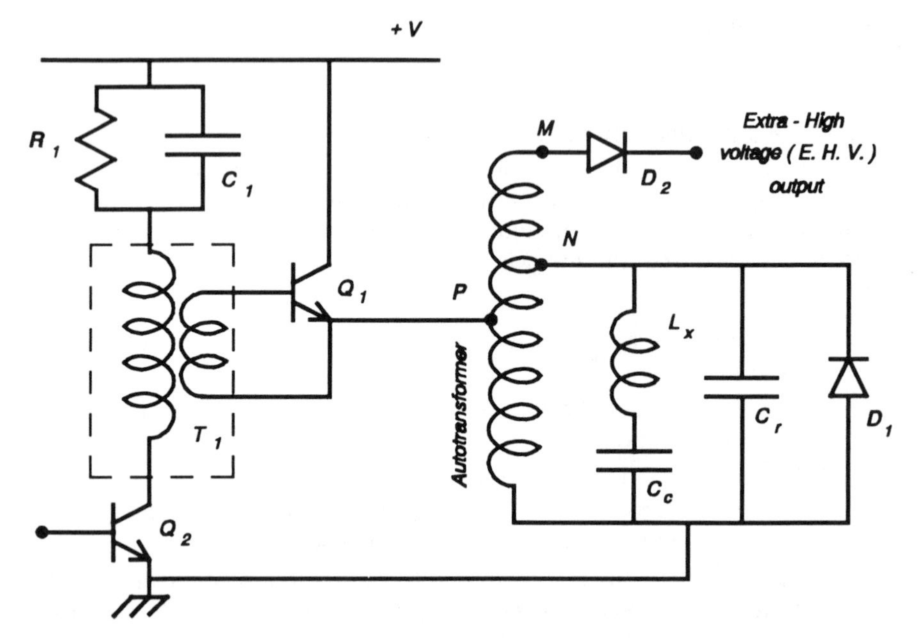

Figure 7.10. The driver of the horizontal deflection yoke with the extra-high-voltage (EHV) generator required for the anode of the picture tube.

the inductances. However, the diode D_1 conducts providing a path for the current to flow into the tuned circuit made up of L_x and C_r. L_x and C_r resonate ("ring") at approximately 60 kHz (four times the horizontal deflection frequency). If the timing is set up correctly, the energy stored in the resonant circuit will be flowing from the capacitor into the yoke just as the next current ramp is starting. This arrangement substantially reduces the amount of power taken from the dc power supply to drive the yoke. Since no diode is connected across the *M-N* portion of the autotransformer, a large voltage spike occurs across it. This is rectified and used to bias the anode of the picture tube.

7.2.8 Picture Tube

The picture tube is an example of a cathode ray tube (CRT) used in a wide variety of display systems. It consists of a glass cylinder that flares into a cone with a nearly flat base. Figure 7.11 shows a cross section of the typical CRT used for the display of television images.

At the end of the glass cylinder, there is an electron gun. The electron gun is made up of a cylinder, closed at one end and coated with oxides of barium and strontium, which give off electrons when heated. The CRT has control

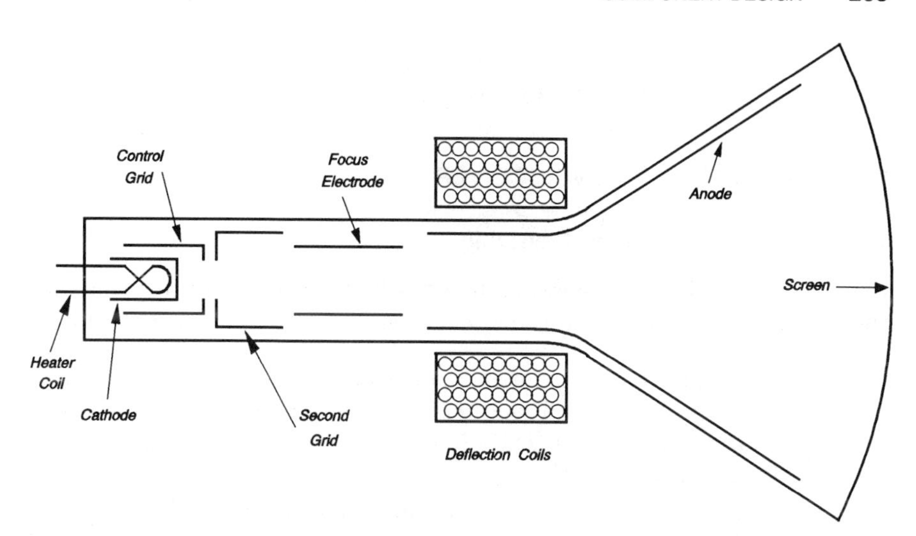

Figure 7.11. A cross-sectional view of the cathode ray tube (CRT). Reprinted with permission from T. Soller, M. A. Starr, and G. E. Valley, *Cathode Ray Tube Displays*, Boston Technical Publishers, Lexington, MA, 1964.

grid(s), focusing, and deflection electrodes or coils. An electric current is passed through the heater coil to bring the temperature of the cathode to the appropriate value; electrons are given off, and, under the influence of a positive voltage placed on the anode, the electrons travel down the tube and strike the screen at high velocity. The grid(s), carries a negative voltage, which can be varied to control the number of electrons that can leave the cathode and eventually reach the screen. This grid(s) therefore controls the intensity of the light forming the image. The electron lens is either an electrostatic or a magnetostatic means of focusing the electrons, usually to produce a small spot on the screen.

The rest of the tube is made up of the deflection electrodes or coils that cause the electron beam to deflect on the screen on the application of the appropriate voltage or current. The anode is a conductive thin-film on the inside wall of the flare as shown and requires several thousand volts positive for proper operation. The screen is a coating of certain silicates, sulfides, fluorides, and alkali halides that emit light when bombarded by electrons. The color of the light emitted is determined by the characteristic wavelength associated with each chemical element present in the screen material. For example, copper-activated 85% zinc sulfide and 15% cadmium sulfide gives off yellow light, whereas 93% zinc sulfide and 7% cadmium sulfide produces a blue-green phosphor. The persistence (or afterglow) of the phosphor can be varied to suit different conditions.

Most modern CRTs used in television receivers have an electrostatic focusing system, because it is simpler and less expensive to manufacture. The deflection system is, however, magnetostatic because of the tendency to have shorter tubes and larger screen areas.

7.3 COLOR TELEVISION RECEIVER

7.3.1 Demodulation and Matrixing

In Section 6.3.8, it was established that black-and-white and color television signals had many common features. Their basic differences are as follows:

(1) In addition to the luminance information (L_w), the chrominance information [$(L_r - L_w)$ and $(L_b - L_w)$] was transmitted using a quadrature modulation (DSB-SC) scheme with a (color) subcarrier at approximately 3.58 MHz.
(2) On the back porch of the horizontal blanking pulse, an eight-cycle burst of the color subcarrier was superimposed.

One scheme used for the recovery of the original red, blue, and green signals is shown in Figure 7.12. It is described as prepicture tube matrixing.

From the output of the video detector, the color burst separator picks up the eight cycles of the 3.58-MHz subcarrier and passes them onto the color subcarrier regenerator. The regeneration circuit is essentially a very high Q factor circuit tuned to 3.58 MHz. The eight-cycle burst of signal causes the high Q circuit to go into a free-running oscillation mode. The subcarrier must have limited variation in both amplitude and phase, for proper synchronous detection, and, therefore, the decay of the amplitude during the free-running period must be controlled. To maintain the amplitude variation to 90% of the initial value for the time required to sweep one horizontal line (63 μsec) a Q factor of 7000 is required. Crystals are used in the regenerator because only crystals have the necessary high Q factor.

The regenerated 3.58-MHz signal is split into two branches. The first branch goes through a 90-degree phase shifter to produce the quadrature signal required for the synchronous demodulation of the $(L_r - L_w)$ and $(L_b - L_w)$ signals contained in the output of the color bandwidth filter. After demodulation, the matrix reproduces the third difference signal, namely $(L_g - L_w)$. The composite video signal containing the luminance information L_w, is now combined in the adders to give the original red, green, and blue information. After suitable amplification, they drive the picture tube cathodes. A variation on this scheme called picture tube matrixing, produces the $(L_g - L_w)$ signal and the addition of the L_w is carried out in the picture tube by driving the grids and the cathodes differentially. This is shown in Figure 7.13.

When the I and Q signal scheme is used in the transmission, it must be decoded into $(L_r - L_w)$ and $(L_b - L_w)$ after the synchronous detectors and before the signal is applied to the matrix.

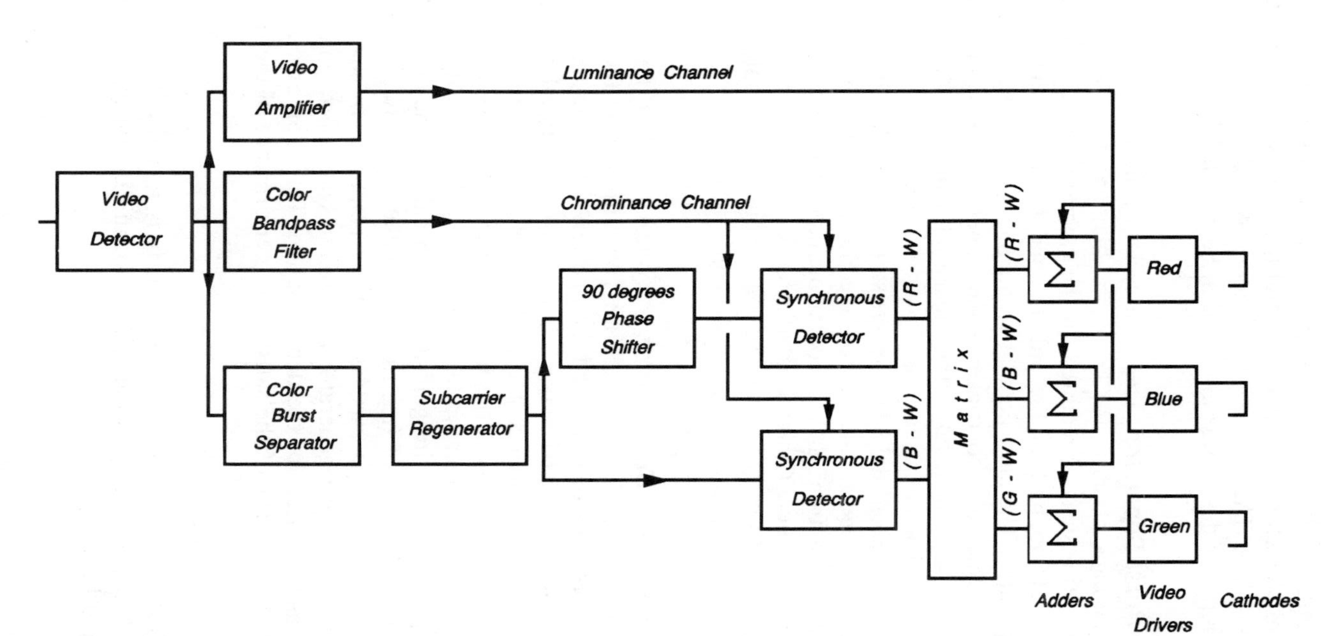

Figure 7.12. The schematic diagram for color separation in the receiver. This is described as prepicture tube matrix, because the colors are separated outside the tube.

7.3.2 Component Circuit Design

The 90-degree phase shift circuit was discussed in Sections 4.4.3a and 4.4.3b. The synchronous detector (four-quadrant multiplier) was discussed in Sections 2.6.4 and 4.4.3c. The design of the adder was described in Section 4.4.3e. Video amplifier design was discussed in Section 6.3.4b.

7.3.2a Color Burst Separator. The eight cycles of the color burst subcarrier riding on the back porch of the horizontal blanking pulse are shown in Figure 6.24. Since this signal is required for the synchronous detection of the chrominance signal, no part of the video signal must be allowed to get through the separator. The most popular technique is to use the falling edge of the horizontal sync-pulse to trigger a gate open and to close the gate immediately after the time it takes the eight cycles of 3.58 MHz to come through (2.23 μsec). The circuit used is shown in Figure 7.14.

Q_2 is biased in the off state by the resistors R_3 to R_6. Its emitter is coupled through the tuned circuit LC to the collector of Q_1. The base of Q_1 is driven by the output from the horizontal oscillator (see Figure 7.3). The resonant frequency of the LC circuit is adjusted so that the rise-time of the rectangular waveform from Q_1 is delayed long enough to let only the eight cycles of the 3.58 MHz signal through to the collector of Q_2.

7.3.2b Color Subcarrier Regenerator. A typical circuit of the color subcarrier regenerator is shown in Figure 7.15. The output from the transistor Q_1 is taken from both the collector and the emitter. C_2 is adjusted to cancel the shunt capacitance of the crystal (see Figure 2.18). The crystal is chosen so that its series resonant frequency is 3.58 MHz and it resonates when it is excited by the subcarrier. In the series mode resonance, the crystal impedance is essentially a short circuit (ideal voltage source). The losses in R_e and R_3 in parallel with R_4 are therefore insignificant. The circuit has the high Q factor to perform about 217 cycles of free-running oscillations, between bursts of color subcarrier signals. Q_2 is used as a buffer amplifier. In some earlier models, the signal from the regenerator was used to synchronize an oscillator whose output was then used for the demodulation. This is an expensive way to solve the problem in view of the satisfactory performance of the free-running system.

7.3.2c The Matrix. The function of the matrix is to reproduce the third difference signal $(L_g - L_w)$. Equation (6.3.29) for this operation was derived and it is repeated here.

$$(L_g - L_w) = -\frac{0.30}{0.59}(L_r - L_w) - \frac{0.11}{0.59}(L_b - L_w) \qquad (7.3.1)$$

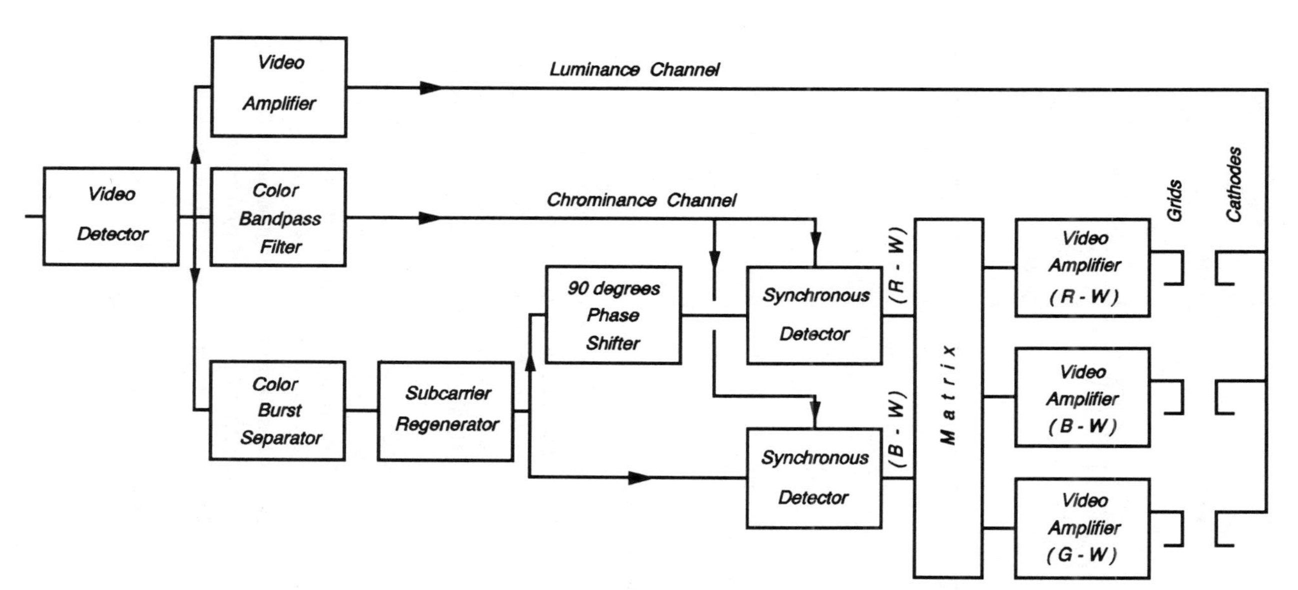

Figure 7.13. The block diagram for color separation in the tube. Note that the color differences are applied to the tube directly.

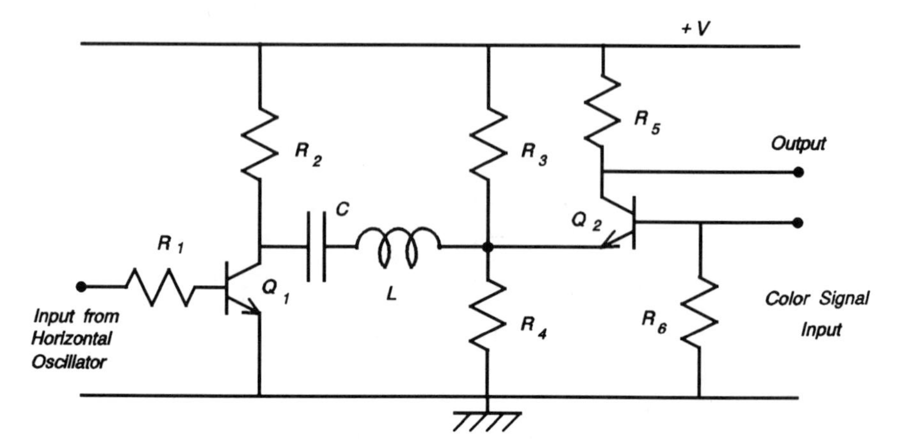

Figure 7.14. The color subcarrier burst separator. The output is used to synchronize the color subcarrier regenerator.

It is clear that this can be achieved with a simple resistive network and a phase-reversing buffer amplifier.

7.3.3 Color Picture Tube

In Section 7.2.8, it was stated that difference phosphor materials on the screen of the picture tube give different colors of light when bombarded by electrons. A number of different techniques for producing images in color on the CRT exist; all of them rely on this phenomenon. The most successful of these uses the shadow mask technique. Figure 7.16 illustrates the basic concept.

Three electron guns representing the three primary colors are mounted symmetrically on a plane perpendicular to the axis of the tube. Electrons from each gun will therefore bombard different areas of the screen after going through the holes in the mask. If the area of the screen corresponding to the blue gun is coated with a phosphor that produces blue light, and so on, then it is clear that by depositing the different phosphor materials in dots of triplets at the appropriate points on the screen, all the different colors can be reproduced by modulating the relative saturation of each component.

The shadow mask technique has the following disadvantages:

(1) The shadow mask has to be very carefully aligned for good color resolution.

(2) The presence of the mask reduces the average brightness (luminance) of the image but this can be compensated for by increasing the electron emission of the guns.

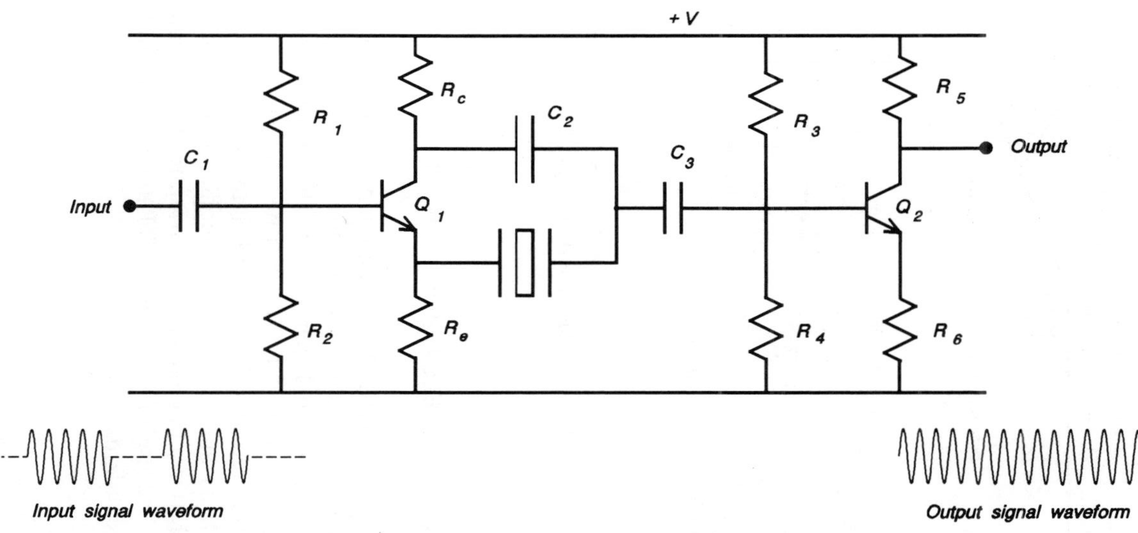

Figure 7.15. The color subcarrier regenerator. Note that because of the high Q factor of the crystal in the circuit, it is virtually free-running between bursts of the subcarrier.

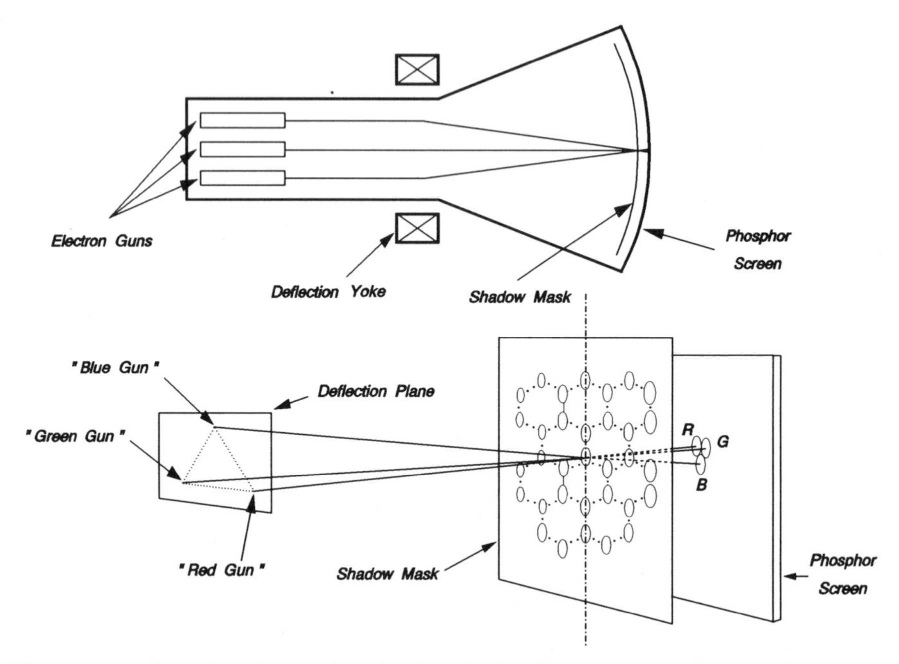

Figure 7.16. The color picture tube showing details of the shadow mask. Reprinted with permission from H. B. Law, "Shadow Mask," *Proceedings of the IRE*, © IRE (now IEEE), **39**(10), 1951.

7.4 HIGH-DEFINITION TELEVISION

High-definition television (HDTV) is an attempt to produce images of quality comparable to that of a 35-mm motion picture film. The proposed system has 1125 lines of scan with 60 fields per second. The required bandwidth is 20 to 25 MHz. The development of HDTV has been plagued by many technical problems. However, recent gains in bandwidth due to the development of technologies such as optical fiber and satellite transmission have helped toward making HDTV a real possibility. Currently, it is the subject of intense political interest because of the commercial implications involved in establishing common standards.

BIBLIOGRAPHY

G. Kennedy, *Electronic Communication Systems*, McGraw-Hill, New York, 1970.

H. B. Law, "Shadow Mask," *Proc. IRE*, 39(10), 1187, 1951.

N. W. Parker, "Television Image-Reproducing Equipment," in *Electronics Engineers' Handbook*, D. G. Fink and D. Christiansen, Eds, 2nd ed., McGraw-Hill, New York, 1982.

S. Sherr, *Fundamentals of Display System Design*, Wiley-Interscience, New York, 1970.

J. Smith, *Modern Communication Circuits*, McGraw-Hill, New York, 1986.

REFERENCES

1. T. Soller, M. A. Starr, and G. E. Valley, *Cathode Ray Tube Displays*, Boston Tech., Lexington, MA, 1964.

PROBLEMS

7.1 Give two advantages of a vestigial sideband modulation system. What are its disadvantages? A transmission consists of a carrier and its upper sideband. Show that when the modulation index is small compared to unity, an ordinary envelope detector can be used to demodulate it. Calculate the signal power in a vestigial sideband signal if the peak value of the carrier voltage is 5 V when the load resistor is 2.5 kΩ and the modulation index is 0.15. You may assume that the diode is ideal.

7.2 A television receiver is tuned to receive a transmission with a carrier frequency of 67.25 MHz (channel 4). If the intermediate frequency is 45.75 MHz, what is (are) the possible frequency(ies) at the local oscillator terminals of the mixer? If there are two possible frequencies, does one of them offer an advantage over the other? The local oscillator signal for the television receiver is generated at a lower frequency and multiplied by a factor of 512. Calculate the frequency of the oscillator and design a Hartley-type oscillator, given that the tuning capacitor has the value 144 pF. Discuss the precautions you would take to minimize changes in the frequency of oscillation.

7.3 Using suitable block diagrams and derived equations, explain how quadrature multiplex is used to transmit and receive the chrominance information in a color TV system. State the conditions that must be satisfied for demultiplexing to be possible. How are these conditions met in the domestic TV receiver?

7.4 Describe the construction and operation of a color TV picture tube. Explain why a color tube requires a higher dc voltage (extra-high-voltage—EHV) than the monochrome for equal luminance.

7.5 Describe with the help of a suitable diagram the operation of a color subcarrier regenerator. Show that for a series-tuned *RLC* circuit at resonance,

(1) The total energy stored in the circuit is a constant

(2)

$$Q_o = 2\pi \frac{\text{Energy stored}}{\text{Energy dissipated per cycle}}$$

In a crystal-controlled color subcarrier regenerator operating at 3.58 MHz, the total energy remaining in the circuit after 218 cycles of free-running oscillation is 90% of its original value. Calculate the peak value of the voltage across the capacitor at the end of the free-running cycle and estimate the Q_o of the circuit.

PART 3

TELEPHONE

8

THE TELEPHONE NETWORK

8.1 INTRODUCTION

The early history of the telephone system has been outlined in Chapter 1. The growth of the telephone system has been truly phenomenal and forecasts show a continuing growth as new services such as data transfer, facsimile, and mobile telephone are added.

The telephone differs from the broadcasting system in two basic ways:

(1) In broadcasting, a few people who, in theory, have information send it out to the many who are presumed to want the information; it is one-way traffic. The communication link provided by the telephone is usually two-way traffic.
(2) The basic idea of broadcasting is to make the message available to anyone who has the equipment and the interest to tune in. This is in contrast to the norm in the telephone system where the privacy of the message is guaranteed by law.

Because of these differences, the two systems tend to handle very different types of information—public versus private—and their patterns of development have been different.

8.2 TECHNICAL ORGANIZATION

For a telephone system to work, there must be a minimum of two people who wish to communicate. It is then possible to install the circuit shown in Figure 8.1. This would be quite adequate, except for the fact that these two people

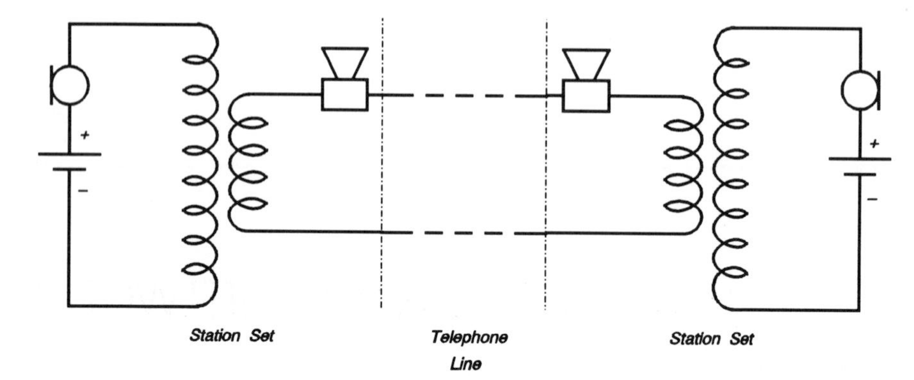

Figure 8.1. The basic telephone system.

may not want to talk to each other all the time and therefore some additional system has to be set up for either person to indicate to the other that they wish to talk. What was added was a bell on the called party's premises that can be rung from the calling party's premises.

Presumably the success of this prototypical communication would soon attract the attention of other people who would want to set up similar systems. It is clear that soon the situation depicted in Figure 8.2 would develop, where every subscriber would have to be wired up to every other subscriber. That would be prohibitively expensive and quite impractical.

Evidently, the way to deal with the situation is to connect every subscriber to a central location and arrange to have an attendant to interconnect the various subscribers in whatever combination is required. That central location is, of course, the central office, which has and continues to have a central role in the telephone system. The system configuration would then be as shown in Figure 8.3.

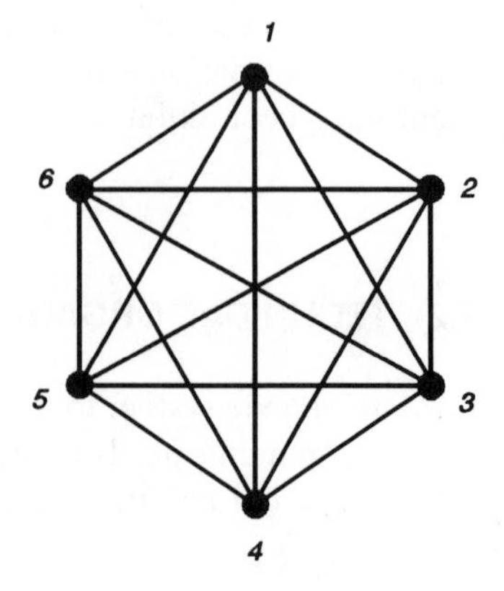

Figure 8.2. The connection diagram for six subscribers showing all the 15 possible telephone lines.

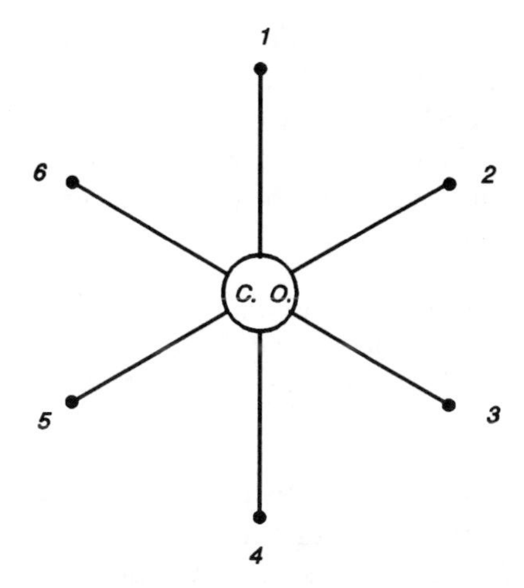

Figure 8.3. The introduction of a central office reduces the number of lines to six.

Assuming that the system has six subscribers, then each subscriber has access to the other five subscribers. But in the meantime, a group also of six in the next town have heard of the success of the system and have set up a similar system of their own. Now there is a possibility of reaching eleven other subscribers if a connection can be made between the two systems. This brings up two very important points:

(1) The greater the number of people on the communication network, the more attractive it is for other people to join.
(2) There has to be some level of compatibility between the two systems.

The system would have evolved, as shown in Figure 8.4.

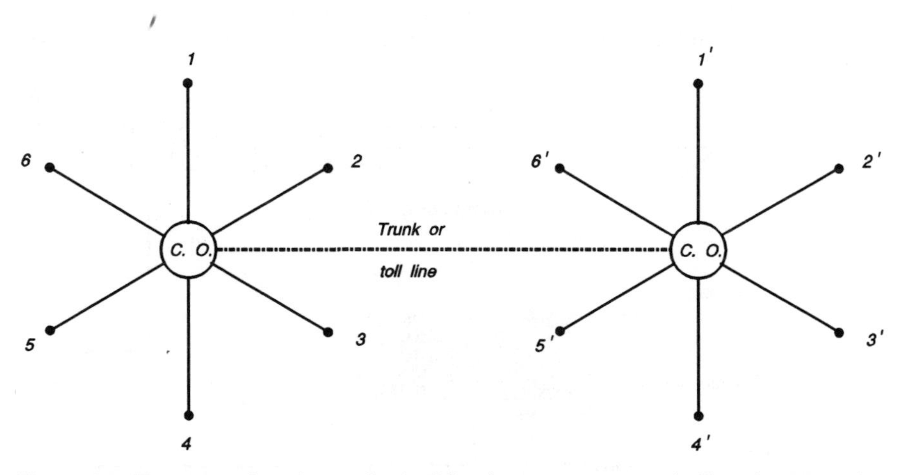

Figure 8.4. The connection of a trunk of toll line between two central offices increases the number of subscribers that can be reached from 5 to 11.

Continuing with the story, the distance between the two towns is quite long and the initial cost and upkeep are high, but, if this line can be made to carry more than one conversation simultaneously, the cost per conversation will be substantially reduced. It is also very likely that because of the length of the line, the quality and the reliability of the system may be degraded. This brings up an important point:

> The greater the number of communication channels that can be established over the same link, the less the cost per message. Multiplex and the conservation of bandwidth will become goals of several generations of communications engineers.

The telephone system that started with people talking to each other has acquired more than people for clientelle. Increasingly, the network is being used to supply services to machines, such as computers, facsimile devices, and security guard services.

8.3 BASIC TELEPHONE EQUIPMENT

The basic telephone has surprisingly very few parts. These are shown in Figure 8.1. When the microphone is connected in series with the battery, it produces a current proportional to the pressure of the sound impinging on it. The transformer eliminates the dc and sends the ac portion of the current through the line. The earphone at the receiving end changes the variation of the current into sound. Evidently, the system works in the reverse direction.

8.3.1 Carbon Microphone

Figure 8.5 shows a cross section of the carbon microphone [1]. It has a lightweight aluminum cone with a flexible support around the periphery, so

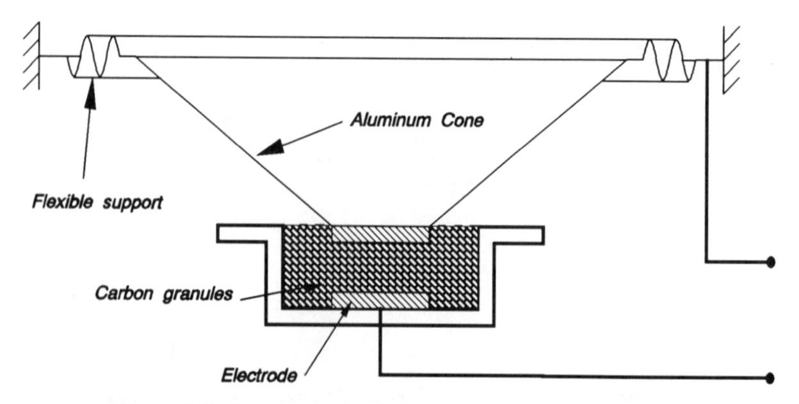

Figure 8.5. A cross-sectional view of the carbon microphone.

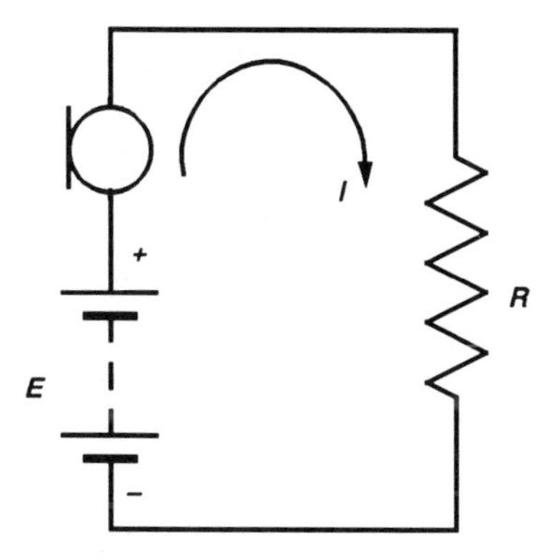

Figure 8.6. The carbon microphone with dc supply E and load resistance R.

that it will deflect (vibrate) to the changing sound pressure level. Attached to the apex is a disc that acts as a piston when the cone deflects. A plastic housing with an electrode attached to the bottom contains a loose pile of carbon granules. When the pressure on the cone is increased, the carbon granules become compressed, the resistance goes down and more current flows. The opposite happens when the pressure is released.

Assuming that the sound pressure level on the carbon microphone is a sinusoid, then the resistance of the device is

$$r(t) = r_0(1 + k \sin \omega t) \tag{8.3.1}$$

where r_0 is the mean resistance, k is a coefficient less than unity, and ω is the frequency of the sound pressure. When the microphone is connected to a battery of electromagnetic force E volts in series with a load R (as in Figure 8.6), we have

$$I = \frac{E}{R + r_0(1 + k \sin \omega t)} \tag{8.3.2}$$

$$= \frac{E}{R + r_0} \cdot \frac{1}{1 + \dfrac{kr_0}{R + r_0} \sin \omega t} \tag{8.3.3}$$

$$I = \frac{E}{R + r_0}\left(1 + k\frac{r_0}{R + r_0} \sin \omega t\right)^{-1} \tag{8.3.4}$$

Using the binomial expansion,

$$I = \frac{E}{R + r_0}\left[1 - \left(\frac{kr_0}{R + r_0}\right)\sin \omega t + \left(\frac{kr_0}{R + r_0}\right)^2 \sin^2 \omega t - \cdots\right] \tag{8.3.5}$$

Let

$$I_0 = \frac{E}{R + r_0} \tag{8.3.6}$$

then

$$I = I_0 - I_0\left(\frac{kr_0}{R + r_0}\right)\sin \omega t + \frac{I_0}{2}\left(\frac{kr_0}{R + r_0}\right)^2 - \frac{I_0}{2}\left(\frac{kr_0}{R + r_0}\right)^2 \cos 2\omega t + \cdots \tag{8.3.7}$$

Because $kr_0/(R + r_0)$ is smaller than unity, higher order terms can be ignored. If it is desirable to reduce the second harmonic distortion, $kr_0/(R + r_0)$ can be reduced, but in doing so the amplitude of the fundamental will be reduced as well. A compromise between distortion and signal amplitude has to be made.

The carbon microphone has the following attractive properties:

(1) It is simple and therefore inexpensive to manufacture.
(2) It is robust; it is not likely to need attention even in the hands of the public.
(3) It acts as a power amplifier; under normal bias conditions; the electrical power output far exceeds the acoustic power input. It does not normally require addition amplification.
(4) Its input–output characteristics are shown in Figure 8.7. The nonlinearity at low input levels helps to suppress background noise and that at high levels acts as an automatic gain control.

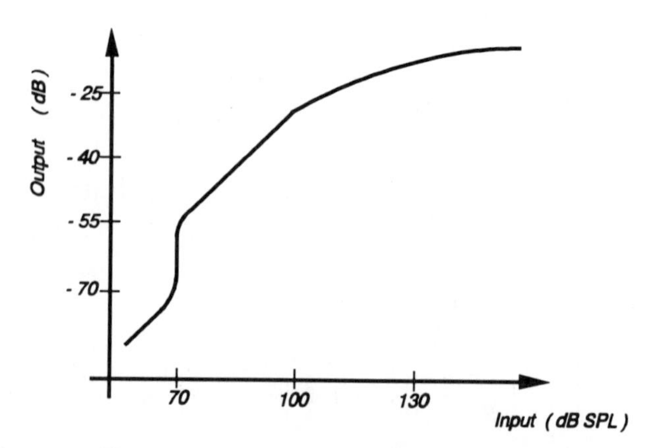

Figure 8.7. The input – output characteristics of the carbon microphone.

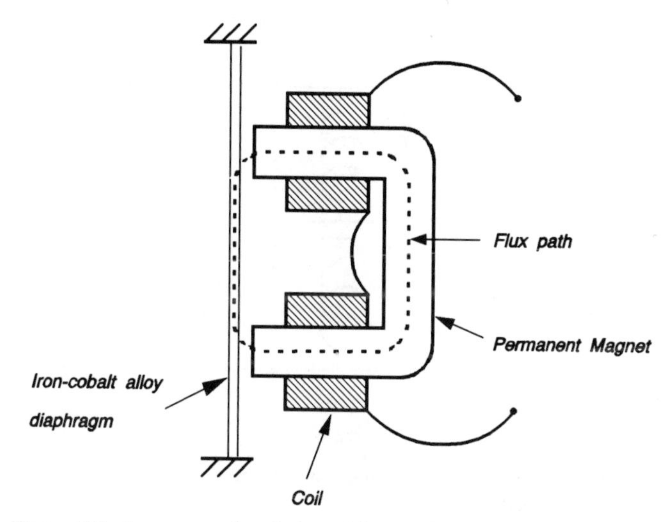

Figure 8.8. A cross-sectional view of the moving-iron telephone receiver.

8.3.2 Moving-Iron Telephone Receiver

A cross section of the moving-iron telephone receiver is shown in Figure 8.8. It consists of a U shaped permanent magnet that carries a coil as shown. In front of the open face of the U, a thin cobalt-iron diaphragm is held by an annular ring support with a short distance between them.

The signal current is passed through the coil; assuming that it is sinusoidal, the flux generated by the current will aid the pull of the permanent magnet on the diaphragm for one-half of the cycle, and it will deflect accordingly. During the other half of the cycle, the coil flux will oppose that of the magnet and the diaphragm will deflect much less.

The force between two magnetized surfaces is given by

$$F = \frac{B^2}{2\mu_0} \qquad [\mathrm{N/m^2}] \tag{8.3.8}$$

where B is the flux density in teslas, μ_0 is the permeability of free-space, that is, $4\pi \times 10^{-7}$. Let A be the area of the pole face (in square meters), B_0 be the flux due to the permanent magnet (in teslas), and $b_0 \sin \omega t$ be the flux due to the current (in teslas). The force in newtons

$$F = \frac{2A}{2\mu_0}(B_0 + b_0 \sin \omega t)^2 \tag{8.3.9}$$

$$= \frac{A}{\mu_0}\left(B_0^2 + 2B_0 b_0 \sin \omega t + b_0^2 \sin^2 \omega t\right) \tag{8.3.10}$$

$$= \frac{A}{\mu_0}\left(B_0^2 + \tfrac{1}{2}b_0^2 + 2B_0 b_0 \sin \omega t - \tfrac{1}{2}b_0^2 \cos 2\omega t\right) \tag{8.3.11}$$

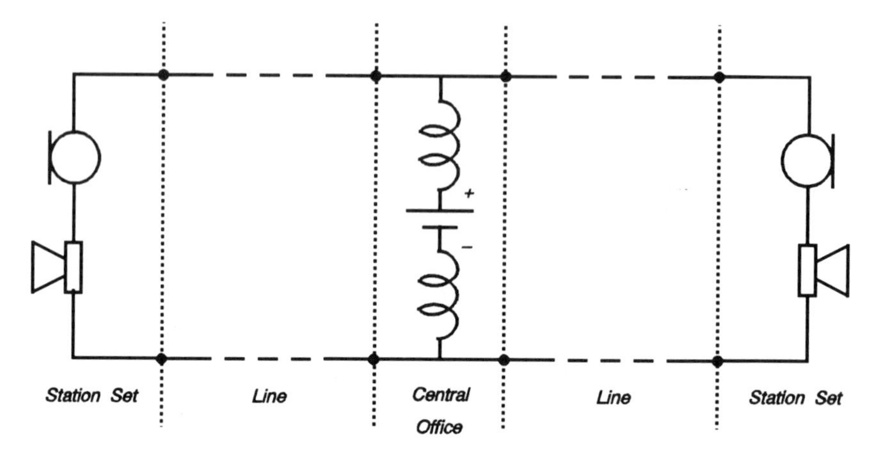

Figure 8.9. The local batteries are replaced by a central power supply.

The second harmonic component can be reduced by making B_0 large compared to b_0. This will increase the dc component of the force, which is likely to cause the diaphragm to touch the magnet. Note that when B_0 is zero (no permanent magnet), the device produces only the second harmonic. This is to be expected since both the positive and negative halves of the sinusoid will exert an equal force of attraction on the diaphragm.

8.3.3 Local Battery — Central Power Supply

The system as depicted in Figure 8.1 is powered from the batteries located on the customer's premises. The batteries are of interest because they are a hazard to the customer and they pose a very difficult problem for the maintenance staff. Furthermore, their reliability is questionable because of their location among other considerations.

The solution to the problem is to have a common power supply located at the central office (out of the way of the telephone subscriber) and readily available to the maintenance personnel. The reliability of the service can then be improved by installing a backup power supply. The scheme for achieving this end is illustrated in Figure 8.9. The central office battery in series with two inductors is connected to the lines of the calling and called party as shown. The inductors have a high inductance and therefore appear to be open circuits at audio frequency but short circuits at direct current. Every call requires two such inductors to complete the connection.

8.3.4 Signaling System

The signaling system consisted of a magneto and a bell, which responded to high ac voltage input. The magneto is a hand-operated alternator whose flux

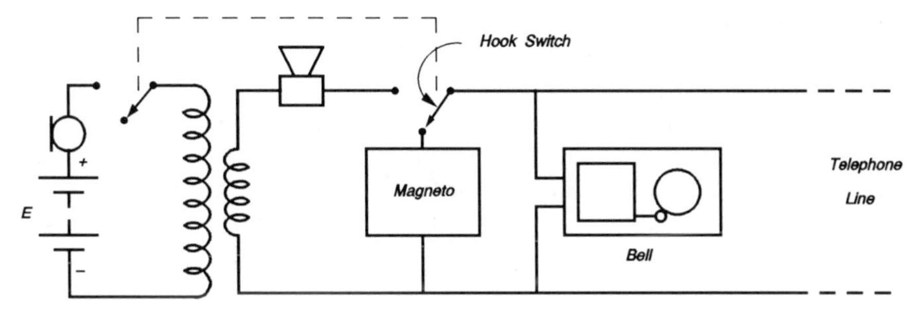

Figure 8.10. The elemental telephone with signaling devices (magneto and bell) shown.

is produced by a permanent magnet. The calling party turns the crank to produce approximately 100 V ac. The current travels down the telephone line and causes the bell at the called party's end to ring. To avoid damage to the telephone receiver and to conserve battery power, the hook switch disconnects them from the line when the telephone is not in use. A simplified diagram of the signaling system is shown in Figure 8.10.

8.3.5 Telephone Line

Physically, the telephone line consists of a pair of copper wires supported on glass or porcelain insulators mounted on wooden poles. Electrically, an infinitesimally short piece of line can be modeled as shown in Figure 8.11 [2, 3]. The elemental series resistance and inductance are represented by δR

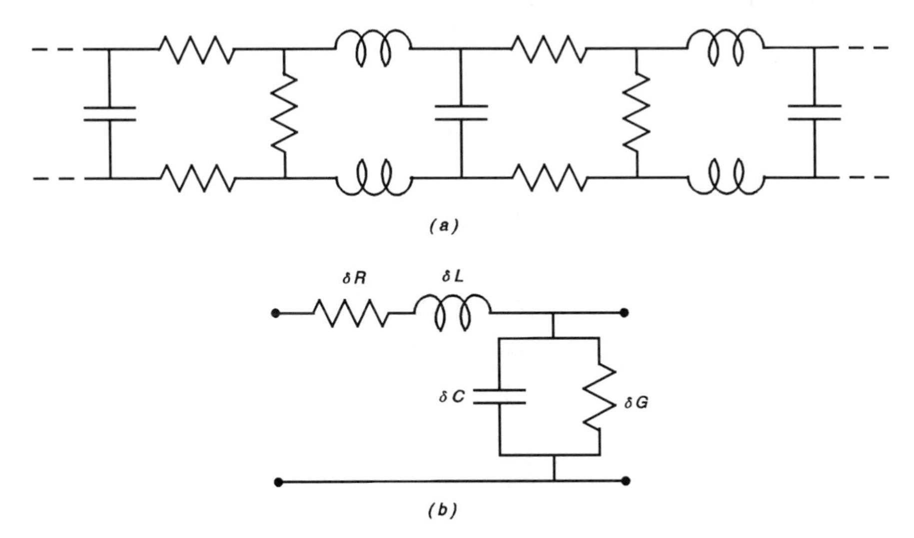

Figure 8.11. (*a*) The equivalent circuit of the telephone line showing series resistance *R*, inductance *L*, shunt capacitance *C*, and conductance *G*. (*b*) An elemental equivalent circuit of the telephone line.

and δL and the elemental shunt capacitance and conductance are represented by δC and δG, respectively.

The analysis of the model is beyond the scope of this book. However, the analysis shows that the telephone line, at voice frequencies can be approximated by an *RC* low-pass filter whose cut-off frequency is a function of its length. The longer the line is, the lower the cut-off frequency. The characteristics of a typical telephone line is shown in Fig. 10.1.

8.3.6 Performance Improvements

From Figure 8.9, it can be seen that the dc required to power the carbon microphone has to flow through the receiver. This is not a good idea, since it will make B_0 in Eq (8.3.11) larger or smaller than it should be. A second disadvantage is that all the ac current generated by the carbon microphone has to flow through the receiver. This will produce a very loud reproduction of the speaker's own voice. The psychological effect is that the speaker lowers his voice, making it difficult for his listener to hear what he is saying. This phenomenon is called sidetone. The two problems can be solved by using the circuit shown in Figure 8.12. It is an example of a hybrid.

8.3.6a The Hybrid. The carbon microphone is connected to the center-tap of the primary of the transformer. One end is connected to the telephone line and the other is connected to an *RC* network, which approximates the impedance of the line. The secondary is connected to the receiver. There is still a path for direct current from the central office battery to flow through the carbon microphone.

In the transmit mode, the ac produced by the microphone divides equally with one half flowing through the telephone line and the other half in the

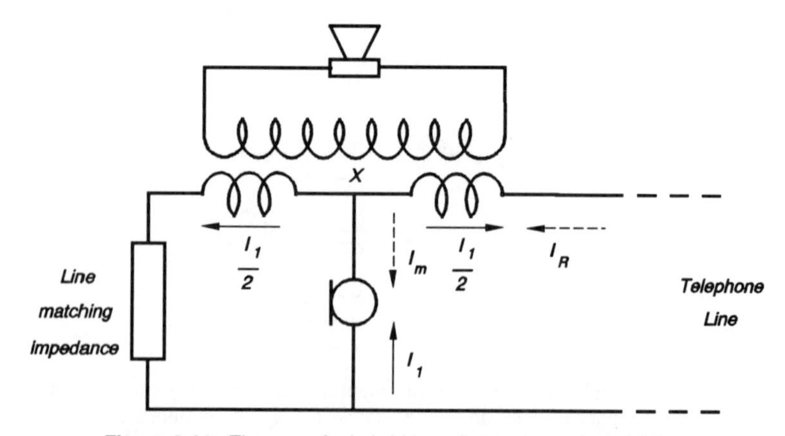

Figure 8.12. The use of a hybrid transformer to control sidetone.

line-matching impedance. Because these currents are in opposite directions in the primary of the transformer, no net voltage appears across the secondary. The speaker cannot hear himself. In the receive mode, the current I_R flows through the first half of the primary winding and then splits at node X with I_m flowing through the carbon microphone where the energy is dissipated. The remainder $(I_R - I_m)$ flows through the second half of the primary into the line-matching impedance. This time, the directions of the two currents in the primary are the same; a net voltage appears across the receiver.

In practice, the level of sidetone fed back to the speaker has to be carefully controlled. When it is too low, the telephone appears "dead" to the speaker and his normal reaction is to raise his voice. When the sidetone is too high, it has the opposite effect.

8.3.6b Rotary Dial. The rotary dial came with the invention of the automatic central office. Evidently, the automatic central office offered several advantages over the manual office. There was increased security of the messages since there was no human interface in setting up calls. The time for setting up and releasing a call was substantially reduced by reducing the possibility of operator errors. It guaranteed 24-hr service with fewer more highly trained personnel.

The dial is simply a method of issuing instructions to the central office, and it does this by producing a binary coded message by mechanically opening and closing a switch. The basic dial is as shown in Figure 8.13. It has a finger wheel with ten finger holes and it is mechanically coupled by a shaft to a second wheel, which has ten cam lobes as shown. The shaft is mounted so that both wheels can rotate about the axis. The wheel assembly is spring loaded so that, when it is rotated in a clockwise direction, it will return at a constant speed under the control of a mechanical governor.

To operate the dial, the caller inserts his or her index finger into the hole corresponding to the number and pulls the finger wheel to the finger stop and then releases it. While the finger wheel is rotating in the clockwise direction, the lever X is free to move out of the way of the cam lobes without disturbing the switch lever Y. When the wheel assembly is rotating in the counter-clockwise direction, every cam lobe that passes X will cause the switch lever Y to open the switch. If current is flowing through the switch, the current flow will be disrupted the number of times corresponding to the number of the finger hole. The current pulses can be used to operate a device (to be discussed later) at the central office to effect the required connection. The return spring, cam lobes, and mechanical governor are designed to produce 10 pulses per second with about equal mark-to-space ratio. Since it is bound to take the subscriber much longer then 1/10 sec to rotate the finger wheel again, a pause longer than 1/10 seconds can be recognized by the central office as an inter-digit pause. It is then possible to send a second and subsequent strings of pulses to effect a connection that

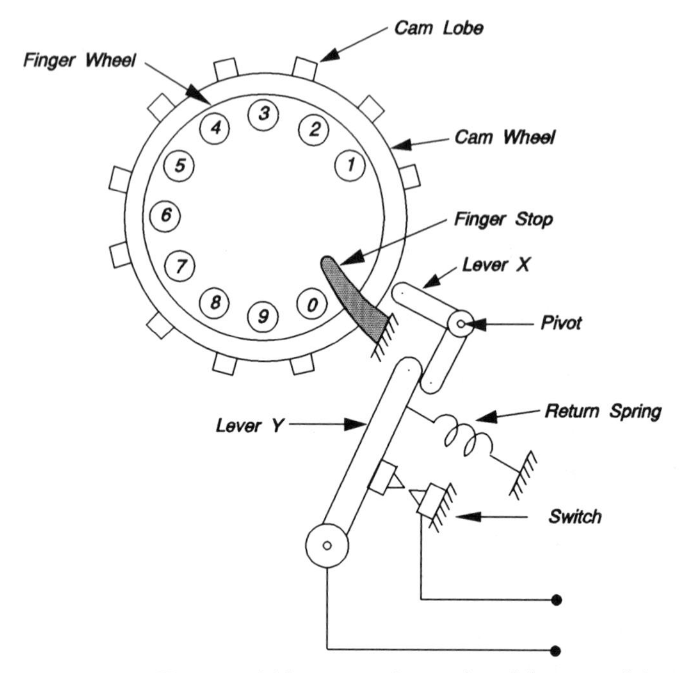

Figure 8.13. The essential features and operation of the rotary dial.

requires a multidigit code. Note that when the digit "0" is dialed, ten pulses are produced.

8.3.6c Telephone Bell. The telephone bell has two brass gongs with a clapper, which is operated by an electromagnet. It is mechanically and electrically tuned to respond optimally (resonance) to current at 20 Hz. It is left connected to the telephone line at all times, but the high impedance of its electromagnetic coil ensures minimal effect at voice frequencies. Also the 10-Hz pulse from the rotary dial has no significant effect on it. Nominally, it operates on 88 V, 20 Hz ac supplied to it from the central office in the ring-mode.

8.3.7 Telephone Component Variation

The telephone components described in this section are meant to be a representative sample of what can be found within the territory of any telephone operating authority or company. For each component, there are several possible variations, some made to get around patent rights granted to others and some to lower cost and improve reliability.

The physical appearance and electrical characteristics of the subscriber telephone instrument have changed since it was first put into service. However, in broad terms, it remained basically the same until the introduction of electronics in the form of semiconductor devices. The availability of amplifiers at very low cost, offered various options, such as new microphones, electronic sidetone control, tone ringers, and tone dialing. Some of these will now be discussed.

8.4 ELECTRONIC TELEPHONE

By the late 1960s, a number of electronics research and development organizations were working on the development of electronic telephone sets. Manufacturing cost reduction, improved performance, and the possibility of offering subscribers new uses for their telephone were incentives for this development. The first truly electronic component to emerge was the tone dialer more popularly known by its trade name TOUCH-TONE™. Telephone sets with other electronic components were not far behind, so that by the early 1980s there were a number of telephones on the market with none of the well established components.

8.4.1 Microphones

The features of the carbon microphone that made it indispensable in the telephone set for a very long time were low cost, acoustic-to-electrical power amplification, suppression of low-level background noise, and high-level compression. Its drawbacks were distortion, high dc requirements, and changes in its acoustic sensitivity due to dc flowing in it. All of its advantages can be obtained with none of the disadvantages by using other microphones in conjunction with a suitable amplifier. The new microphones could be made considerably smaller than the carbon microphone. Examples of such microphones include the following

(1) *The Electret Microphone* [4]. This is a variation on the capacitance microphone. The incident sound causes the distance between two plates of a capacitance to change, resulting in a change of voltage. The output voltage and power are very low. The output impedance is very high (10 pF). The normal biasing of capacitor microphones is averted by a built-in charge that is placed on capacitor during the manufacturing process. It has an excellent frequency response, and it is normally used in acoustical measuring instruments.

(2) *Ribbon Microphone*. A thin aluminum ribbon is suspended in the field of a small powerful magnet. The incident sound causes the ribbon to vibrate in the field causing a voltage proportional to its velocity to be induced in it. It has a very good frequency response but very low output power.

(3) *Crystal Microphone*. A crystal of Rochelle salt (sodium potassium tartrate), quartz, and other piezoelectric materials produce a voltage when subjected to mechanical deformation. The crystal is cut into a

thin layer with suitable conductors connected to the faces. The incident sound causes a voltage to appear across the conductors. The microphone is very often made up of several layers of crystals.

8.4.2 Receiver

The receiver is one of the few components that has successfully resisted change since Alexander Graham Bell patented it in 1876. The materials used for making the magnet and diaphragm and the actual mechanical construction have changed, but the basic principle of operation remains the same.

8.4.3 Hybrid

The function of the ideal hybrid is to direct the signal from the microphone on to the telephone line without loss and to direct the incoming signal on the line to the receiver with no loss. The operation of the hybrid is therefore similar to the operation of a circulator—a well known device in microwave engineering. The two devices are compared in Figure 8.14. There are two major differences:

(1) There are two critical paths in the hybrid (transmit and receive) but three critical paths in the circulator. In a normal circulator, the transmitter cannot have a path to the receiver since this will cause instability.

(2) Circulators are realizable in reasonable physical dimensions at microwave frequencies. A direct application of circulator theory at audio frequency predicts a device several kilometers in diameter.

The problem of realizing a circulator at audio frequency can be solved by using a *gyrator* [5], better known for its ability to invert impedances [6, 7]. Gyrators were the subject of intense interest at a time when the microelectronics industry was looking for a microminiaturized version of the inductor.

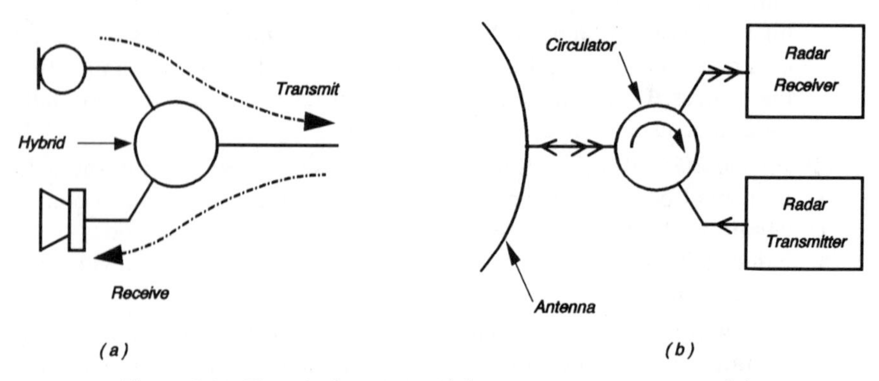

Figure 8.14. The telephone hybrid (*a*) compared to the circulator (*b*).

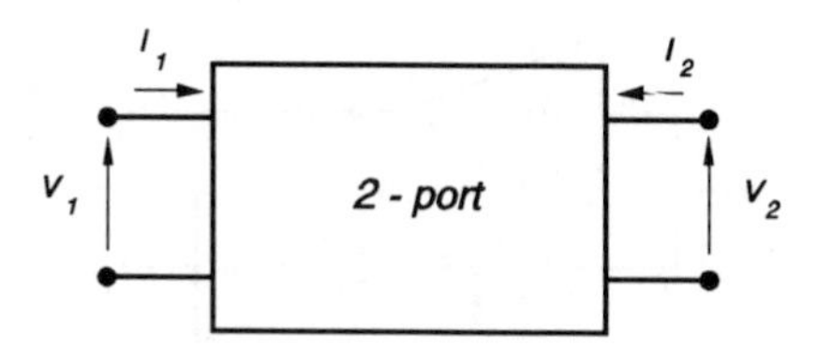

Figure 8.15. A two-port with its defining voltages and currents.

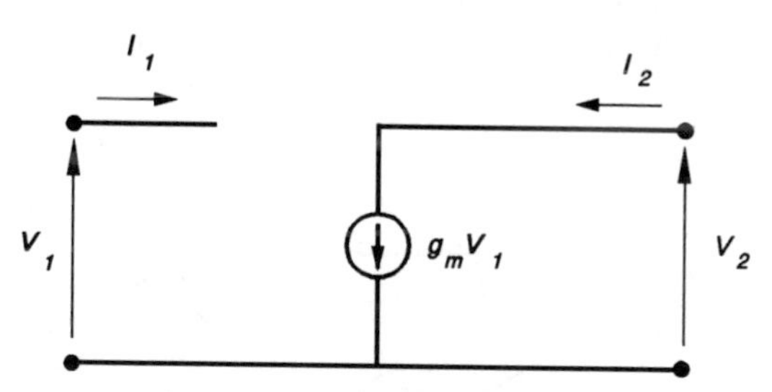

Figure 8.16. The ideal voltage-controlled current source.

Consider the two-port shown in Figure 8.15 with the port voltages and currents as indicated. The two-port is described by the matrix

$$\begin{bmatrix} I_1 \\ I_2 \end{bmatrix} = \begin{bmatrix} Y_{11} & Y_{12} \\ Y_{21} & Y_{22} \end{bmatrix} \begin{bmatrix} V_1 \\ V_2 \end{bmatrix} \tag{8.4.1}$$

Applying this description to the ideal voltage-controlled current source shown in Figure 8.16 gives

$$\begin{bmatrix} I_1 \\ I_2 \end{bmatrix} = \begin{bmatrix} 0 & 0 \\ g_m & 0 \end{bmatrix} \begin{bmatrix} V_1 \\ V_2 \end{bmatrix} \tag{8.4.2}$$

where g_m is the transconductance.

Consider a second ideal voltage-controlled current source with a 180-degree phase shift. The matrix equation is

$$\begin{bmatrix} I_1 \\ I_2 \end{bmatrix} = \begin{bmatrix} 0 & 0 \\ -g_m & 0 \end{bmatrix} \begin{bmatrix} V_1 \\ V_2 \end{bmatrix} \tag{8.4.3}$$

When the two ideal voltage-controlled current sources are connected back to back as shown in Figure 8.17, the matrix equation is

$$\begin{bmatrix} I_1 \\ I_2 \end{bmatrix} = \begin{bmatrix} 0 & -g_m \\ g_m & 0 \end{bmatrix} \begin{bmatrix} V_1 \\ V_2 \end{bmatrix} \tag{8.4.4}$$

The two-port can be converted into a three-terminal element by lifting the ground. Its matrix equation is then

$$\begin{bmatrix} I_1 \\ I_2 \\ I_3 \end{bmatrix} = \begin{bmatrix} 0 & g_m & -g_m \\ -g_m & 0 & g_m \\ g_m & -g_m & 0 \end{bmatrix} \begin{bmatrix} V_1 \\ V_2 \\ V_3 \end{bmatrix} \tag{8.4.5}$$

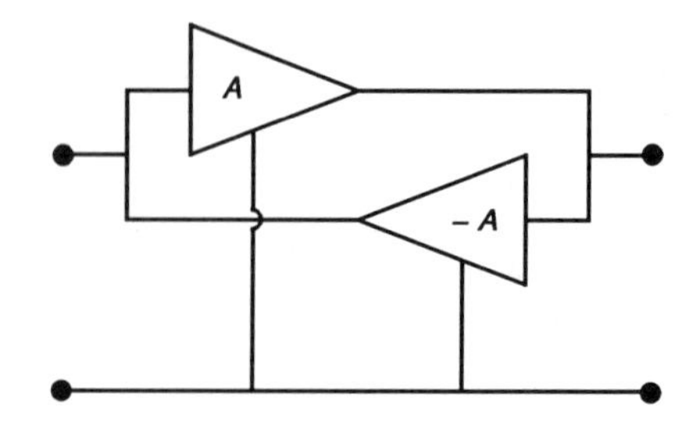

Figure 8.17. Two ideal voltage-controlled current sources connected back to back to form a gyrator.

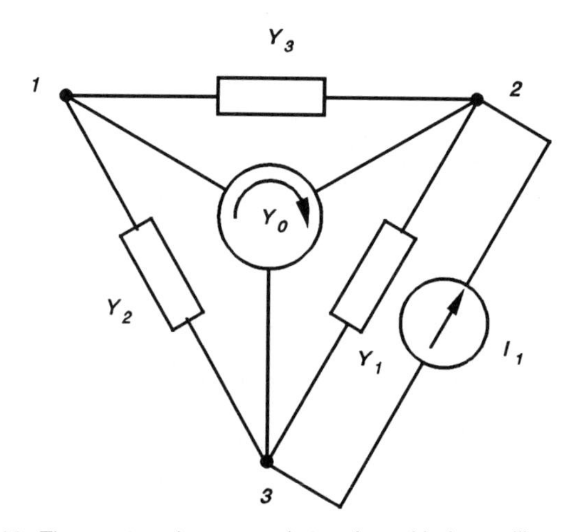

Figure 8.18. The gyrator when properly terminated behaves like a circulator.

Figure 8.18 shows the three-terminal circuit in which the transconductance g_m has been replaced by the more general transadmittance Y_0, and port 1 has been terminated in an admittance Y_1, port 2 in Y_2 and port 3 in Y_3.

Consider that an ideal current source I_1 is connected across port 1, so that a voltage V_1 appears across it. This will induce voltages V_2 and V_3 across ports 2 and 3, respectively. The voltage ratio

$$\frac{V_2}{V_1} = \frac{Y_1(Y_3 - Y_0)}{Y_2Y_3 + Y_0^2} \tag{8.4.6}$$

Similarly,

$$\frac{V_3}{V_1} = \frac{Y_1(Y_2 + Y_0)}{Y_3Y_2 + Y_0^2} \tag{8.4.7}$$

When $Y_1 = Y_2 = Y_3 = Y_0$, Equation (8.4.6) gives

$$\frac{V_2}{V_1} = 0 \tag{8.4.8}$$

and Equation (8.4.7) gives

$$\frac{V_3}{V_1} = 1 \tag{8.4.9}$$

This means that voltage across port 1 appears across port 2, but no voltage appears across port 3, that is,

$$V_2 = V_1 \qquad \text{and} \qquad V_3 = 0 \tag{8.4.10}$$

When the process is repeated with the ideal current source connected across port 2,

$$V_3 = V_2 \qquad \text{and} \qquad V_1 = 0 \tag{8.4.11}$$

Similarly with the ideal current source across port 3,

$$V_1 = V_3 \qquad \text{and} \qquad V_2 = 0 \tag{8.4.12}$$

It is clear that the circuit is behaving like a circulator oriented in a clockwise direction.

To get the desired effect, the transadmittance of the gyrator amplifiers have to be adjusted to fit the line admittance. The admittances of the microphone and receiver circuits have to be tailored so that the gyrator sees an admittance equal to its own admittance connected to each port. In practice, less than a perfect match can be achieved and therefore some sidetone is obtained. This property can be exploited to adjust the level of the sidetone to a comfortable level.

Several electronic telephones use the audio frequency circulator concept and its various manifestations as hybrids.

8.4.4 Tone Ringer

In most electronic telephone sets, the bell has been replaced by some kind of tone ringer, which usually emits an attractive musical note or notes to signal the presence of a call. Quite often amplitude and frequency modulation are used to enhance the tone. The tone ringer must satisfy the following conditions:

(1) The input impedance must be high so as not to interfere with the signal on the line to which it is permanently connected.
(2) It has to operate on the 20 Hz, 88 V ac ringing signal.

In terms of circuit design, what is required is one or two audio frequency oscillators, a modulator, a power supply fed from the 20-Hz ring signal, an audio frequency amplifier, and a loudspeaker. The design of all these items have been discussed earlier. Oscillators were discussed in Section 2.4, modulators in Section 2.6, audio frequency amplifiers in Section 2.7, and loudspeakers in Section 3.4.8. The power supply for the ringer would require a rectifier and a capacitor.

8.4.5 The Tone Dial

Instead of producing current pulses to signal to the central office the number dialed, the tone dial produces a pair of audio frequency tones. The frequencies of these tones are carefully chosen so they are not harmonically related. This reduces the probability of other tones or signals being recognized as dialed numbers. The dial pad and the corresponding frequencies produced are shown in Figure 8.19.

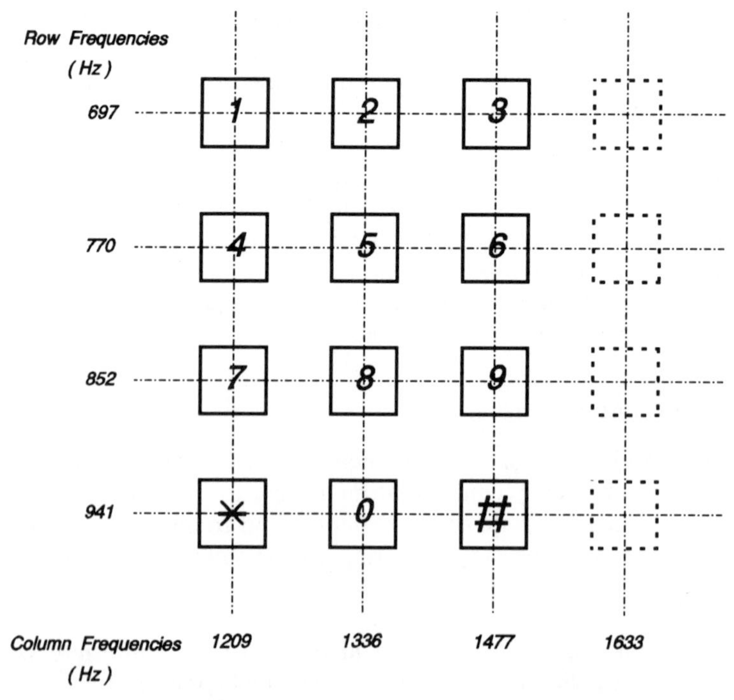

Figure 8.19. The push-button dial and its corresponding frequencies.

8.4.5a TOUCH-TONE™ — DIGITONE™ Dial. The dial consists of two essentially identical oscillators one of which produces the high-frequency tones and the other the low-frequency tones. The change in frequency is achieved by switching in capacitors of appropriate values when the dial push button is depressed. Each button has a unique pair of tones associated with it. The central office equipment recognizes the number dialed by the two frequencies present.

The early versions of the TOUCH-TONE™ dial used discrete bipolar transistors in conjunction with an inductor and capacitances in a Colpitts configuration. Both the low- and high-frequency groups were produced by a single oscillator. One of the capacitances in each group was changed as the various buttons were depressed to produce the required frequencies. Later versions used an integrated amplifier with an *RC* twin-tee feedback circuit to produce the tones. The basic configuration of the circuit is shown in Figure 8.20.

Two conditions have to be met for oscillations to occur according to the Barkhausen Criterion:

(1) The closed loop gain must be equal to unity. In practice, the loop gain must be slightly larger than unity for sustained oscillation.
(2) The change in phase around the loop must be an integer multiple of 2π rad.

The classical *RC* twin-tee filter has values of *R*s and *C*s as shown in Figure 8.21*a* and the amplitude and phase responses in Figure 8.21*b*. Its transfer function is given by

$$\frac{v_2}{v_1} = -\frac{1 - \omega^2 C^2 R^2}{1 - \omega^2 C^2 R^2 + j4\omega CR} \tag{8.4.13}$$

Rationalizing and equating the imaginary part to zero gives the frequency at which the phase angle is zero or 180 degrees

$$\omega_0 = \frac{1}{RC} \tag{8.4.14}$$

Under this condition, the output voltage $v_2 = 0$; the circuit has a null in its frequency response with a very high Q factor. The high Q factor can be exploited for high stability of the oscillating frequency if the oscillator is designed to operate at the frequency of the null. However, using the classical twin-tee values of *R*s and *C*s in oscillator design will be self-defeating, because an amplifier with infinite gain will be required.

Departure from the standard ratios of *R*s and *C*s produces lower values of Q factors. The amplitude-frequency responses of the modified twin-tee filter (Figure 8.20) with different values of a are shown in Figure 8.22*a*. The

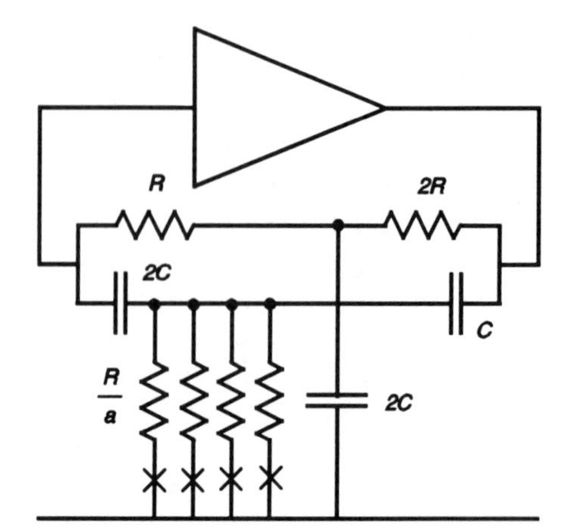

Figure 8.20. The modified twin-tee feedback oscillator. The closing of one of the four switches connects a different R/a to produces the tones. Two such circuits are used in the dial; one for the low frequencies and the other for the high frequencies.

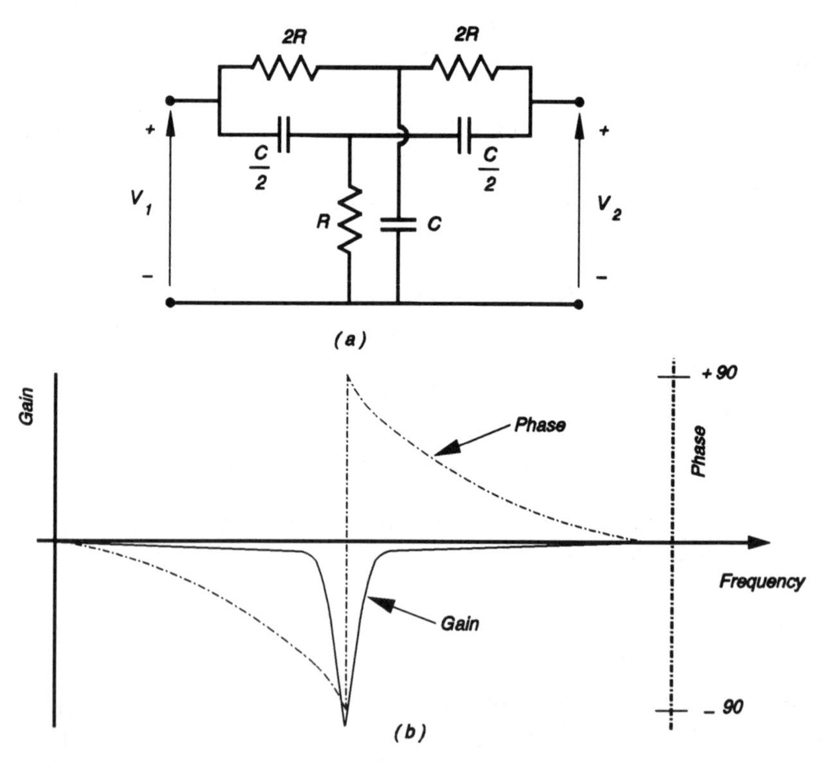

Figure 8.21. (*a*) The classical twin-tee notch filter showing its configuration and circuit element ratios. (*b*) The gain and phase characteristics of (*a*).

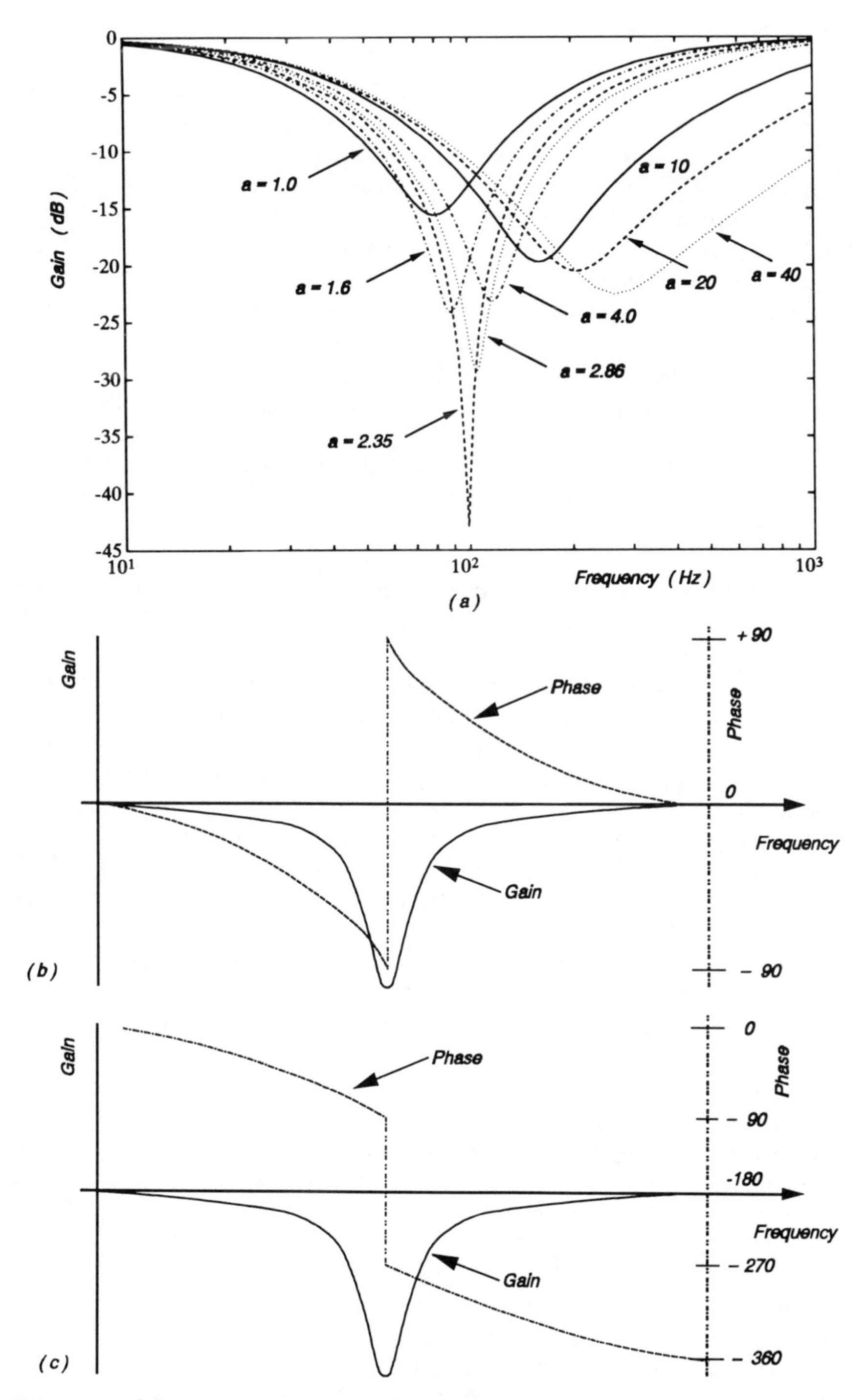

Figure 8.22. (*a*) The gain response of the modified twin-tee for various values of *a*. (*b*) The gain and phase characteristics of the modified twin-tee when *a* is less than 2.25. (*c*) The gain and phase characteristics of the modified twin-tee when *a* is greater than 2.25.

frequencies at which the phase shift of the filter is zero or 180 degrees coincides with the frequency at which the output voltage is a minimum (null). It can be seen that the notch frequency for the particular configuration and circuit element ratios in Figure 8.20 changes with the parameter a. It can also be observed that the depth of the notch varies as a changes; the deepest notch occurring when a is equal to approximately 2.25. For values of a less than 2.25, the modified twin-tee circuit has gain and phase characteristics as shown in Figure 8.22*b*. When a is equal to or greater than 2.25, the gain and phase response is as shown in Figure 8.22*c*. The oscillator design technique described here therefore works only when a is greater than 2.25.

The oscillator design exercise consists of identifying the value of a, which has its notch at the required frequency. The gain needed from the amplifier at the null or where the 180-degree phase shift occurs can be identified. The amplifier can then be designed to have the required gain and a phase shift of 180 degrees.

8.4.5b Digital Tone Dial. The digital tone dial attempts to exploit the low cost of digital integrated circuits. The design of the system is shown in Fig. 8.23.

The crystal-controlled oscillator generates a signal at 3.58 MHz. This frequency was chosen because a crystal designed to operate at that frequency was readily available and it was inexpensive. (It is used in the color burst carrier of television sets; see Section 7.3.2b.) The actual frequency is not important, so long as it is sufficiently high that division by an integer will produce frequencies that lie within the permitted error margin of the tone frequencies. The oscillator frequency is first divided by 16 to give a clock frequency of 223.75 kHz. The push-button pad has a number of contacts that are used to send a binary logic statement of 1s and 0s to the N coder. The function of the N coder is to generate its own set of 1s and 0s as input for the divide-by-N. The divide-by-N consists of a set of resetable binary counters with additional logic circuitry to reset the counters so that the number N can be changed according to the output of the N coder. The function of the eight-stage shift register is to produce eight sequential pulses at the output that have $\frac{1}{8}$ the period of the required sinusoidal signal. These pulses are used to drive the digital-to-analog converter to produce a crude eight-step approximation to the sinewave. The low-pass filter attenuates the unwanted harmonics before the signal drives the telephone line through the line driver. Except for minor differences in the N coder, the two halves of the circuit are the same. The basic circuit blocks used are the NOR, NAND, NOT or inverter, and the resetable bistable multivibrator, known collectively as logic gates. These gates are used in large quantities but each gate occupies such a small area on an integrated circuit chip that the cost is minimal. For example, this circuit had 10 NOR, 4 NAND, 44 NOT, and 23 resetable bistable multivibrators.

Logic gates can be realized in different forms, each with its own mode of operation. They are grouped into families such as *complementary metal*

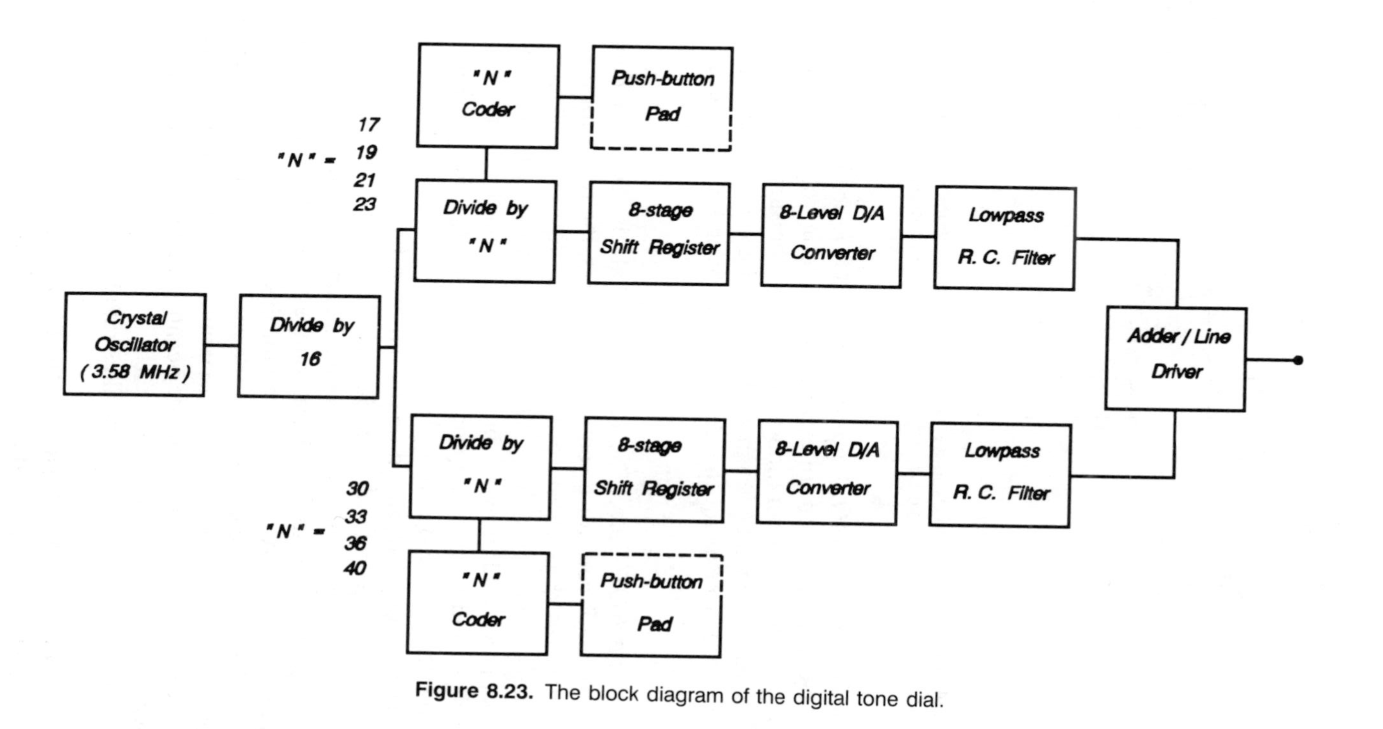

Figure 8.23. The block diagram of the digital tone dial.

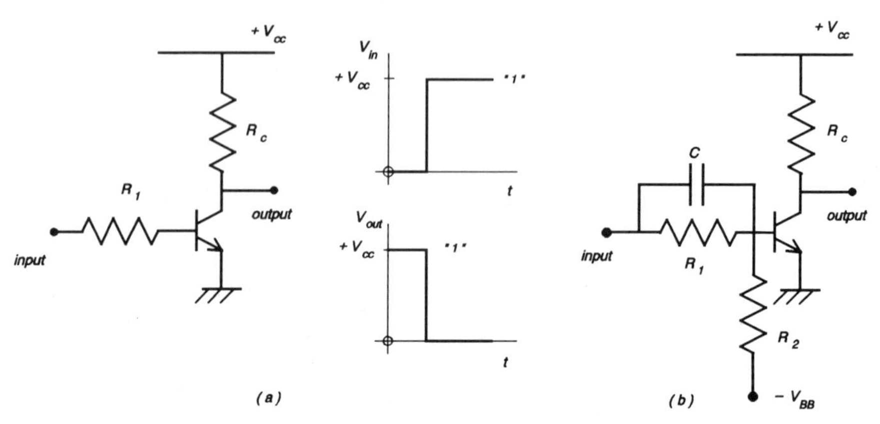

Figure 8.24. (*a*) The basic NOT gate and its transition diagram. (*b*) The NOT gate with speed-up capacitor and negative biasing to improve noise immunity.

oxide semiconductor (CMOS), *transistor-transistor logic* (TTL), *resistor transistor logic* (RTL), and so on. In explaining the design of logic gates, the logic family whose mode of operation appears to be the simplest is chosen.

Before discussing the design of the digital tone generator it is best to digress in order to explain the design of the basic building blocks shown in Figure 8.23.

Design of the NOT Gate. The basic NOT gate is shown in Figure 8.24*a*. When the input is zero (or essentially zero) volts, the transistor is cut off, and no current flows through the collector resistor R_c. The collector is then 1, that is, the system voltage V_{CC}. When the input is 1, current flows through the resistor R_1 and forward biases the base-emitter junction. Collector current flows and given an appropriate value for R_c, the transistor goes into saturation, that is, the output is 0.

Two components are added on, as shown in Figure 8.24*b* to improve the performance of the gate. A capacitor is connected across R_1. Because the voltage across a capacitor cannot change instantaneously, the leading edge of a positive pulse will cause the base voltage to rise immediately, causing the transistor to conduct. A capacitor used for this function is called a "speed-up" capacitor. The second component is the resistor R_2, which is connected to a dc source $-V_{BB}$. The purpose of this circuit is to ensure that the base-emitter voltage is normally kept at a slightly negative value so the probability of the gate switching due to noise is reduced.

The design of this circuit is best illustrated by example:

Example 8.4.1 The NOT gate. A NOT gate drives a load that requires $I_L = 1$ mA. The dc supply voltage $V_{CC} = 10$, $V_{BB} = 5$ V, and the transistor is a silicon NPN bipolar with $\beta = 100$. Determine suitable values for R_1, R_2, and R_c.

Solution. To prevent the load current from interfering with the operation of the gate, it is necessary to make the collector current about 10 times the load current

$$I_c = 10 \times I_L = 10 \text{ mA}$$

The transistor goes into saturation when the voltage drop across R_c is equal to $V_{CC} = 10$ V

$$R_c = V_{CC}/I_c = 10 \text{ V}/10 \text{ mA} = 1 \text{ k}\Omega$$

Base current required for 10-mA collector current is

$$I_b = I_c/\beta = 10 \text{ mA}/100 = 100\ \mu\text{A}$$

$$R_1 = (V_{CC} - V_{BE})/I_b = (10 - 0.7)/100\ \mu\text{A} = 93 \text{ k}\Omega$$

In practice, the values of R_1 may be reduced to half the calculated value to increase the margin of safety.

To improve the noise immunity, let the base voltage be reverse biased by -3 V when the input is 0 V. This means that R_2 has to be chosen so that

$$R_2/R_1 = 2/3$$

$$R_2 = 62 \text{ k}\Omega$$

It is necessary to verify that when the input is 10 V (V_{CC}), R_2 and V_{BB} will not hold the base voltage below 0.7 V. This step may be carried out by considering the transistor to be disconnected. The voltage at the node of R_1 and R_2 can then be calculated. Any voltage greater than 0.7 V ensures that the transistor will indeed be biased on. In this case the base voltage would have been 1.0 V but the base-emitter diode will hold it at 0.7 V.

The value of the speed-up capacitor depends on the frequency of operation of the gate. A reasonable choice is 50 to 100 pF.

Design of the NOR Gate. Figure 8.25 shows a two-input NOR gate with its truth table. When both inputs are 0s, no current flows in either transistor and the output is 1. When either A or B is a 1, current flows in the corresponding transistor which goes into saturation and the output is 0. Finally when both inputs are 1s, both transistors conduct and the output is a 0.

The NOR-gate may be viewed as two NOT gates sharing a common collector resistor. The design of the NOR gate is the same as that of the NOT gate.

Design of the NAND Gate. A modification of the NOT gate by the addition of an extra resistor and two diodes gives a NAND gate, as shown in Figure 8.26 together with its truth table. When either A or B or both of them are 0s,

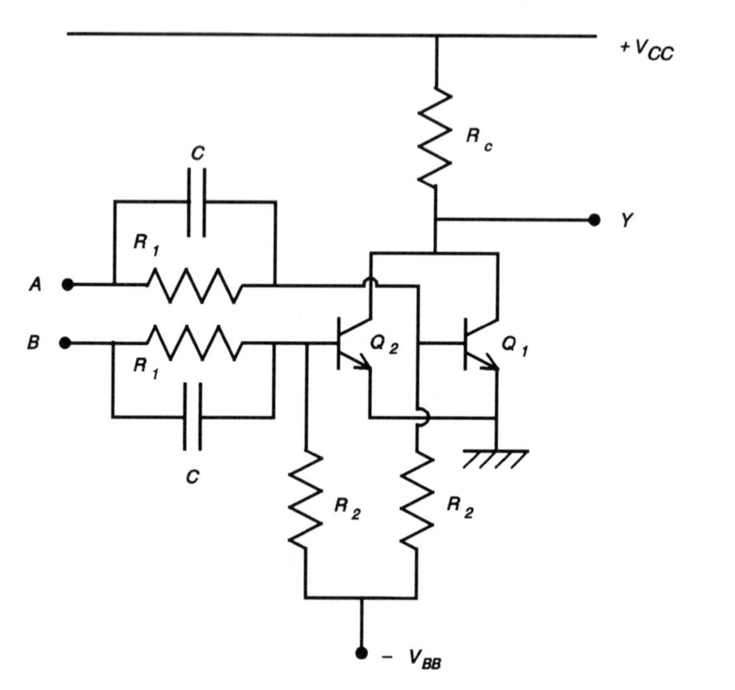

Truth Table

Inputs		Outpu
A	B	Y
0	0	1
1	0	0
0	1	0
1	1	0

Figure 8.25. The NOR gate with its truth table.

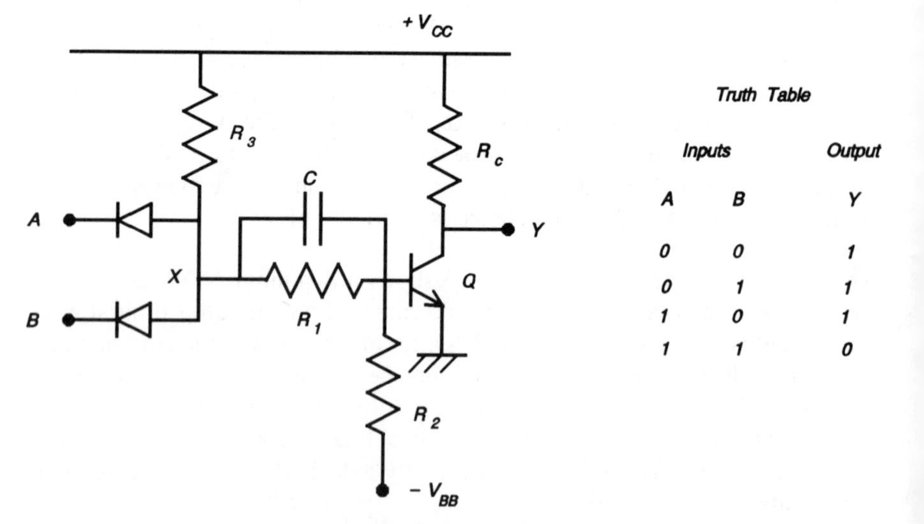

Truth Table

Inputs		Output
A	B	Y
0	0	1
0	1	1
1	0	1
1	1	0

Figure 8.26. The NAND gate and its truth table.

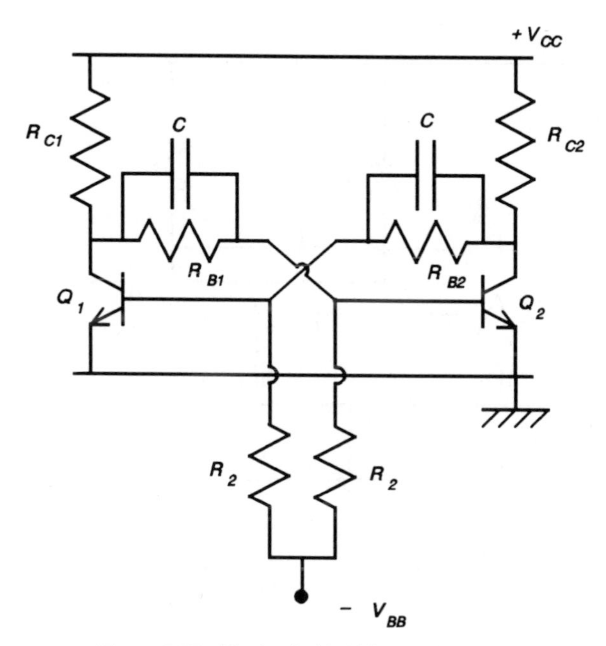

Figure 8.27. The basic bistable multivibrator.

current flows from V_{CC} through R_3 and the appropriate diode(s) to ground. The voltage at node X is held at 0.7 V and Q remains in cut off. Only when both A and B are 1s can base current flow through R_3 and R_1 and cause Q to go into saturation producing a 0 at the output.

The design of the NAND-gate follows from that of the NOT gate. The procedure for calculating the values of R_c, R_2, and C remain the same. R_1 from the NOT gate example has to be split in two to form R_3 and R_1 as shown in Figure 8.26. A reasonable compromise might be to make R_3 = 33 kΩ and R_1 = 60 kΩ. The circuit that drives the NAND gate will then not have to sink large currents through the diodes.

Design of the Bistable Multivibrator. The easiest way to understand the operation of the bistable multivibrator is to consider it to be two NOT gates connected back to back, as shown in Figure 8.27. When the dc power is first switched on, suppose Q_1 draws more current than Q_2. The collector voltage of Q_1 will therefore start to drop faster than that of Q_2. The voltage drop at the collector of Q_1 will be passed on by the capacitor C to the base of Q_2. The base-emitter junction of Q_2 will be a little less forward biased, and its collector current will be reduced at a rate determined by the current gain of Q_2. Its collector voltage will therefore tend to rise, pulling up the base of Q_1 and thereby causing the base-emitter junction to become more forward biased. More current will then flow through the collector of Q_1 because of

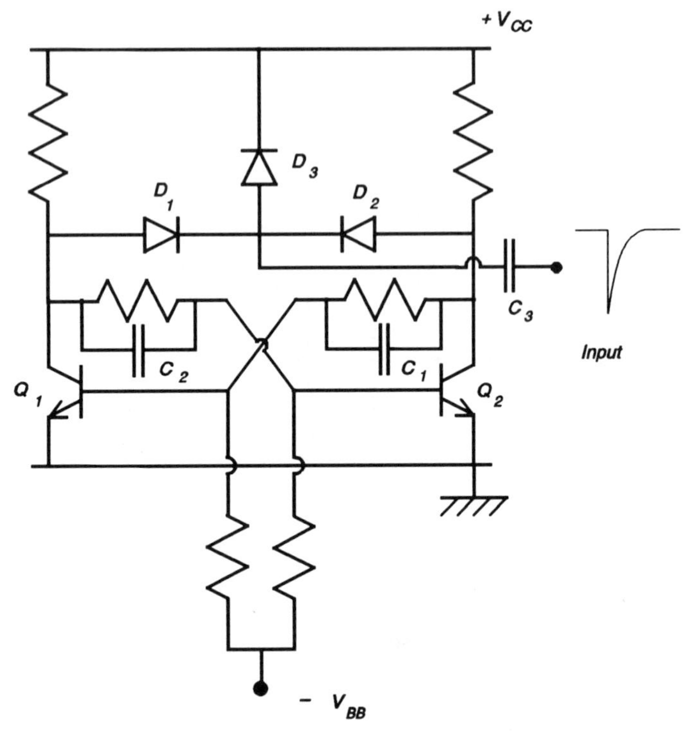

Figure 8.28. The bistable multivibrator with steering diodes, called a flip-flop.

the current multiplication in the transistor, causing the collector voltage to drop faster than before. This is the end of one cycle and the event that started the process has become magnified. This is a regenerative process and comes to stop when Q_1 goes into saturation and Q_2 is cut off. Unlike the astable multivibrator described in Section 7.2.7, there is no mechanism to cause the two transistors to switch back and forth. Q_1 stays in saturation and Q_2 is cut off until some external event causes them to change states. This event could be a small positive-going pulse or trigger applied to the base of Q_2. If the trigger is sufficient to bring Q_2 into partial conduction momentarily, this will cause current to flow in the collector of Q_2. Its collector voltage will drop, causing the base-emitter junction of Q_1 to be less positively biased. Less collector current will flow and the collector voltage of Q_1 will start to rise causing the base-emitter junction of Q_2 to become even more forward biased. Eventually, Q_2 goes into saturation and Q_1 is cut off. This is the end of the regenerative cycle in the opposite direction and the circuit will remain in the current state until another external event causes it to change states.

The bistable multivibrator can be made much more interesting and useful by the addition of the "steering circuit" shown in Figure 8.28. The diodes D_1, D_2, and D_3 are connected and their common node is coupled to a source of negative-going clock pulses by the capacitor C_3. Assuming that Q_1 is con-

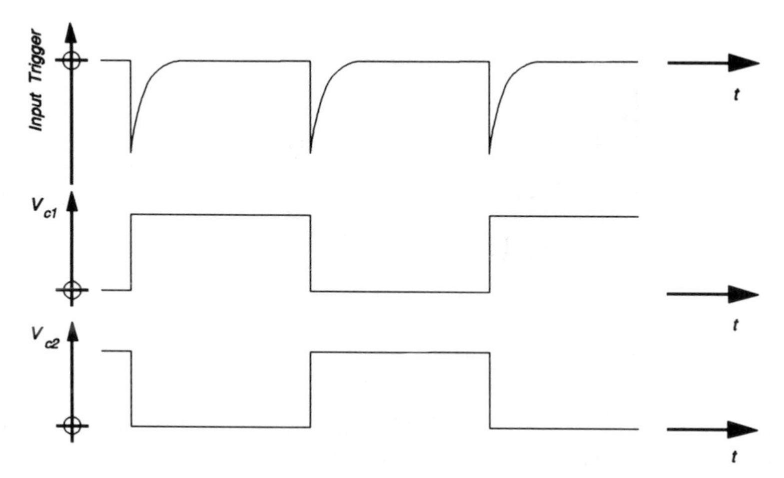

Figure 8.29. A timing diagram for the flip-flop showing the input trigger and the collector voltages.

ducting, its collector voltage will be essentially at ground potential, whereas that of Q_2 will be at V_{CC}. When the negative-going pulse arrives, D_2 will conduct momentarily causing the voltage at the collector of Q_2 to drop. The regenerative cycle is set in motion and the cycle ends with Q_1 off and Q_2 on. The next negative-going pulse will cause the transistors to change states yet again. This is illustrated in Figure 8.29. It should be noted that the diode D_3 guarantees that static charge built-up on the capacitor C_3 cannot exceed $V_{cc} + 0.7$ V.

Two very important conclusions can be drawn from Figure 8.29:

(1) The frequency of the output signal taken from either transistor is one-half of the input frequency—this is a *divide-by-two* circuit.

(2) The circuit has a *memory*—if the current state is known, one can deduce the previous state.

In terms of the current discussion, the first conclusion is the more relevant. By using two or more of these bistable multivibrators in cascade, it is possible to get a *divide-by-four* (2^2) circuit, three of them will perform a *divide-by-eight* (2^3) circuit and so on. In Figure 8.23, a cascade of four bistables are used to realize the divide-by-sixteen circuit. The bistable multivibrator with its steering circuit is popularly known as a *flip-flop*.

A set and a reset feature can be added to the circuit, whereby the state of the flip-flop can be set before the arrival of a chain of pulses. For example a 1 can be applied momentarily to the base of Q_1 to ensure that it is on. It is then possible to predict the state it will be in after any given number of input pulses. A flip-flop with the set and reset features is called an S–R flip-flop.

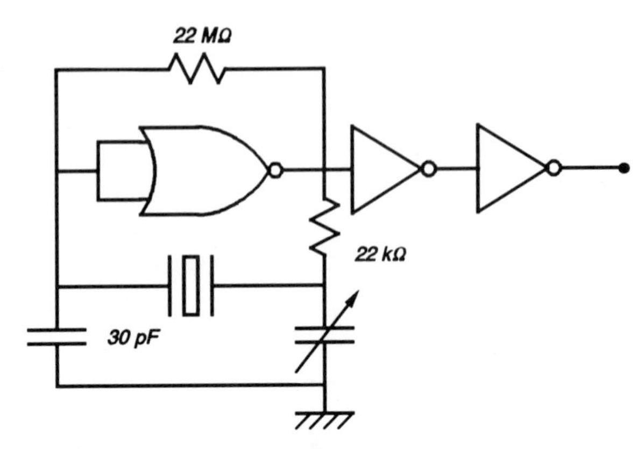

Figure 8.30. The Complementary Metal-Oxide Semiconductor (CMOS) version of the crystal-controlled oscillator. The output is a squarewave.

Yet another useful feature can be added so that when the clock pulse arrives, the flip-flop switches states only when a 1 or a 0 is present at a designated terminal. The flip-flop is then called a J–K flip-flop. Flip-flops are discussed in almost any textbook on digital circuits and the interested reader is advised to refer to such texts for more details.

The Crystal-Controlled Oscillator. The crystal-controlled oscillator is shown in Figure 8.30. With the two inputs to the NOR gate tied together, it becomes a NOT gate which is in effect a high-gain amplifier with phase inversion. The crystal together with the resistors and capacitors provide the feedback path. The output waveform is not quite a squarewave. The two inverters in cascade amplify and clip the signal to give a good square waveform.

The Divide-by-Sixteen Circuit. As mentioned earlier, this is a cascade of four flip-flops connected in a binary "ripple counter" arrangement. It divides the crystal oscillator frequency by 16 to give an output of 223.75 kHz.

The N*-Coder Circuit*. The N-coder uses the information from the dial pad to set up a pattern of 1s and 0s and sends these to the divide-by-N circuit. The particular pattern decides what the value of N has to be. The N-coder uses only NOR and NOT gates.

*The "Divide-by-*N *Circuit*. A combination of the set and reset features and the exploitation of other logic circuits such as NOR, NAND, and NOT gates provides the possibility to reset a cascade of flip-flops after n pulses of input, where n takes on values other than 2^m where m is an integer. This is the

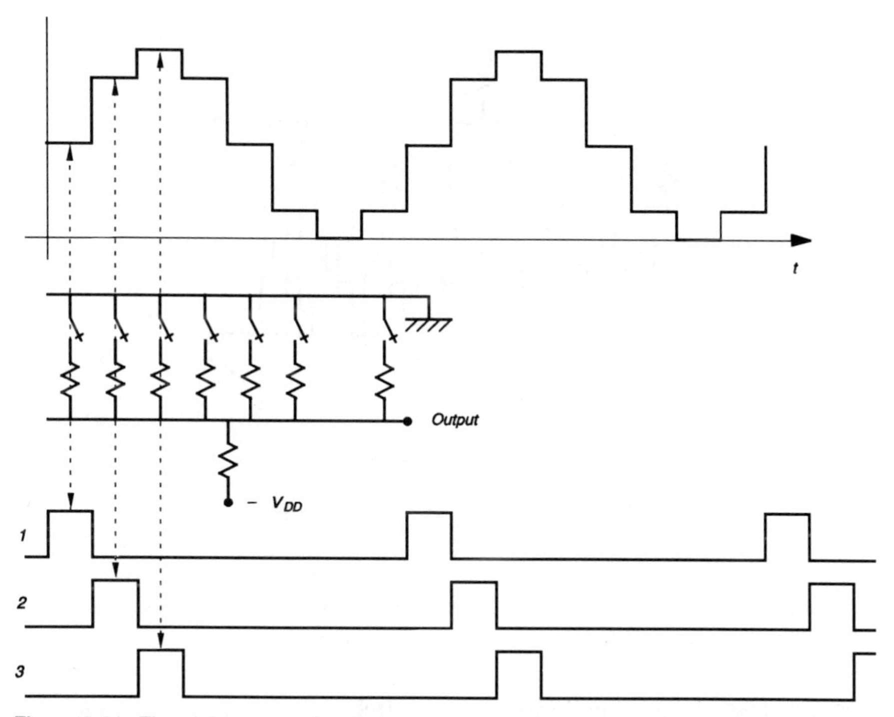

Figure 8.31. The eight-stage shift register closes each switch in turn. The resistor values are chosen to produce eight current levels which approximate a sinewave.

scheme used in the divide-by-N circuit with the N-coder supplying the necessary 1s and 0s to the appropriate flip-flops to set and/or reset them.

Using the lowest frequency (697 Hz) as an example, the input frequency to the divide-by-N circuit is 223.75 kHz. The selected value of N is 40. This gives an output frequency of 5.594 kHz, which is used to clock the eight-stage shift register.

The Eight-Stage Shift Register. The eight-stage shift register is made up of a cascade of flip-flops, NOR gates, and NOT gates to set up a sequence of pulses that have a width equal to $\frac{1}{8}$ the period of the required frequency. The fraction $\frac{1}{8}$ is used because the required sinewave is to be approximated by an eight-step function as shown in Figure 8.31.

The Eight-Level Digital-to-Analog Converter. The sequential pulses from the shift register are used to drive a bank of eight field-effect transistors switches connected in parallel as shown in Figure 8.32. The output is an eight-level approximation to a sinusoidal current. Going back to the example, the

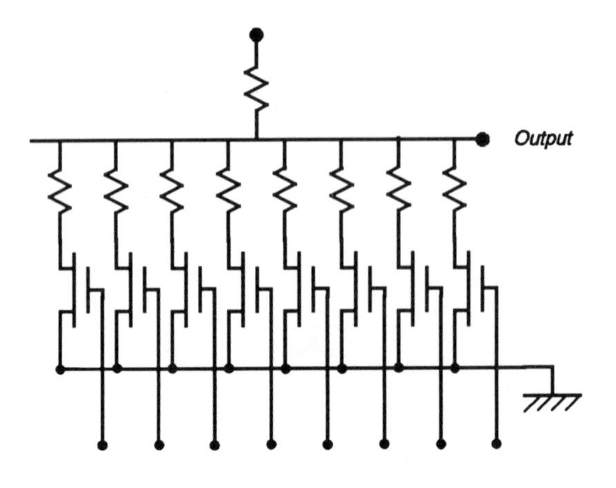

Figure 8.32. The switches in Figure 8.31 are replaced by FETs.

fundamental frequency of the approximation to the sinusoid is $5.594/8 =$ 699.3 Hz. The error is 0.3%

The Low-Pass Filter. The low-pass filter attenuates the higher harmonics of the waveform before the signal reaches the line driver. With the exception of very short lines, this filter is not required; the telephone line is a good enough low-pass filter. The harmonic content can be reduced by using more than eight steps to represent the sinewave.

Table 8.1 shows the target frequency f_0, the required divisor n, the nearest integral divisor N, the actual frequency f_0', and the error for all 12 tone-dial signals.

TABLE 8.1 Target Frequencies for the Digital Tone Dial Design Shown in Figure 8.23

Target Frequency, f_0 (Hz)	$16 \times 8f_0$	Required Divisor, n	Integer, N	Frequency, f_0' (Hz)	Error (%)[a]
1633	209,024	17.12	17	1645.2	+0.7
1477	189,056	18.94	19	1472.0	−0.3
1336	171,008	20.93	21	1331.8	−0.3
1209	154,752	23.13	23	1216.0	+0.6
941	120,448	29.72	30	932.3	−0.9
952	109,056	32.83	33	847.5	−0.5
770	98,560	36.32	36	776.9	+0.9
697	89,216	40.13	40	699.2	+0.3

[a]The errors indicated here are within the ±1.5% allowed.

8.5 DIGITAL TELEPHONE

Until the introduction of the digital telephone, all signals from the station set were transmitted to the central office in analog form over the subscriber loop, which in most cases is the twisted pair of plastic-insulated copper wire. In the central office it may be subjected to a number of sophisticated signal processing techniques, depending on the routing of the message and the medium of transmission. In the digital telephone, the analog signal is converted into digital form by a *codec* (*co*der-*dec*oder) using *pulse-code modulation* (PCM). The pulses are sent along the twisted pair to the central office where they may be further processed before transmission to their destination.

8.5.1 The Codec

The codec is available on an integrated circuit chip, and it is installed in the station set. Figure 8.33 shows a simplified configuration of the system. In this configuration, the analog part of the telephone is left unchanged. The physical distance from the telephone hybrid to the 2-to-4 wire hybrid can vary from essentially zero when the codec is installed in the station set to a few kilometers when the codec is in the central office or other points in between them. In the transmit mode, the analog signal is fed to the input of the analog-to-digital (A/D) converter, which gives a set of binary outputs in PCM in a parallel format. A parallel-to-serial (P/S) converter changes the parallel format to a serial format so it can be sent down the transmit leg of the digital line. In the receive mode, the input to the decoder is in PCM serial mode and the serial-to-parallel converter changes the signal into the parallel mode before going onto the digital-to-analog (D/A) converter. The analog output then goes to the receiver. The design of the circuits in the boxes shown in Figure 8.34 now follow. The order of presentation has been changed to take advantage of the fact that an A/D converter makes use of a D/A converter.

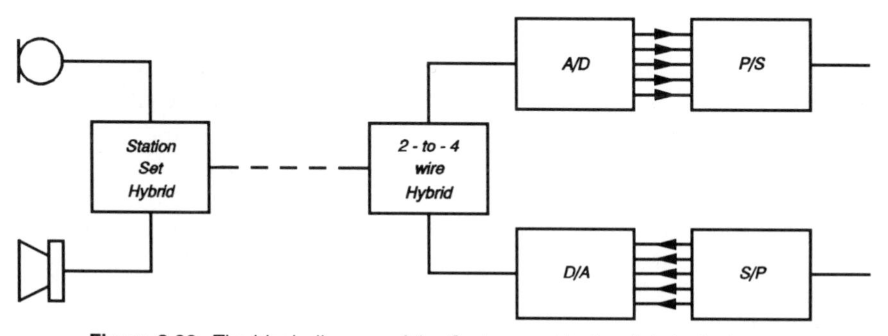

Figure 8.33. The block diagram of the Codec used in the digital telephone.

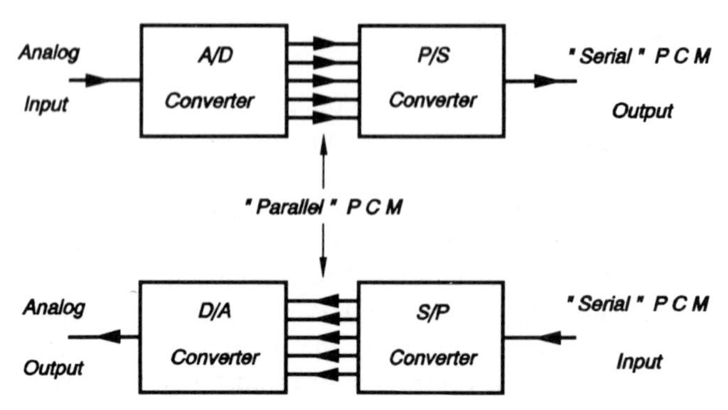

Figure 8.34. Details of the parallel-to-series PCM converter for both transmit and receive modes.

8.5.1a Digital-to-Analog Converter. As the name suggests, this circuit takes a digital input and converts it to the equivalent analog output. The digital input is presented in the form of 1s and 0s from the least significant bit (LSB) to the most significant bit (MSB). For example the number 26 is presented as:

$$\begin{array}{cccccccc} & 2^0 & 2^1 & 2^2 & 2^3 & 2^4 & \\ \hline \text{LSB} & 0 & 1 & 0 & 1 & 1 & \text{MSB} \\ & 0 + & 2 + & 0 + & 8 + & 16 & = 26 \end{array}$$

Consider the operational amplifier connected as shown in Figure 8.35.

This is a summing amplifier (Section 4.4.2e) whose output is given by

$$v_0 = -\frac{V^+ R_2}{R_1}\left[a_0 2^0 + a_1 2^{-1} + a_2 2^{-2} + a_3 2^{-3} + \cdots\right] \quad (8.5.1)$$

The coefficients a_n are digital bits, which are either 0 or 1 when the switch S_n is down or up, respectively. The voltage V^+ is the reference voltage and R_1 and R_2 are chosen to give a suitable level of signal at the output.

One obvious disadvantage of the circuit is that the larger the number of bits presented, the larger is the ratio of the largest resistor to the smallest. In integrated circuits, large resistors occupy large areas of the chip and are therefore undesirable. However, the function of the R_1 resistors is to scale the current flowing into the summing point of the amplifier according to the significance of the bit. Current scaling can be achieved by changing the circuit to that shown in Figure 8.36. The switches are either connected to the inverting input of the amplifier or they are connected to ground; but the

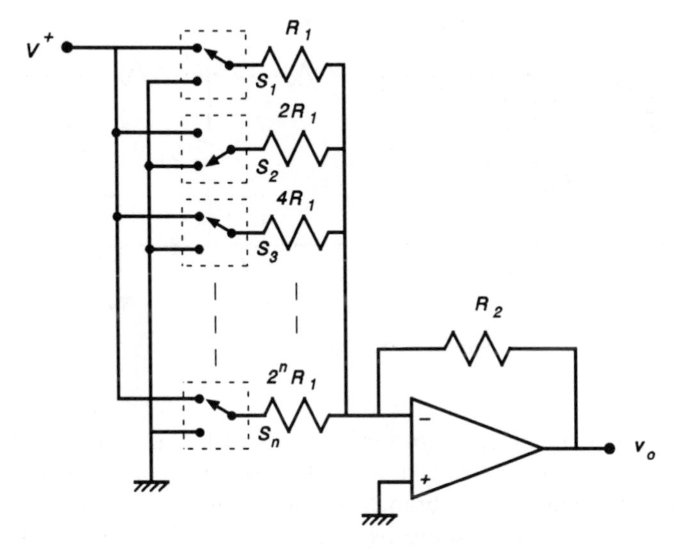

Figure 8.35. The circuit diagram of the digital-to-analog converter.

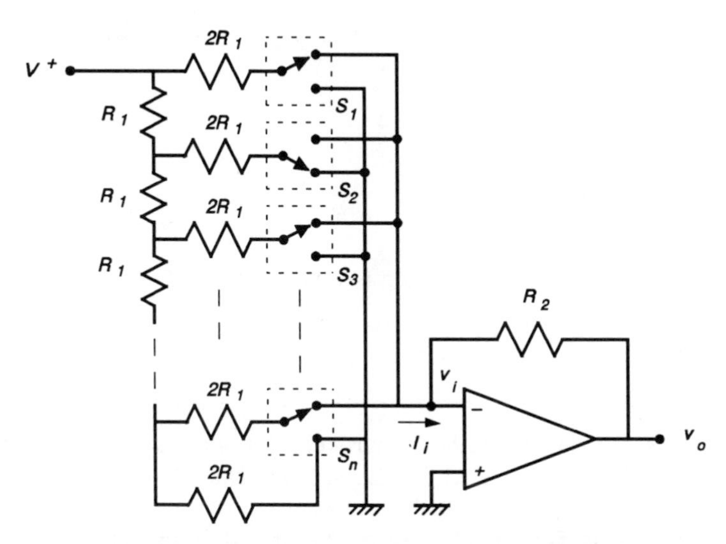

Figure 8.36. An improved version of the digital-to-analog converter which uses the $R-2R$ ladder.

inverting input is a virtual ground, so the resistive network can be redrawn as a ladder, as shown in Figure 8.37.

Looking to the right at point X, the resistance seen is R_1. Again looking to the right at point Y, the resistance seen is still R_1. It follows that looking at all points corresponding to X gives a resistance R_1 and hence the input resistance of the ladder is R_1. This type of ladder is described as an R–$2R$

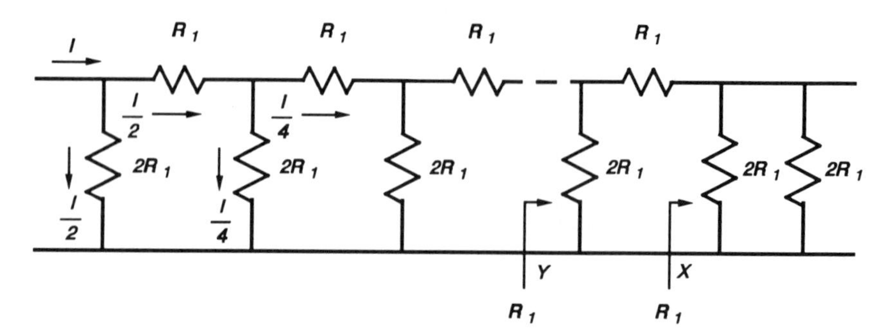

Figure 8.37. The $R-2R$ ladder.

ladder. Consider the current flowing into the ladder, I. It follows that I will divide equally at node 1 as shown. At node 2, the process is repeated with the current $I/4$ flowing in each branch, so that at node n the current in each branch is $I/2^n$. All the currents are summed at the input of the operational amplifier to give

$$I_i = \frac{V^+ - v_i}{R_1}\left[a_1 2^{-1} + a_2 2^{-2} + a_3 2^{-3} + \cdots\right] \tag{8.5.2}$$

But

$$I_i = \frac{v_i - v_0}{R_2} \tag{8.5.3}$$

When the operational amplifier gain is very large,

$$v_0 = -\frac{V^+ R_2}{R_1}\left[a_1 2^{-1} + a_2 2^{-2} + a_3 2^{-3} + \cdots\right] \tag{8.5.4}$$

where the as determine whether a particular bit is a 1 or a 0. Note that the error in the conversion is $\pm \frac{1}{2}$LSB (Least Significant Bit).

8.5.1b Analog-to-Digital Converter. Several different strategies can be used in the design of A/D converters each with its own advantages and disadvantages. For this discussion, the *ramp counter A/D converter* has been chosen, because its operation is relatively easy to understand. The ramp counter A/D converter is shown in Figure 8.38.

Two inputs are applied to the comparator: the analog signal and a voltage from the resistive network. The comparator output is a 0 if the analog signal is greater than the voltage from the resistive network and a 1 if the opposite is true. The comparator output and a clock drive a two-input NAND gate so that the NAND gate permits the clock pulses to reach the resetable binary

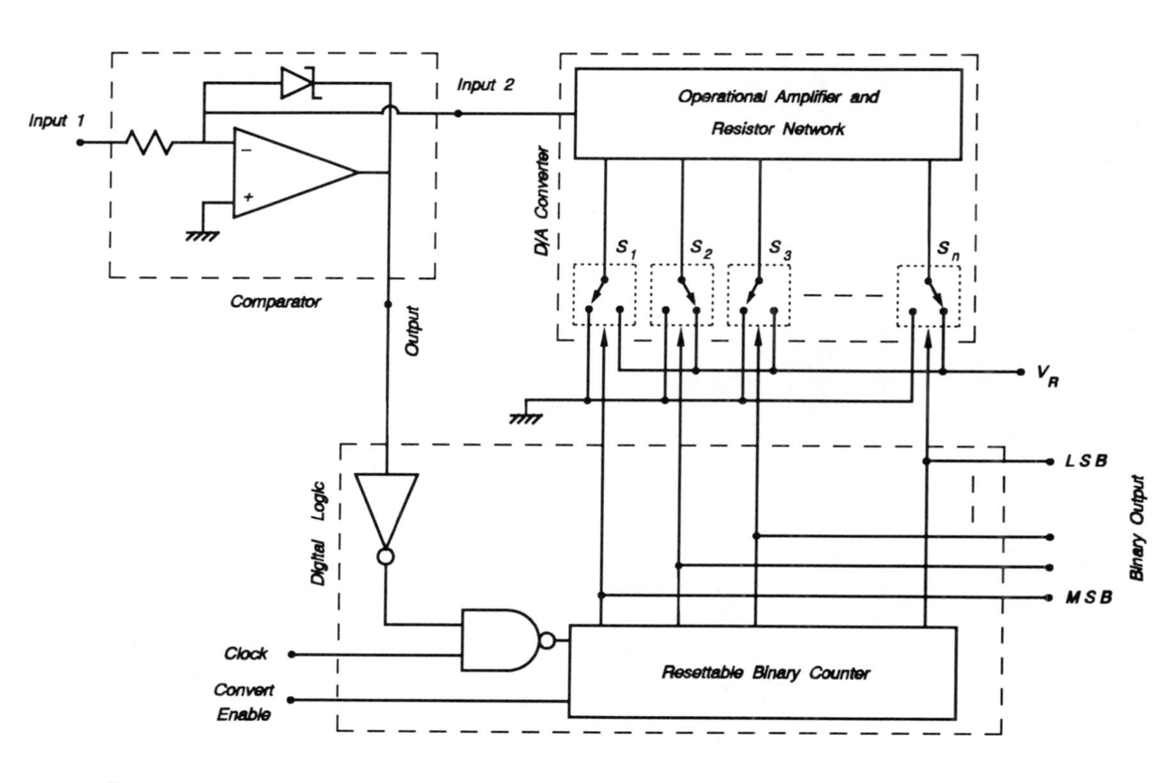

Figure 8.38. The block diagram of the analog-to-digital converter. Note that it contains within it the digital-to-analog converter.

counter only when the comparator output is 0; that is the NAND-gate input is a 1, because of the inverter. The binary counter counts the clock pulses and simultaneously closes or opens the switches S_1 to S_n depending on the bit (1 or 0) that it applies to the switches. Which switches are closed or open determine the output voltage of the resistive network. This is in fact a D/A converter! So long as the output voltage of the resistive network is lower than the input analog voltage, the binary counter continues to count and continues to increase the output voltage of the resistive network. Finally, the voltage from the resistive network exceeds the analog input voltage and the comparator output changes to 1. The NAND gate stops the clock pulses from reaching the binary counter and it therefore stops counting. The output from the binary counter can now be read. The convert enable clears the registers of the binary counter ready for the next sample of the analog input voltage to be converted.

The design of the NOT gate, the NAND-gate, the binary counter, and the switches were discussed in Section 8.4.5. The design of the resistive network plus operational amplifier (D/A converter) was the subject of Section 8.5.1a. The comparator, parallel-to-serial converter, serial-to-parallel converter, and the 4-to-2 wire hybrid will now be discussed.

8.5.1c Comparator. The circuit of the comparator is shown in Figure 8.39*a*. The ideal operational amplifier is assumed to have infinite gain so that if the input varies infinitesimally about 0 V, a positive-to-negative change will

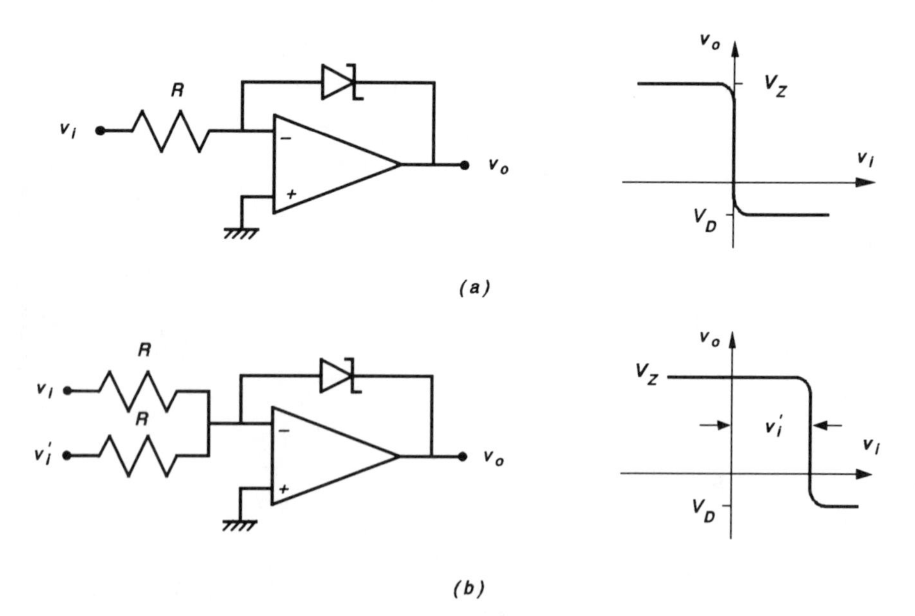

Figure 8.39. (*a*) The zero-voltage comparator and its output characteristics. (*b*) The comparator for v_i and v_i' and its output characteristics.

cause the output to go from saturation at the negative dc supply voltage to the positive. The operational amplifier circuit is therefore a comparator that compares its input to 0 V. To stop the output from saturating at the dc supply rails, a zener diode may be connected as shown in Figure 8.39*a*. For a very small positive voltage input, the output will be negative but it will be clamped when it drops to -0.7 V by the forward-biased zener diode. Equally, a very small negative voltage input will cause the output to become positive, but again it will be clamped, this time by the zener voltage V_z of the diode. Evidently the clamping voltages were chosen so that a "0" ≈ -0.7 V and $V_z \approx V^+ -$ the system voltage. The input–output characteristics are given in Figure 8.39*a*.

Now consider the circuit in Fig. 8.39*b*, where there are two input voltages v_i and v_i' connected by equal resistances R to the inverting input of the operational amplifier. Evidently, the only way to keep the output from going positive or negative (clamped condition) is that the current supplied by v_i must be equal to the current absorbed by v_i'. The two voltages must be equal in magnitude. When they are not equal, which they are not most of the time, the output is clamped at one extreme or the other (1 or 0). The input–output characteristics show the effect of v_i' as an offset voltage. In the context of the A/D converter, the output of the comparator changes state when the output of the resistive network exceeds the analog input voltage by, at the most, the value of the LSB.

8.5.1d Parallel-to-Serial Converter. The output of the A/D converter, which is in pulse-code modulation (PCM parallel) has to be converted into serial form so that it can be sent along a single communication channel. A *universal asynchronous receiver/transmitter* (UART) performs this function. This an elaborate, large-scale-integration (LSI) integrated circuit, which can convert data from parallel to serial and vice versa. However, the basic operation of an eight-digit parallel-to-serial converter is illustrated in Figure 8.40. The parallel information is available at the nine (eight plus the start/stop) terminals. The moving switch sweep past and "reads" the bits in serial form. Note that one or more of the bits can be used for the "stop/start" code to tell the receiver when a segment of the code starts and stops. Other bits may be used as a parity code for correcting the signal code in the presence of noise. Examples of parallel-to-serial shift registers are the MC14014B (Motorola Semiconductor), the MM54HC165 (National Semiconductor Corp.) and the SN54166 (Texas Instruments) integrated circuit and are listed in most integrated circuit handbooks.

8.5.1e Serial-to-Parallel Converter. The operation of the serial-to-parallel converter is much simpler than its opposite number. Figure 8.41 shows an eight-stage serial-to-parallel converter. The stop/start pulse is used to clear the shift register and to trigger the input enable circuit after a specified delay. The serial input data can then start to enter the shift register moved

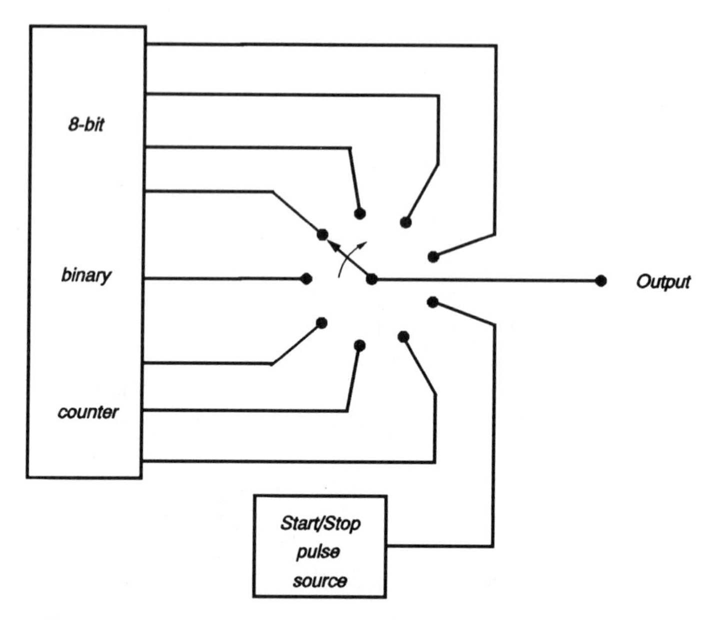

Figure 8.40. A mechanical analog of the parallel-to-serial converter.

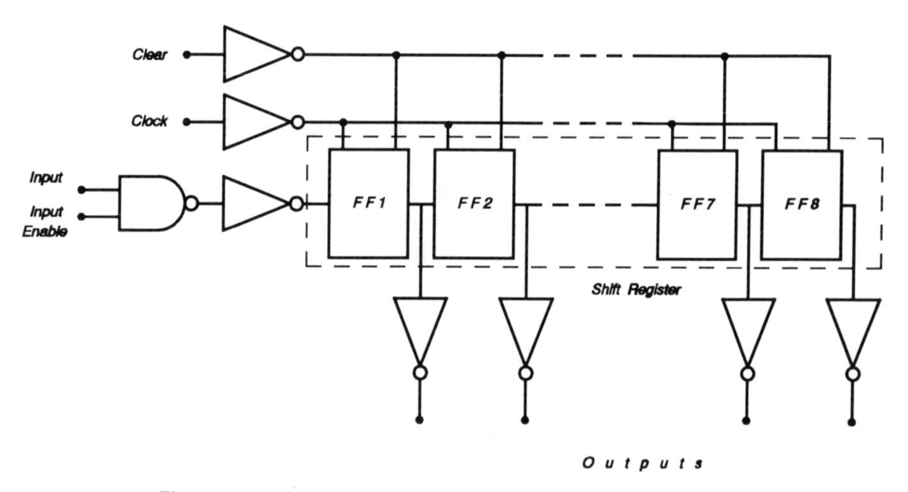

Figure 8.41. A block diagram of the serial-to-parallel converter.

on by the clock. The arrival of the next stop/start pulse signals the end of the datastream. The binary outputs are read and the shift register is cleared for the next stream of data to arrive. An example of an integrated circuit serial-to-parallel converter is the SN54164 (Texas Instruments).

8.5.1f 4-to-2 Wire Hybrid. The basic 4-to-2 wire hybrid is shown in Figure 8.42. There are two identical but separate transformers, T_1 and T_2, con-

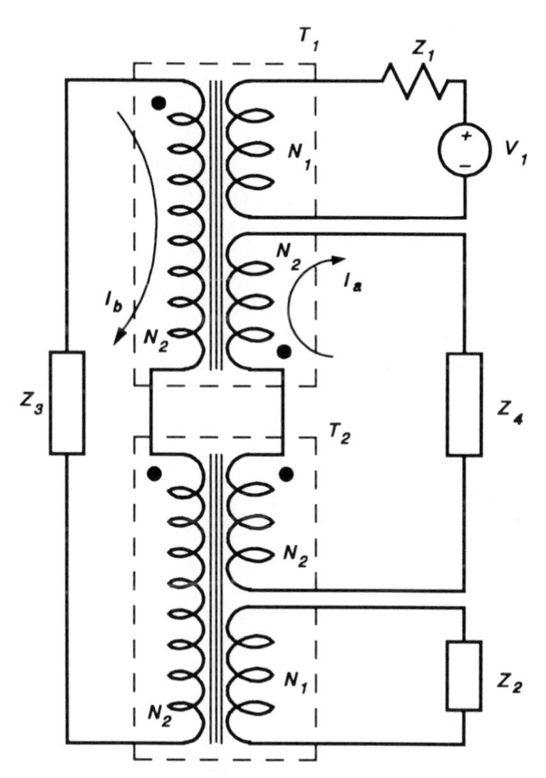

Figure 8.42. The 4-to-2 wire hybrid. (Reprinted with permission from *Transmission Systems for Communication*, 5th ed. AT & T, Bell Labs., 1982.)

nected to conform to the dot notation as indicated. When the number of turns in each coil is as shown, a voltage source V_1 in series with an impedance Z_1 connected to port 1 will cause a voltage equal to

$$\frac{N_2}{N_1}V_1 \tag{8.5.5}$$

to appear across the other two coils of T_1 each of which has N_2 turns. Assuming that $Z_1 = Z_2$ and $Z_3 = Z_4$. It follows that the current I_a will be equal to I_b, and they will flow in the directions indicated. In transformer T_2, the magnetomotive force (mmf) due to the two currents will cancel each other and the net voltage across port 2 will be zero. The signal input power from port 1 is divided equally between ports 3 and 4. If the voltage source had been connected across port 2 instead of port 1, the same effect would be observed, except there will be no voltage across port 1.

Consider the situation when a voltage source is connected across port 3. The voltages across the two halves of port 4 will cancel each other. However, voltages

$$\frac{N_1}{2N_2}V_3 \tag{8.5.6}$$

will appear across ports 1 and 2.

This arrangement can be used as a station-set hybrid, if the transmitter is connected across port 1, the receiver across port 2, and the line across port 3, so long as an impedance equal to that of the line is connected across port 4. Its normal application, however, is as a 4-to-2 wire hybrid; the four wires being connected to ports 1 and 2 and the 2 wires being connected to port 3.

8.6 CENTRAL OFFICE

8.6.1 Manual Office

The manual central office is largely history, although one sometimes hears stories of a few survivors in very remote places. A description of its operation was given in Chapter 1. It is evident that it was just a matter of time before automation would eliminate the telephone operator's job. The surprise was that it was not the human errors due to voice communication problems, delay in getting service, or rising labor cost that was the motivation for automation; it was the lack of security of the message.

8.6.2 Basics of Step-by-Step Switching

As described in Section 8.3.6b, the information from the telephone subscriber to the central office concerning the number to which he wants to be connected comes in the form of a series of current pulses. These current pulses are at a frequency of 10 Hz, with variable and certainly much longer interdigit pauses. The engineering problem was to use these current pulses to move a mechanical switch into the required position to effect the connection.

It is most convenient to start a description of the process with a small and fictitious central office with 10 subscribers each with the pulse dial described in Section 8.3.6b. Under these conditions, switching would take place when a single string of dial pulses arrived from the calling subscriber. Figure 8.43 shows an electromechanical contraption that would perform the switching function. The serrated wheel is attached to the switch wiper arm, and it is spring loaded to keep it in its rest position as shown. When a pulse is applied to the electromagnet A, the pawl pulls the serrated wheel the appropriate distance to cause the switch wiper arm to move from one contact to the next. The wheel is prevented from returning to its rest position after every pulse by the detent and the electromagnet B.

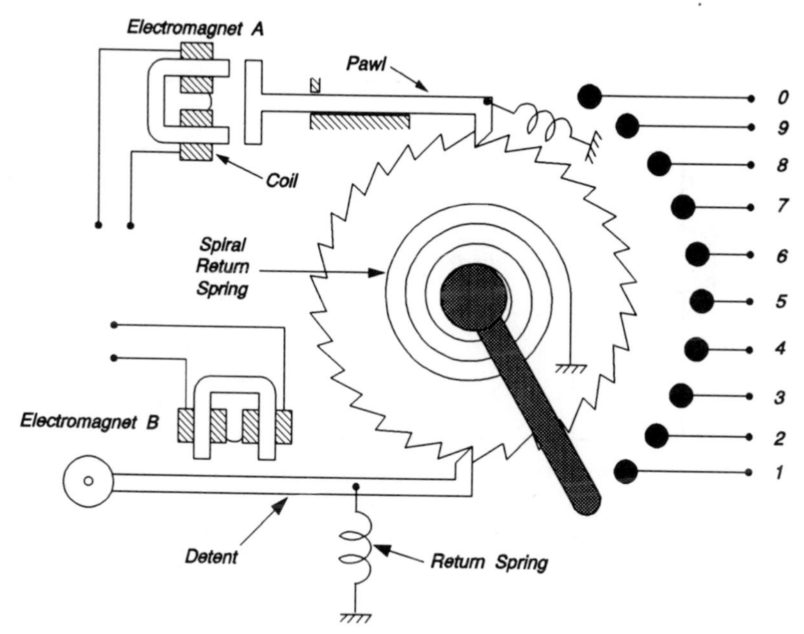

Figure 8.43. An electromechanical switch for a ten-subscriber central office.

Consider the circuit shown in Figure 8.44. To initiate a call, the subscriber lifts the handset, thereby closing the hook switch. Direct current flows in the coil of electromagnet B and operates the detent to stop the serrated wheel from rotating in a clockwise direction; it is however, free to rotate in a counterclockwise direction. The subscriber now operates the dial and short circuits the line seven times, say. The ac current generated flows through electromagnet A and causes it to operate seven times moving the switch wiper arm onto the seventh fixed contact. There are no known telephone system that operates this way! The conversation can now proceed. When the call is over, the replacement of the handset opens the dc current loop, the electromagnet B is deenergized, the detent moves away from the serrated wheel and under the force of its return spring, the switch wiper arm returns to its rest position.

Each of the ten subscribers would be connected to such a switch and all ten switches connected as shown in Figure 8.45. A switching system such as this could be extended to accommodate 100 subscribers by making the following changes:

(1) Dialing will now consist of two trains of pulse with a suitable pause between pulse trains.

(2) Each of the 10 selector switches will be connected to 10 other identical switches.

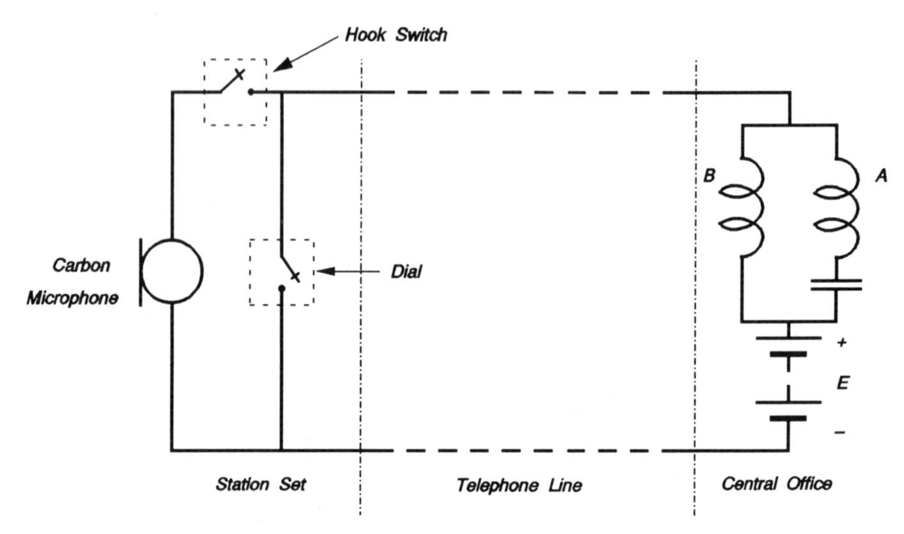

Figure 8.44. The switch shown in Figure 8.43 connected to the station set.

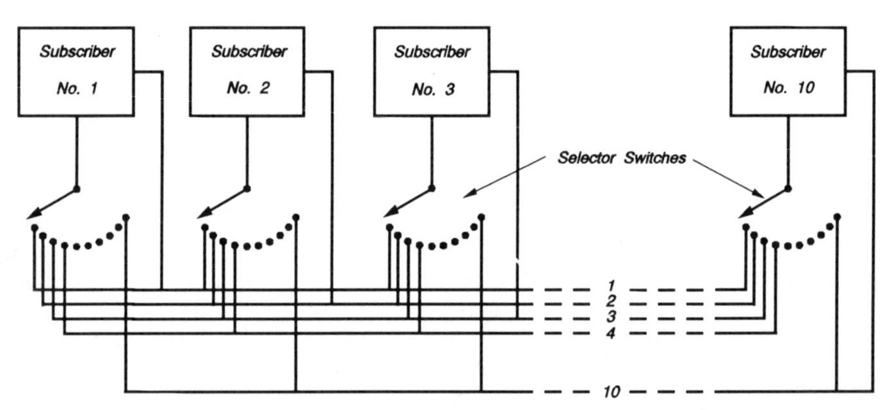

Figure 8.45. The complete ten-subscriber telephone system.

(3) Some electrical or mechanical arrangements will be made to transfer the second train of pulses to the bank of secondary switches.

Further extension to 1000 and 10,000 subscribers is possible by suitable changes in the central office and the introduction of three and four-digit dialing codes, respectively. Such a system is possible and would operate quite well except for the fact that it would be very large and prohibitively expensive to manufacture, install, and operate.

8.6.3 The Strowger Switch

If you can imagine a new version of the fictitious electromechanical contraption shown in Figure 8.43 with movement in two directions, then you have a good idea of how the Strowger switch works. The actual Strowger switch is shown in Figure 8.46.

There are two sets of electromechanical drives: one vertical and the other horizontal. There are ten vertical positions and ten horizontal ones. Such a switch can accommodate up to 100 subscribers and therefore requires two trains of pulse. The first moves the switch wiper upward—to the third level if there were three pulses in the train. During the interdigit pause, control is switched to the horizontal drive. The second train of pulses moves the switch wiper horizontally to the seventh position—if it had seven pulses, to connect to the number 37. At the end of the conversation, the switch must be returned to its initial position.

Replacing the 10-position switches shown in Figure 8.45 with Strowger switches would convert a 10-line office into a 100-line office. It would appear

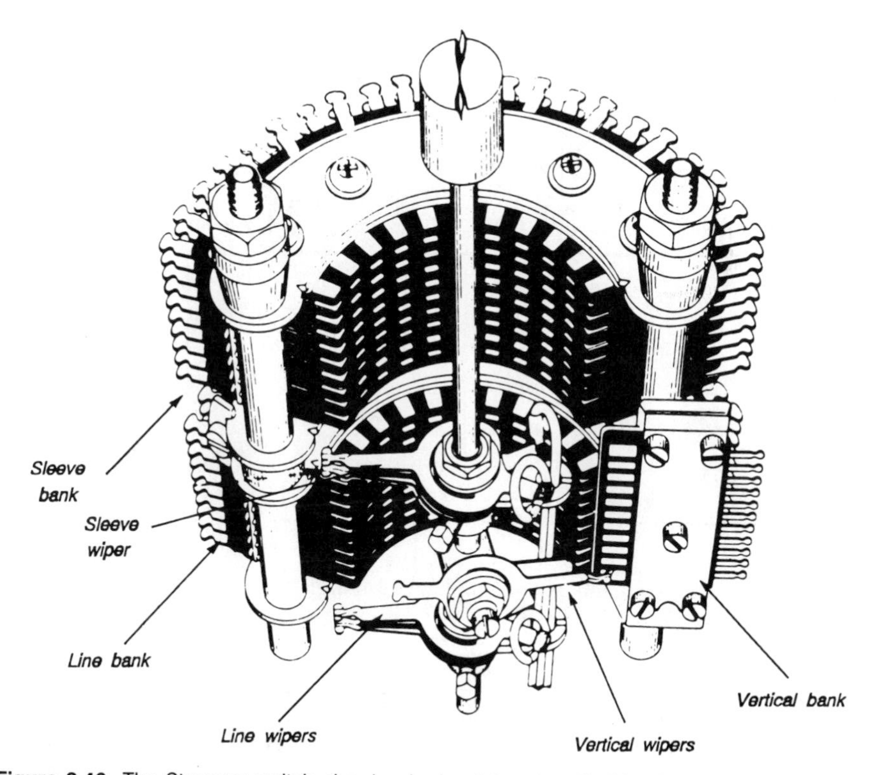

Figure 8.46. The Strowger switch showing horizontal and vertical banks as well as wipers. (Reprinted with permission from B. E. Briley, *Introduction to Telephone Switching*, AT & T, Bell Labs., 1983.)

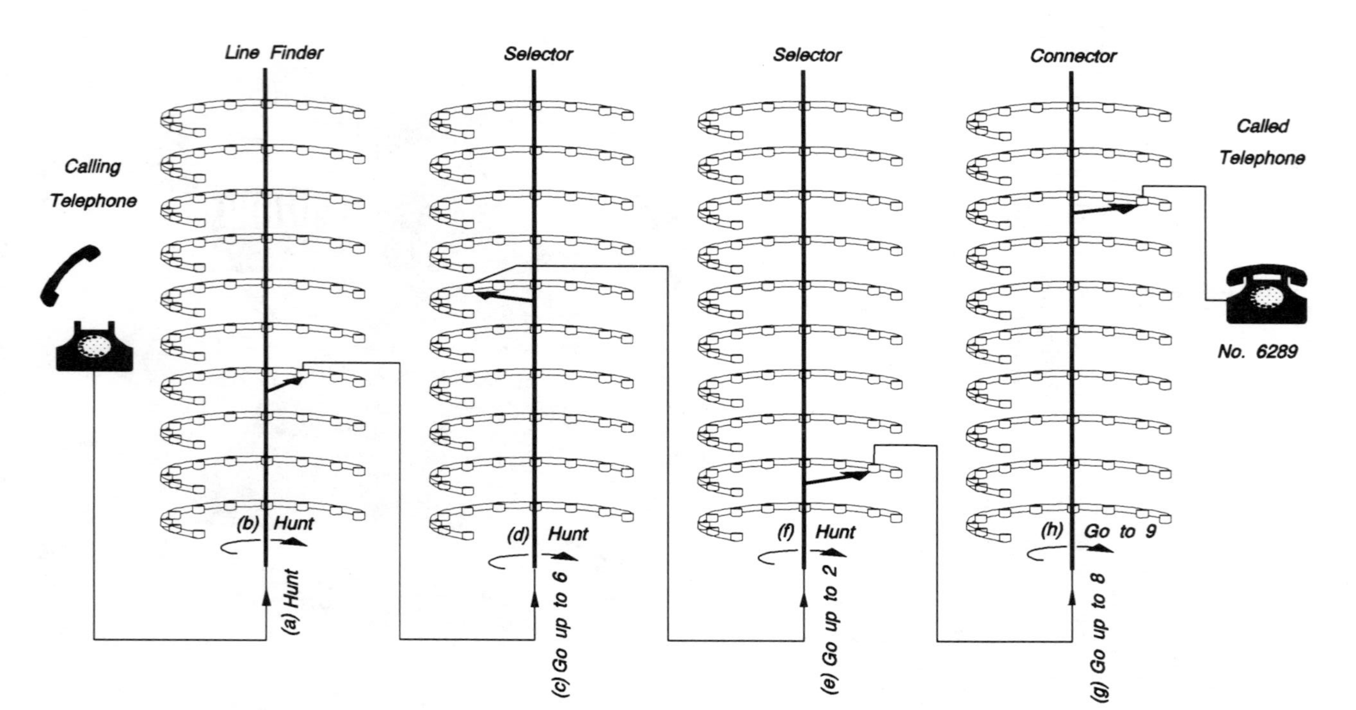

Figure 8.47. A simplified connection through the step-by-step central office. (Reprinted with permission from B. E. Briley, *Introduction to Telephone Switching*, AT & T, Bell Labs., 1983.)

at first that the two-dimensional switch would solve the economic problem, in fact it is more complicated than the 10-position switch and therefore more expensive. The solution to the economic problem was discovered along with the fact that it is most unlikely that in a 100-line office 50 calls will be in progress at the same time. It was therefore not necessary to design the system for maximum capacity operation. The study of traffic statistics revealed that the system could be designed to handle a maximum number of calls (usually about 10% of subscribers) and if it happens that there is one more than this maximum then it would be lost. This led to a redesign of the central office so that the subscribers would share a considerable amount of common equipment. Every new incoming call initiates a hunt for an idle path through the system, and, if no idle paths can be found, a busy signal is returned. The precise details of the inner workings of the central office are not necessary since the object of the exercise is to develop a general appreciation of how a call is connected. Figure 8.47 shows a simplified connection using line finders, selectors, and connectors.

8.6.4 Basics of Crossbar Switching

The basis of the crossbar switching system is a fallback to the manual central office and it is shown in Figure 8.48. In a hypothetical nine-subscriber central office, each subscriber is represented in the vertical and horizontal elements of the matrix. To connect any two of the subscribers, it is necessary to have

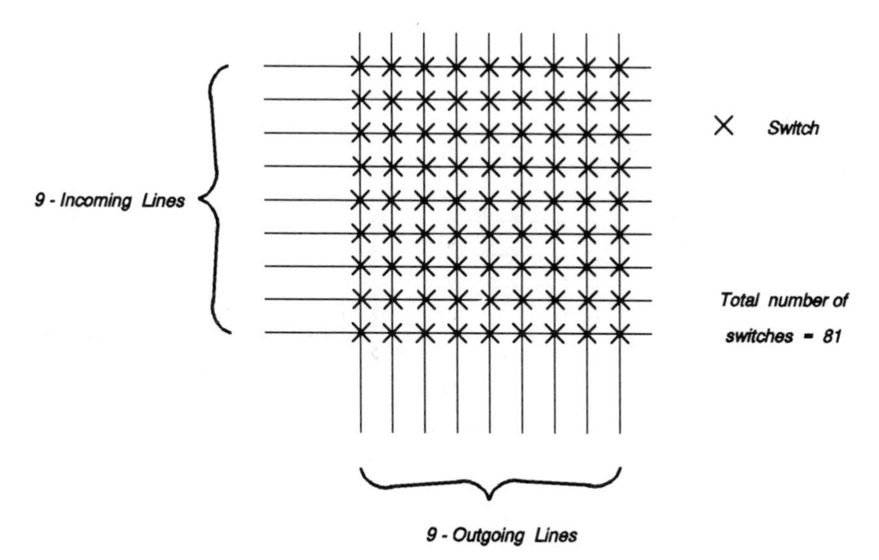

Figure 8.48. A representation of the cross-bar switch for a nine-subscriber central office showing all 81 possible connections. Reprinted with permission from S. F. Smith, *Telephony and Telegraphy A*, 2nd ed., Oxford University Press, Oxford, 1974.

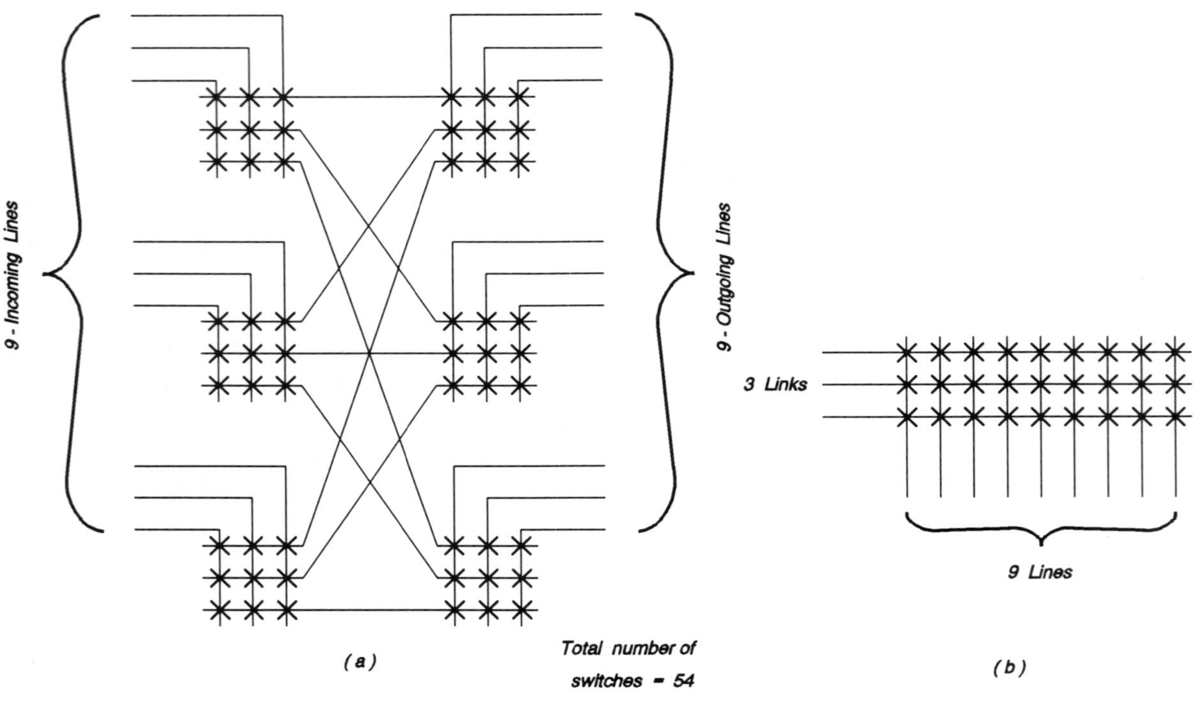

Figure 8.49. (*a*) A switching scheme for a nine-subscriber central office with the number of contacts reduced to 54. (*b*) The use of three links, which reduces the number of conversations that can be carried out simultaneously to three also reduces the number of contacts to 27. (Reprinted from Ref. 9 by permission of Oxford University Press.)

81 contact points. In the crossbar system, it actually takes the operation of two relays to close one contact point, one for the vertical element and the other for the horizontal element. As in the other central office organization schemes, it is possible to reduce the number of contact points:

(1) Reorganizing the interconnection as shown in Figure 8.49*a*. This shows that the number of contacts have been reduced to 54.
(2) Since it is most unlikely that every subscriber will be using her telephone at the same time, let the maximum number of conversations passing through this office be three. By using three links as shown in Figure 8.49*b*, it is still possible to connect every subscriber to every other subscriber so long as no more than three calls are in progress.

The use of such schemes to reduce the number of switches, is referred to as *concentration*.

8.6.5 Central Office Tone Receiver

So far the discussion of switching in the central office has been based on a 10-Hz pulse signal generated by the rotary dial or imitations of it. The use of tone dialing in the telephone system is becoming increasingly more important, because it is faster and it offers the possibility to use the telephone system for purposes other than talking. There are already a number of household appliances on the market which can be connected and controlled from a remote location by dialing special codes.

As discussed earlier, the tone dial generates two distinct frequencies when a single button is pressed. By identifying the two frequencies present the number of the button can be decoded. The system for decoding the signal is shown in Figure 8.50. The tone signal goes through a band split filter, which separates the low frequencies from the high frequencies. Each group is amplified, hard limited to produce squarewaves and filtered again by the channel filters. The output signal from the channel filters are rectified by the detectors, which indicate the presence (1) or absence (0) of the frequencies produced by the dial. For an output to be valid, it has to have two frequencies present, one from each frequency group. The system has several built-in features which prevent mistaken identification of other audio signals as valid dial signals.

8.6.6 Elements of Electronic Switching

The basis of electronic switching is the same as the crossbar technique but with the electromechanical switches replaced by electronic components. The advantages gained are speed of operation, increased reliability, and lower cost. In development, several schemes had to be tried, but none of them

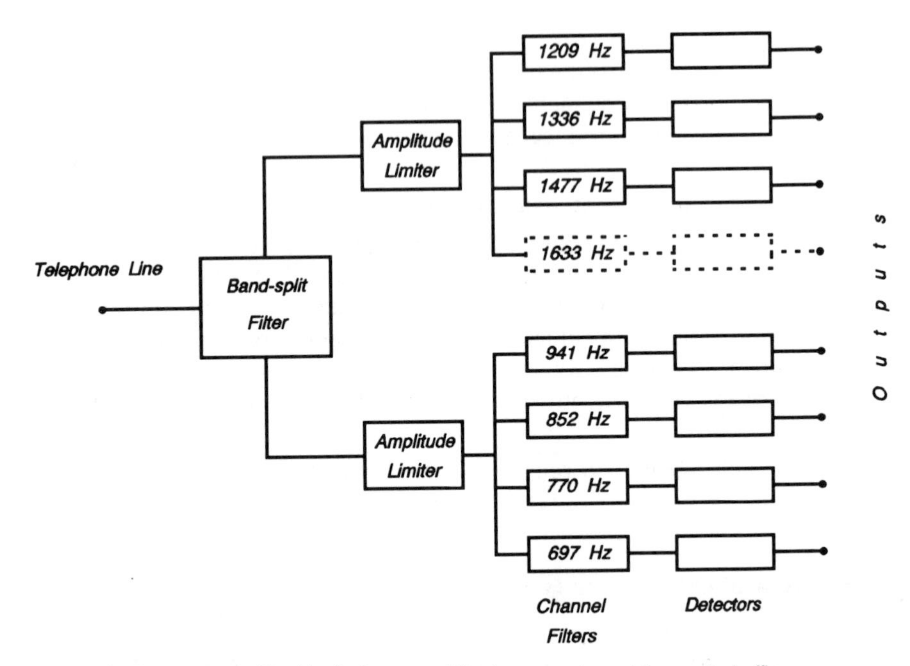

Figure 8.50. The block diagram of the tone receiver at the central office.

worked because the replacement of the metal-to-metal contact in the electromechanical switches with semiconductor switches proved to be unsuitable. The concept of electronic switching had to undergo a fundamental change before success could be achieved. The problem was that the crossbar technique was based on a *space-division multiplex* (SDM), an approach that relied on finding multiple paths for the different conversations passing through the central office by separating them in space, that is, by assigning each path to a space or position in the matrix. *Time-division mutliplex* (TDM), which uses the same path for all conversations but separates them in time, was more successful. Time-division multiplex is discussed in Chapter 9.

So far, in the discussion of what takes place in the central office from the initiation of a call to its end, it has been assumed that the system was under the direct control of the pulses or tones as they arrived. It is certainly advantageous to have a memory in the system so that the information dialed in by the subscriber can be stored, analyzed, and an optimal path determined for the call. In the language of the telephone engineer, the memory device is a *register*. The pulses from the dial are fed to a set of shift registers or decade counters, one set for each decade of the number dialed. With the tone dial, the digits are first decoded and then stored in binary form in a shift register. What we have then are banks of hard-wired switches and a memory in which instructions can be stored, retrieved, and acted upon in due course. To help drive this point home, consider the activities that take place in the central office in order to connect a call. A flow diagram of this is shown in Figure 8.51.

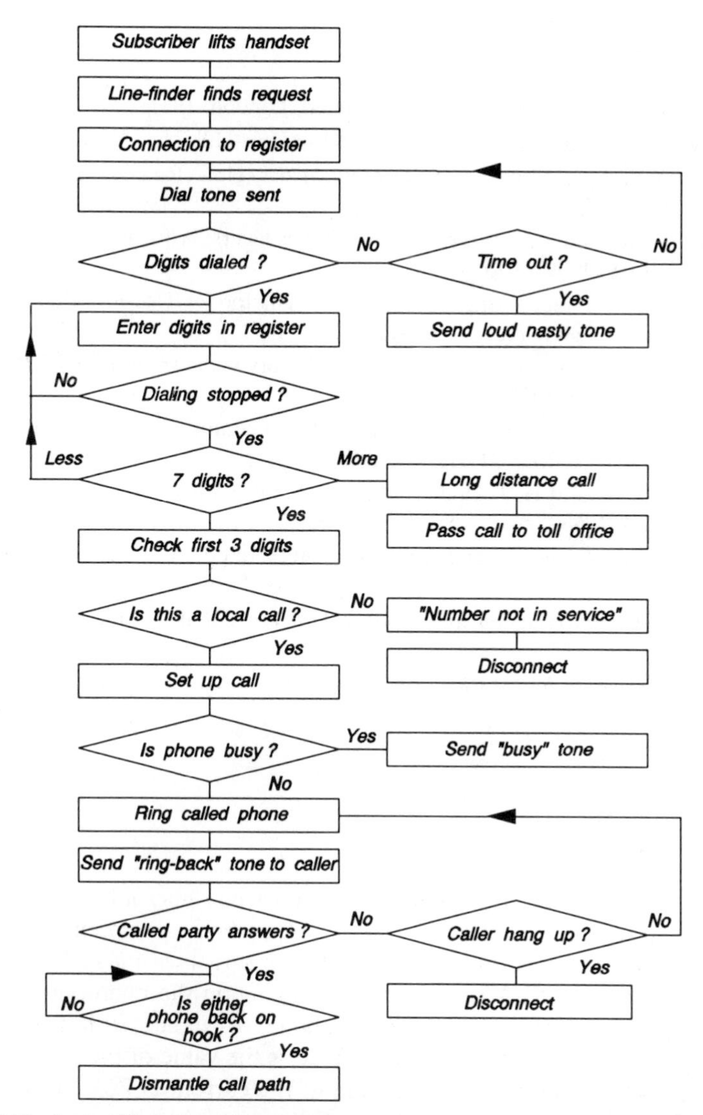

Figure 8.51. A simplified flow chart for setting up a telephone call. The similarity to a computer program flow chart is evident.

The flow diagram looks suspiciously like the flow diagram for a computer program. It is indeed a computer program and instructions can be carried out by a general-purpose computer with the appropriate peripheral equipment. The modern central office has a dedicated computer that controls every aspect of placing a call. Humans simply supervise the machine and intervene when something goes wrong or when they work out a better set of instructions for the machine.

REFERENCES

1. W. Fraser, *Telecommunications*, Macdonald, London, 1957.
2. M. A. Plonus, *Applied Electromagnetics*, McGraw-Hill, New York, 1978.
3. M. N. O. Sadiku, *Elements of Electromagnetics*, Holt, Rinehart & Winston, New York, 1989.
4. G. M. Sessler and J. E. West, "Electret Transducers: A Review," *J. Acoust. Soc. Am.*, 53, 1589–1599, 1973.
5. P. D. van der Puije, "Audio Frequency Circulator for Use in Telephone Sets," *IEEE Trans. Comm. Tech.*, 1267–1271, Dec. 1971.
6. B. D. H. Tellegen, "A General Network Theory with Applications," *Philips Res. Rept.*, 7, 259–269, 1952.
7. B. A. Shenoi, "Practical Realization of a Gyrator Circuit and RC-Gyrator Filters," *Trans. IEEE*, CT-12, 374, 1965.
8. D. F. Sheahan and H. J. Orchard, "High Quality Transistorized Gyrators," *Elec. Lett.*, 2, 274, 1966.
9. S. F. Smith, *Telephony and Telegraphy A*, 2nd ed., Oxford University Press, London, 1974.
10. B. E. Briley, *Introduction to Telephone Switching*, Bell Telephone Laboratories, 1983.
11. Staff, *Transmision Systems for Communication*, 5th ed., Bell Telephone Laboratories, 1983.

PROBLEMS

8.1 Describe the attractive features of the carbon microphone that make it well suited to the telephone system. A carbon microphone has a static resistance of 75 Ω and it is connected in series with a 12-V battery and a load resistor R. A sound wave impinging on the microphone causes a sinusoidal variation of the microphone resistance with a peak value equal to 15% of the static value. What is the value of the load resistor R if the second harmonic component of the current is 7.5% of the current at the fundamental frequency? Calculate the signal power delivered to R at the fundamental frequency and comment on it.

8.2 A successive approximation A/D converter has an output consisting of 8 bits. The input to the converter is $0.82564V_R$ V, where V_R is the reference voltage. The A/D converter is clocked at 1 MHz.

(1) Determine the total error in the conversion

(2) Calculate the acquisition time

If a change is made in the A/D converter so it counts from zero and advances by the value of the LSB for every clock pulse (ramp counter A/D converter), what would be the conversion error in the time given by (2) when the LSB = $0.001\ V_R$? How long will it take the counter ramp A/D to achieve the same conversion accuracy as the successive approximation A/D converter?

8.3 Design a bipolar transistor bistable multivibrator using a 12-V dc power supply. The multivibrator is to drive a load that requires 1 mA. Indicate how you would convert the multivibrator into a flip-flop. Show that the flip-flop performs a divide-by-two function.

8.4 Using standard digital gates, design a circuit that will divide a clock frequency by the following integral numbers: 7, 19, 25, 47, 77, and 92.

8.5 What is a shift register and how does it work? Illustrate your answer with a design of a four-stage shift register. Describe two applications where a shift register may be used.

8.6 What is sidetone and how can it be used to cultivate good speaking habits in telephone users? Does sidetone control offer any other advantages? With the aid of a suitable circuit diagram, describe the operation of a typical sidetone suppression circuit. The circuit of Figure P8.1 shows a transformer hybrid in the "receive-mode" where R_m is the resistance of the microphone. Assuming that $Z_L = Z_B = 900\ \Omega$, $Z_R = 1000\ \Omega$, and $N_1 = N_2 = 200$ turns, calculate the power dissipated in Z_R when $V_{in} = 1$ V rms and the reactance of N_1 is 1200 Ω at the frequency of operation. (*Hint:* Transfer the load Z_R to the primary of the transformer.)

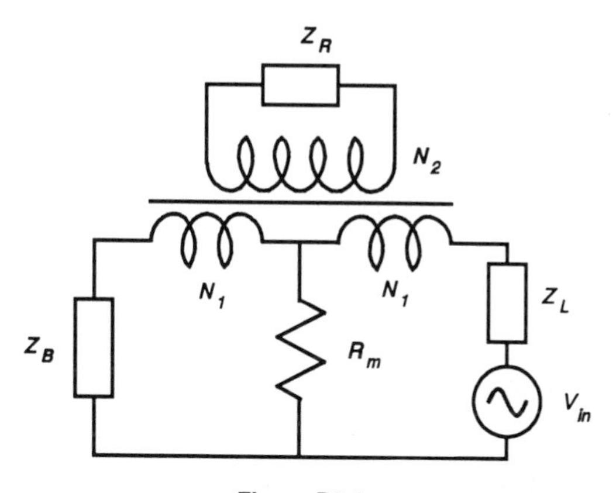

Figure P8.1.

8.7 In the design of digital circuits it is common to find two NOT gates connected in cascade. What is the purpose of such a connection? Using only two-input NOR gates, show how you can realize

(1) A NOT gate

(2) A two-input OR gate

(3) A two-input AND gate

9

SIGNAL PROCESSING IN THE TELEPHONE SYSTEM

9.1. INTRODUCTION

Until the introduction of the digital telephone, there was virtually no signal processing on the subscriber loop. Indeed, there was no need for it. The majority of subscriber loops were able to transmit voice signals with no particular difficulty and in cases where the lines were longer than usual, line "loading" was used with success.

Signal processing has two major aims:

(1) To improve the quality of signal transmission over the telephone communication channels.
(2) To lower the cost of communication by improving the efficiency of channel use.

In general, the quality of a communication channel tends to deteriorate as a function of distance. In addition, long distance channels are expensive to establish and maintain. It follows that the more messages that can be transmitted in a given time, the lower the cost per message. It is therefore on the long-distance channels (trunks or tolls) that signal processing techniques have proven to be most successful. In this chapter, the common signal processing techniques used in the telephone system and some of the circuits employed will be examined.

9.2 FREQUENCY-DIVISION MULTIPLEX

Frequency-division multiplex (FDM) is a technique in which a number of signals can be transmitted over the same channel by modulating each one at a different and appropriate frequency so that they do not interfere with each other. The assignment of specific carrier frequencies to radio stations for broadcasting and other purposes is in fact, FDM. It can be used with amplitude modulation as well as other forms of modulation. In the context of the telephone, FDM is used in conjunction with amplitude modulation.

In normal amplitude modulation, the carrier, upper, and lower sidebands are transmitted. With 100% modulation, the carrier voltage amplitude is twice that of the sidebands. The power in the carrier is therefore $\frac{2}{3}$ of the total. Unfortunately, the carrier has no information content. Each of the sidebands contains $\frac{1}{6}$ of the total power. In radio, 100% modulation is almost never used, so the power content of the sidebands is much less than described. It is noted that the information content is duplicated in the two sidebands. It is clear that one way to beat the corrupting influence of noise on the information content of the transmission is to put as much as possible, if not all of the available power, into one of the sidebands. An added advantage to this scheme is that the required bandwidth is reduced to one-half of its original value. Clearly, this would allow twice as many messages to be sent on the same channel as before. The transmission of only one sideband in an AM scheme is called single-sideband (SSB) modulation. The price paid for this advantage is that it is necessary to reinstate the carrier at the receiver to demodulate an SSB signal. The reinstated carrier has to be in synchronism with the original carrier, otherwise demodulation yields an intolerably distorted signal. Providing a synchronized local oscillator requires complex equipment at the transmitter as well as at the receiver. In SSB radio, an attenuated form of the carrier is transmitted with the signal. This is used to synchronize the local oscillator in the receiver. In the telephone system, a centrally generated pilot signal is distributed to all offices for demodulation purposes. In some cases, a local oscillator without synchronization is used. If the frequency error is small (approximately ± 5 Hz), successful demodulation can be achieved [7].

9.2.1 Generation of Single-Sideband Signals

A block diagram of the SSB generator is shown in Figure 9.1. The signal and the carrier are essentially multiplied by the balanced modulator to give a DSB-SC output. The bandpass filter removes either the lower or upper sideband.

$$f(t) = A \cos \omega_s t \cos \omega_c t \tag{9.2.1}$$

$$f(t) = \frac{A}{2}\left[\cos(\omega_c + \omega_s)t + \cos(\omega_c - \omega_s)t\right] \tag{9.2.2}$$

Assuming the upper sideband is eliminated, we get

$$f_1(t) = \frac{A}{2} \cos(\omega_c + \omega_s)t \tag{9.2.3}$$

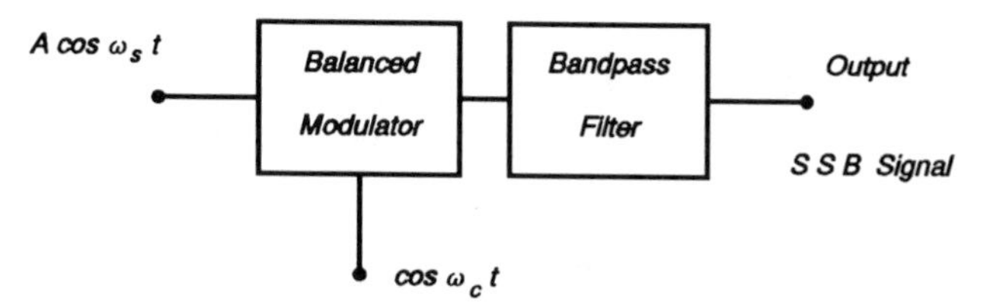

Figure 9.1. The block diagram for SSB modulation.

To eliminate the upper sideband, it is necessary to have a filter with a very sharp cut-off frequency at the carrier frequency. This is not easy to achieve in practice, but the task is made simpler when the modulating signal ω_s has no low-frequency components. Under this condition, crystal and electromechanical filters can be designed to suppress the upper sideband. This is the case for a telephone voice channel, which is nominally from 300 to 3000 Hz.

From Equation (9.2.3), only the upper sideband was transmitted. At the receiving end, the signal is demodulated (multiplied) by a (the) carrier $\cos \omega t$. The result is

$$f_2(t) = \frac{A}{2}\cos(\omega_c + \omega_s)t \cos \omega_c t \tag{9.2.4}$$

$$f_2(t) = \frac{A}{4}\cos \omega_s t + \frac{A}{4}\cos(2\omega_c + \omega_s)t \tag{9.2.5}$$

A low-pass filter is used to separate the required signal at frequency ω_s from that at $(2\omega_c + \omega_s)$.

9.2.2 Design of Circuit Components

The balanced modulator was discussed in Section 4.4.2c. Filter design is outside the scope of this book but a representative list of books on filters is provided in the bibliography at the end of Chapter 3.

9.2.3 Formation of a Basic Group

In the trunk or toll system, 12 channels form a basic group. The basic group is formed by SSB modulation of 12 subcarriers at $64, 68, 72, \ldots, 108$ kHz. These carriers are generated from a 4 kHz crystal-controlled oscillator and multiplied by a suitable factor. The upper sidebands are removed and they are added together to form the group. Figure 9.2*a* shows a block diagram for channel 1. Figure 9.2*b* shows the spectrum of the basic group.

For a small-capacity trunk, the basic group may be transmitted without further processing. The transmission channel can be a twisted pair or coaxial cable.

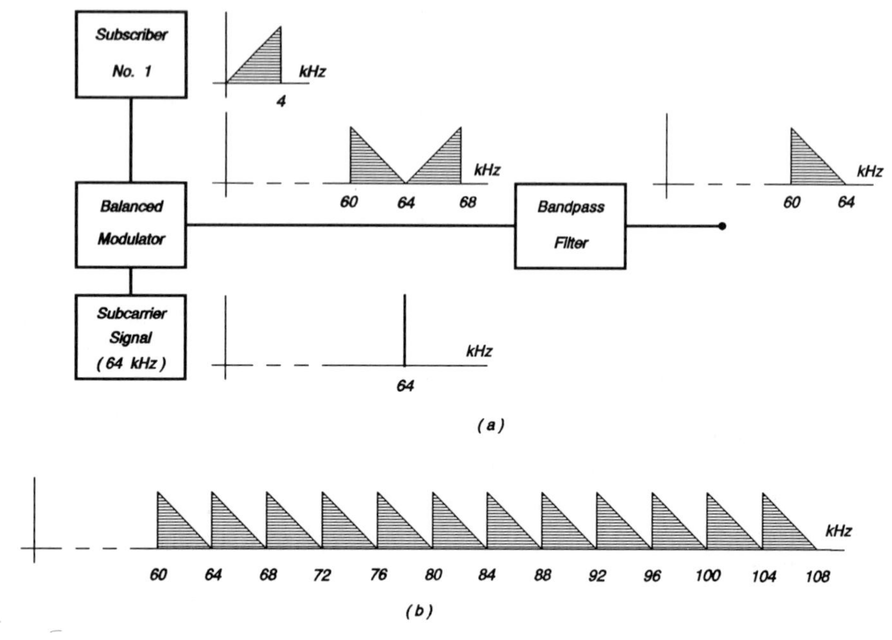

Figure 9.2. Formation of the basic group with spectra. Reprinted with permission from *Transmission Systems for Communications*, 4th ed., AT&T, 1970.

9.2.4 Formation of a Basic Supergroup

For higher capacity channels, five basic groups are combined to form a basic supergroup. Figure 9.3*a* shows the block diagram of the basic supergroup 1. Note that to make the filtering problem easier, the carrier frequency is chosen to be 420 kHz. Figure 9.3*b* shows the frequency spectrum of the basic supergroup.

Table 9.1 shows the carrier frequencies and bandwidths for each basic supergroup. For a 60-channel trunk, the signal can be transmitted in this form. Again, a twisted pair with coil loading or amplification and coaxial cable, may be the media of transmission.

By organizing the 12 basic groups into 5 basic supergroups it is clear that the subcarrier frequencies, the balanced modulators, and bandpass filters can all be duplicated five times over. If the basic group had been made larger, new subcarrier frequencies would have had to be generated and bandpass filters of different characteristics would have been necessary.

9.2.5 Formation of a Basic Mastergroup

To create a 600-channel trunk, 10 basic supergroups are combined to form a basic mastergroup. The frequency spectrum of the basic mastergroup is shown in Figure 9.4. Note that there are gaps of 8 kHz between each basic

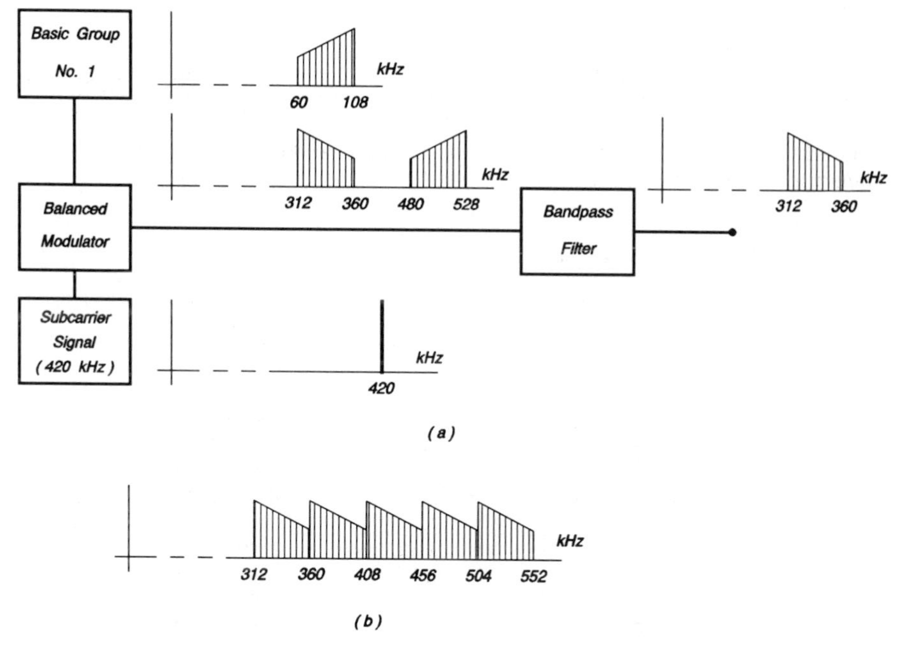

Figure 9.3. Formation of the basic supergroup with spectra. Reprinted with permission from *Transmission Systems for Communications*, 4th ed., AT&T, 1970.

supergroup spectrum. These gaps are designed to make the filtering problem easier.

The carrier frequencies and bandwidths of the 10 basic supergroups are given in Table 9.2. The basic supergroup can be transmitted over coaxial cable or it can be used to modulate a 4 GHz carrier for terrestrial microwave transmission or even sent over a satellite link. Other larger groups can be formed, for example, six mastergroups may be combined to form a *jumbo-group* with 3600 voice-channels.

To recover the original baseband signals from the various groups, the appropriate number of filtering/demodulation processes will have to be

TABLE 9.1 Basic Supergroups

Supergroup Number	Carrier Frequency (kHz)	Bandwidth (kHz)
1	420	312 – 360
2	468	360 – 408
3	516	408 – 456
4	564	456 – 504
5	612	504 – 552

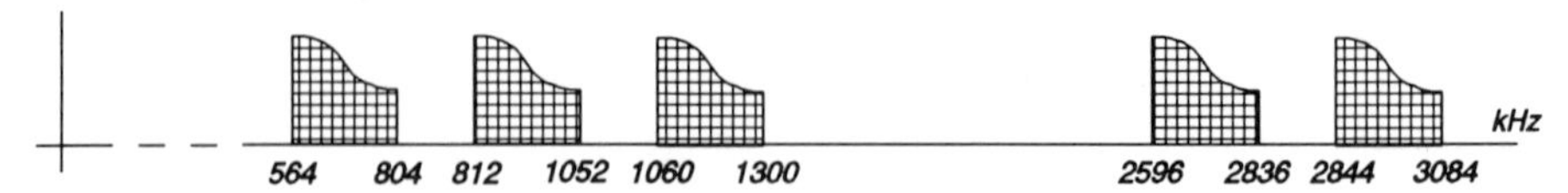

Figure 9.4. Formation of the basic mastergroup with spectra. Reprinted with permission from *Transmission Systems for Communications*, 4th ed., AT&T, 1970.

TABLE 9.2 Basic Mastergroup

Basic Supergroup Number	Carrier Frequency (kHz)	Bandwidth (kHz)
1	1116	564 – 804
2	1364	812 – 1052
3	1612	1060 – 1300
4	1860	1308 – 1548
5	2108	1556 – 1796
6	2356	1804 – 2044
7	2652	2100 – 2340
8	2900	2348 – 2588
9	3148	2596 – 2836
10	3396	2844 – 3084

carried out. At each stage, of the demodulation process, the correct carrier will have to be reinstated for this to be possible.

9.3 TIME-DIVISION MULTIPLEX

In FDM, voice signals were "stacked" in the frequency spectrum so that many such signals could be transmitted over the same channel without interference. In time-division multiplex (TDM), each voice signal is assigned the use of the complete channel for a very short time on a periodic basis. The theoretical basis of this technique is the *Sampling Theorem*. An informal statement of the sampling theorem is:

> If the bandwidth of a signal is B Hz, then the signal can be reconstructed from samples taken at a minimum rate of $2B$ samples per second (*Nyquist sampling rate or frequency*).

The proof of this theorem is beyond the scope of this book. However, there are a number of practical problems which arise in the application of the theorem:

(1) The theorem assumes that the samples have infinitesimally narrow pulse widths. This is clearly not so in a practical circuit. The sampling

rate is usually chosen to be higher than the Nyquist frequency, because it is the minimum; it is discrete to avoid extreme conditions when dealing with an imperfect situation.

(2) The theorem assumes that an ideal low-pass filter is used to bandlimit the signal ahead of the sampler. When using a practical filter, it is necessary to sample the signal at a slightly higher rate (oversampling) to avoid distortion due to aliasing.

A TDM system with two input signals is illustrated in Figure 9.5. The samplers or commutators are shown here as switches that are driven in synchronism.

The TDM system shown in Fig. 9.5 is an example of pulse-amplitude modulation (PAM) system. Practical TDM systems based on PAM have been built and used in the telephone system (No. 101 ESS—PBX, [8]).

9.3.1 Pseudodigital Modulation

To code an analog signal in pulsatile form one can use the height of the pulse, the width (or duration), or the position of the pulse relative to a standard position. When the height is used, it is called *pulse-amplitude modulation* (PAM). When the coding is in terms of the width it is called *pulse-width modulation* (PWM) and when the position is used it is called *pulse-position modulation* (PPM). Pulse height, width, and position are analog quantities, which in turn can be quantized and represented by a binary code where the digits are present (1) or absent (0). When this has been done the modulation scheme is called *pulse-code modulation* (PCM). Although PCM is qualitatively different from the other modulating schemes, they are compared in Figure 9.6.

These schemes would work equally well in a noiseless environment. When noise is present, and it always is, PCM has a clear advantage over the others. In the case of PAM, PWM, and PPM the receiver has to determine what the original amplitude, width, and position were, respectively, in order to reconstruct them. In PCM, the decision is simplified to whether the digit sent was a 1 or a 0. In all cases, it is necessary to transmit timing information with the signal, so that the receiver knows where the bit stream starts and stops.

9.3.2 Pulse-Amplitude Modulation Encoder

To illustrate the design principle of a PAM communication channel, a four-channel PAM system has been chosen. The coder or commutator is shown in Figure 9.7. The master clock drives the four-phase ring counter. The ring counter drives four sampling gates on and off in the correct sequence. When one of the four outputs is on (1) all the others are off (0) so only the sampling gate with the 1 is connected to the adder. Note that both

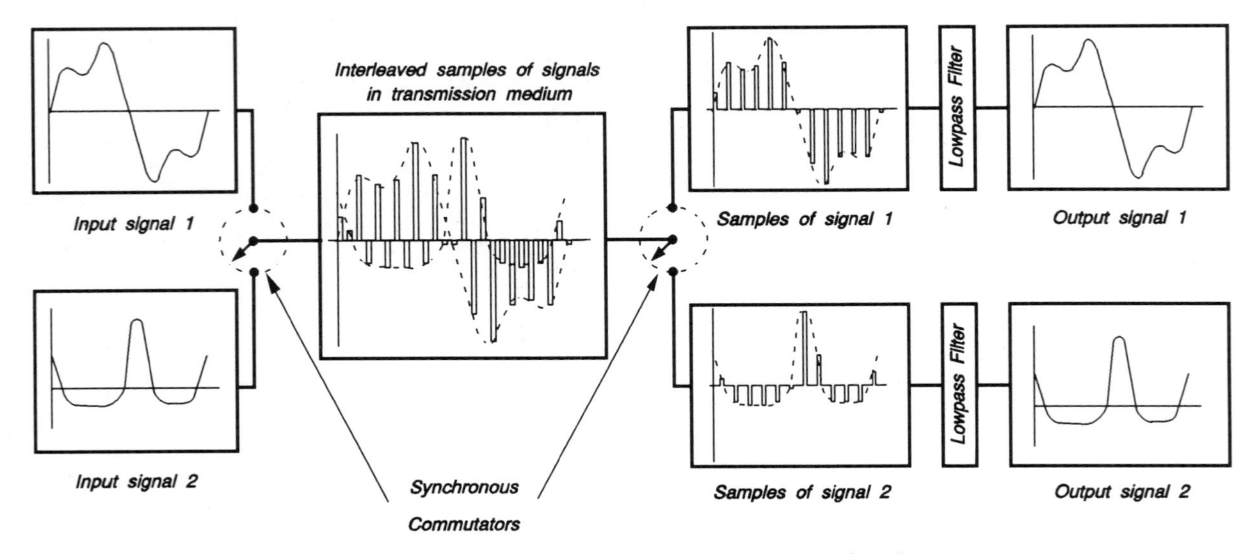

Figure 9.5. A mechanical illustration of time-division multiplex (TDM) with pulse amplitude modulation (PAM). Reprinted with permission from B. P. Lathi, *Modern Digital and Analog Communication Systems*, CBS College Publishing, New York, 1983.

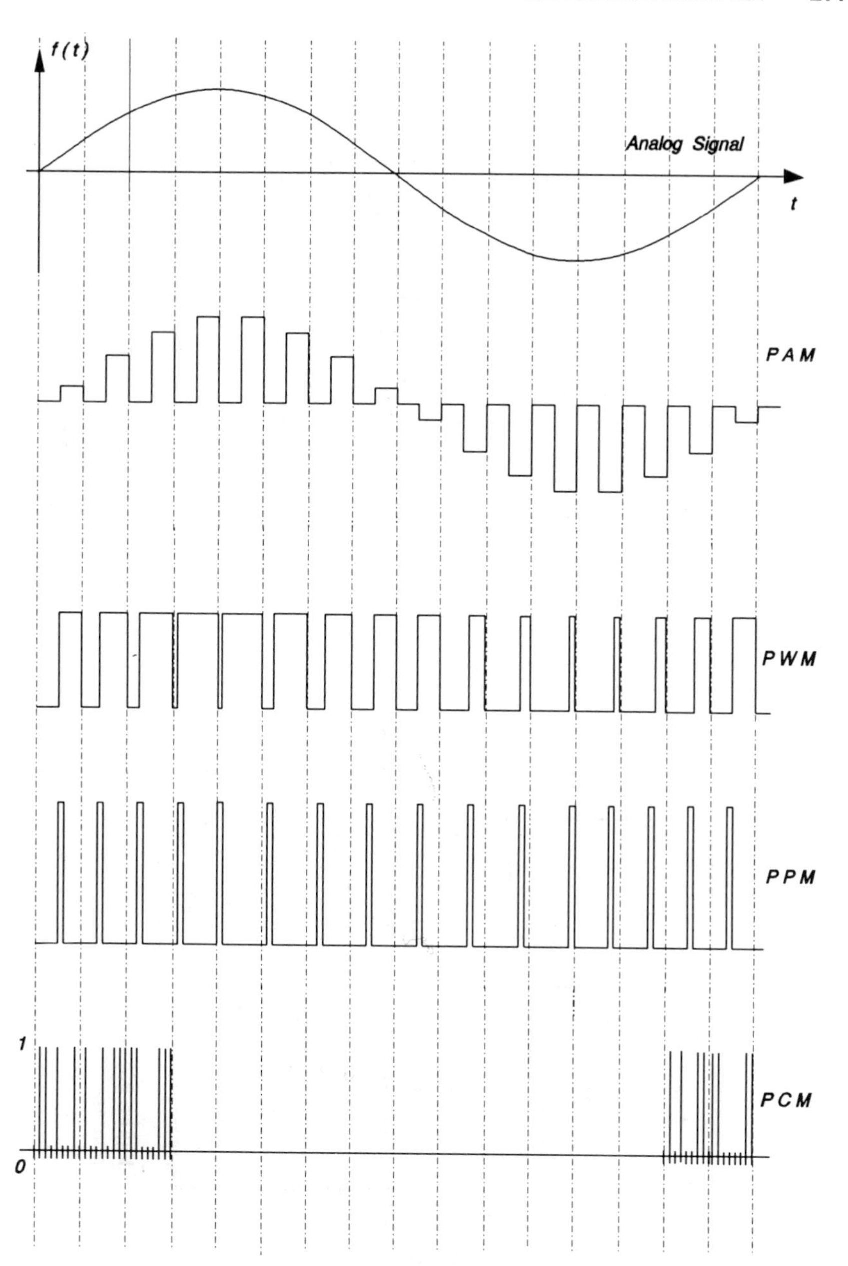

Figure 9.6. A comparison of PAM, PWM, PPM, and PCM. Note that PAM, PWM, and PPM are not truly digital since they convey information by the variation of analog quantities, that is amplitude, duration, and position in time. Reprinted with permission from B. P. Lathi, *Modern Digital and Analog Communication Systems*, CBS College Publishing, New York, 1983.

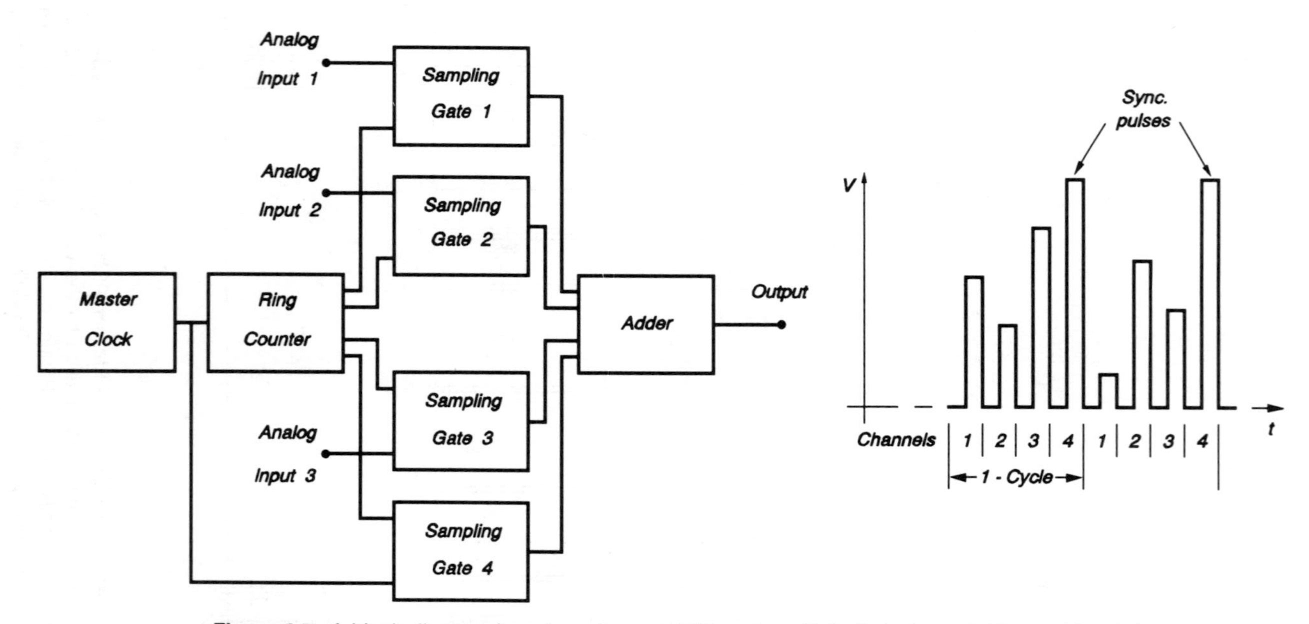

Figure 9.7. A block diagram for a four-channel PAM system. Note that channel 4 is used for timing purposes.

inputs to the fourth sampling gate are connected to the ring counter. This means that channel 4 will always produce a positive pulse. The amplitude of this pulse is adjusted to be higher than the most positive value of the analog input voltage. This is called the *synchronization pulse*, or *sync-pulse* for short. It used to identify and time the other channels.

9.3.2a Four-Phase Ring Counter. The four-phase ring counter and its timing diagram are shown in Figure 9.8. It can be seen from the diagram that in the time taken by one frame, the output pulses go through one cycle. The outputs are used to drive the sampling gates.

9.3.2b Series Sampling Gate. The configuration of the series sampling gate is shown in Figure 9.9. The transistor is an open circuit when the control signal is a 0 and a short circuit when it is a 1. The output is as shown.

9.3.2c Shunt Sampling Gate. The shunt sampling gate is shown in Figure 9.10. The transistor acts as a switch and short circuits the output when the gate voltage is a 1. When the gate voltage is a 0, it is an open-circuit and a path exits between the input and the output.

9.3.2d Series-Shunt Sampling Gate. The two circuits shown above have an inherent deficiency, because its source-to-drain impedance is low but not equal to zero when the transistor is on. To improve the performance, the action of the two gates can be combined, as shown in Figure 9.11.

9.3.2e Operational Amplifier Sampling Gate. The circuit is shown in Figure 9.12. The operational amplifier is connected to give a gain of R_2/R_1 when the transistor is in the off state. When the transistor is on, R_2 is short-circuited, and the gain is reduced to unity.

9.3.2f Multiplier Sampling Gate. A PAM sampler can be seen as a multiplication of the analog signal and a train of pulses. The process is illustrated in Figure 9.13. One of the best methods for accomplishing analog multiplication is to use the four-quadrant analog multiplier. This circuit was described in Section 2.6.3. A practical integrated circuit realization of this is the MC1595 (four-quadrant multiplier manufactured by Motorola Semiconductor Products Inc.).

9.3.2g The Adder. The adder is discussed in Section 4.4.2e.

9.3.3 Pulse-Amplitude Modulation Decoder

The first step in the recovery of the original three signals is to reverse the action of the commutator by separating them into their respective channels. Low-pass filters are then used to reconstruct the analog waveform from the PAM pulses. The PAM decoder system is shown in Figure 9.14.

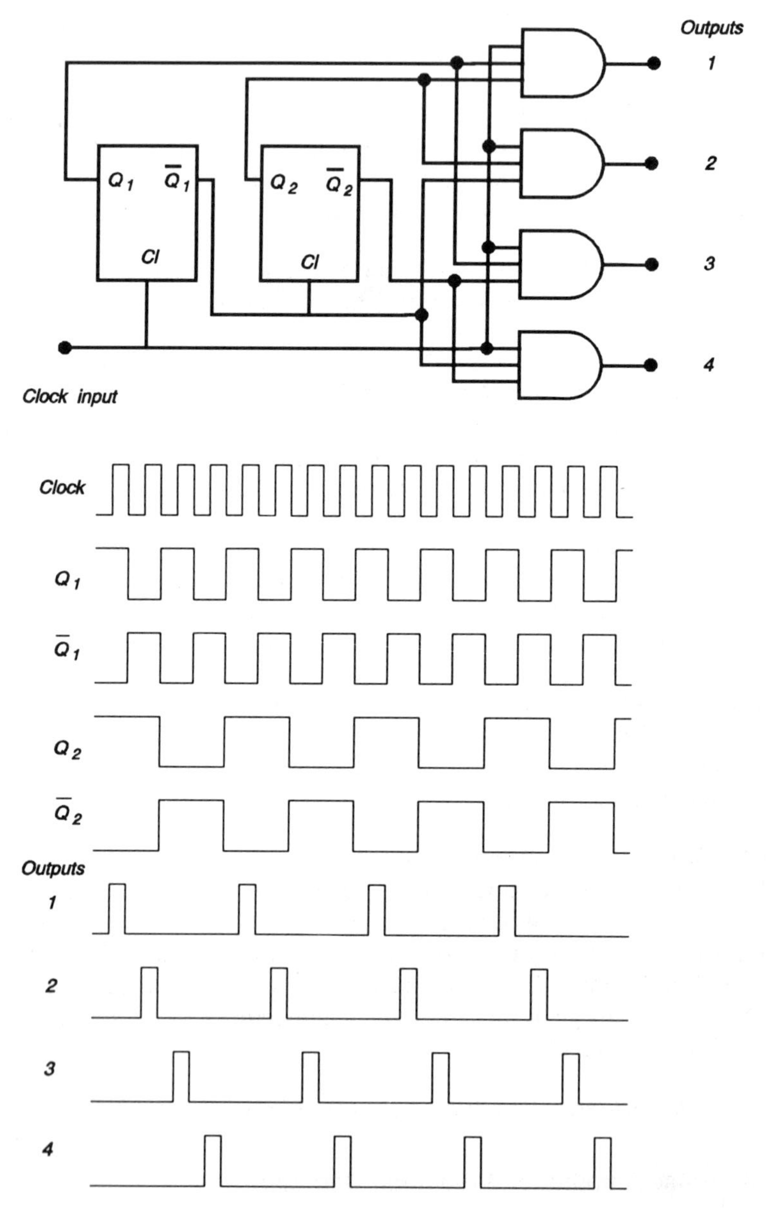

Figure 9.8. A block diagram of a four-phase ring counter with its timing diagram.

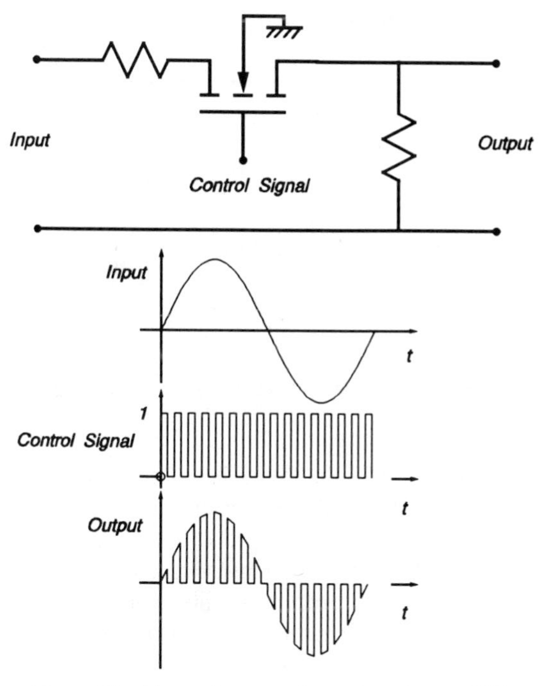

Figure 9.9. The series sampling gate using an FET.

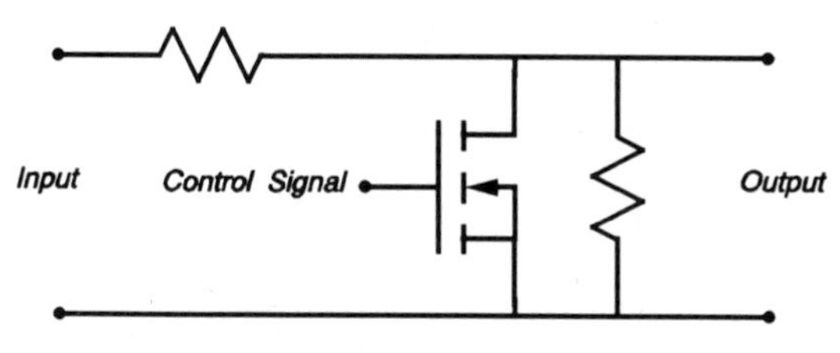

Figure 9.10. The shunt sampling gate.

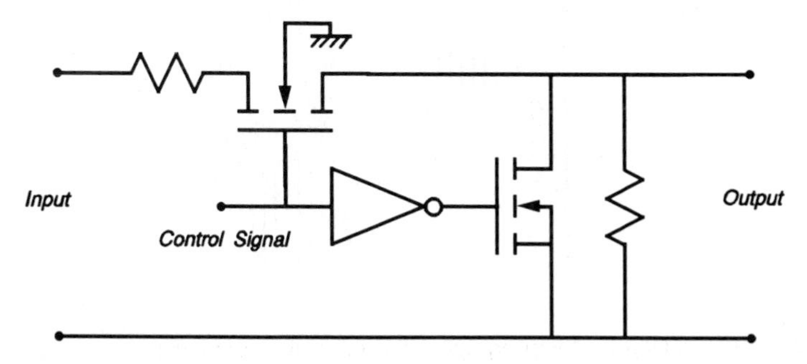

Figure 9.11. A combination of the series and shunt sampling gates improves performance.

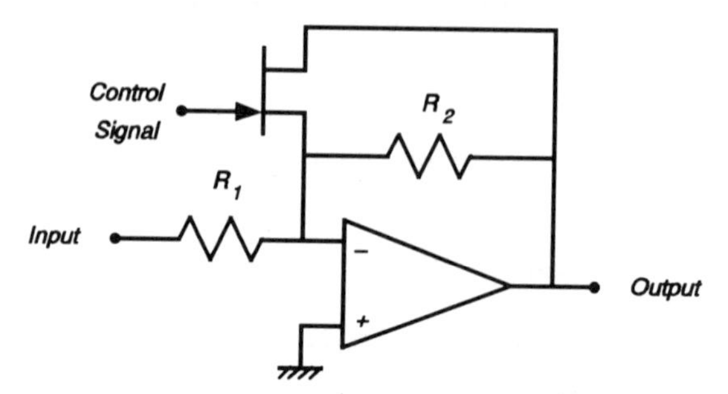

Figure 9.12. A sampling gate using an operational amplifier.

The incoming signal is fed into the Schmitt trigger. The trip level of the Schmitt trigger is set so that only the large sync-pulse will trigger it. The output of the Schmitt trigger is then used to synchronize an *astable multivibrator*. The astable multivibrator then runs in synchronism with the master clock in the PAM coder. The output of the astable multivibrator drives a four-phase ring counter, which produces four sequential output pulses in synchronism with the ring counter in the encoder. These pulses are used to

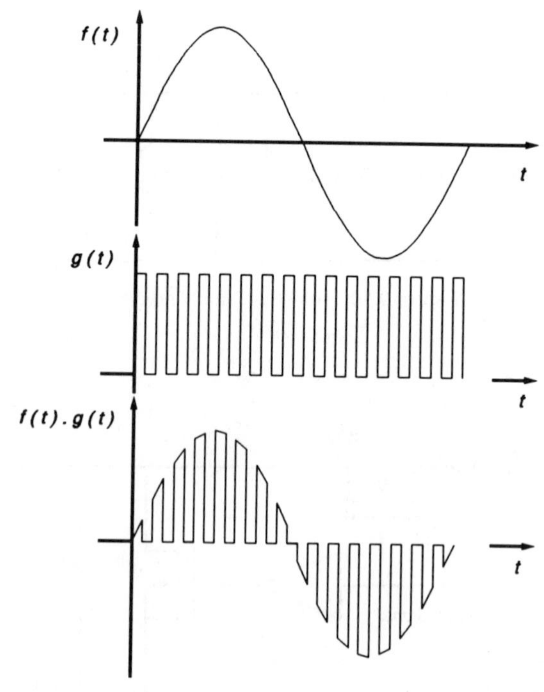

Figure 9.13. The multiplier sampling gate illustrates that sampling is equivalent to the multiplication of two signals.

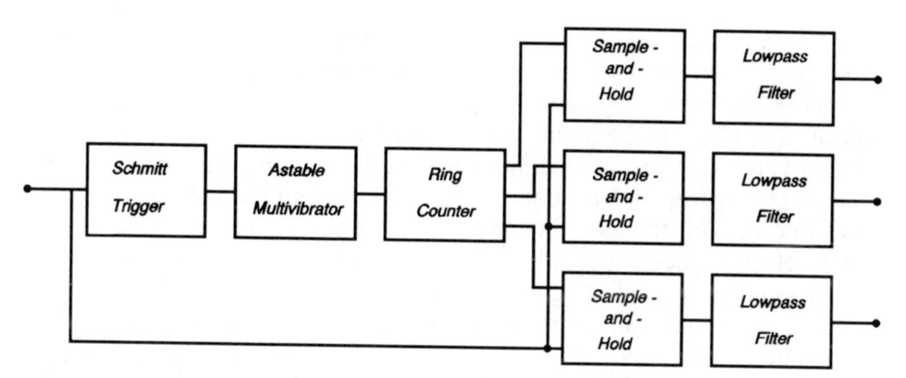

Figure 9.14. A block diagram of the PAM decoder. Note that in Figure 9.7, the first three channels had a signal and the fourth was used for timing purposes.

drive the control gates of the *sample-and-hold* (S/H) circuits. The incoming signal is also fed to the analog inputs of the S/H circuits. Since the control gate of *only* the S/H-1 is open during the period allotted to channel 1, the pulse amplitude of the signal in channel 1 is passed onto the S/H-1. The other channels follow in sequence. The low-pass filters remove the high-frequency components of the PAM signals, producing a replica of the original analog signals.

9.3.3a Schmitt Trigger Design. The circuit diagram of the Schmitt trigger is shown in Figure 9.15. The circuits is designed so that Q_1 has no base current with no input, and it is therefore off. Q_2 is supplied with base current

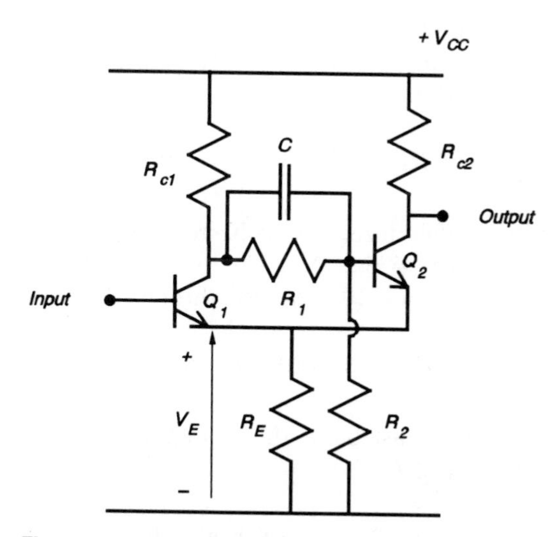

Figure 9.15. The circuit diagram of the Schmitt trigger.

from the resistive chain R_{c1}, R_1, and R_2, and it is therefore on. Current flows in Q_2, which will be in saturation with the correct choice of R_{c2}. The emitter current of Q_2 flows in R_E and sets up a voltage V_E. If an increasing positive voltage is applied to the input, when it reaches the value $V_E + V_{BE}$, Q_1 will start to conduct. Current is drawn through R_{c1} causing the voltage on the base of Q_2 to drop. Q_2 conducts less vigorously and the voltage across R_E tends to drop. But this drop in voltage at the emitter causes the base-emitter voltage of Q_1 to increase rapidly. This is a form of regeneration and proceeds very fast ending with Q_1 conducting and in saturation and Q_2 cut off.

When the input voltage is decreasing, there comes a point when it is slightly below the value $V_E + V_{BE}$. Q_1 conducts less current causing its collector voltage to tend to rise and its emitter voltage to tend to drop. The rising trend at the collector of Q_1 is passed onto the base of Q_2 by R_1 and C. The combined effect of a decreasing V_E and a rising base voltage causes Q_2 to switch on regeneratively and go into saturation. The Schmitt trigger reacts to a slowly changing input voltage by producing a voltage step when its trip level is exceeded. This happens for increasing as well as decreasing voltages. The design of the Schmitt trigger is best illustrated by an example.

Example 9.3.1 Schmitt Trigger Design. Design a Schmitt trigger circuit so that it triggers when a voltage in excess of 3.0 V is applied at the input. The dc supply is 10 V and two NPN silicon bipolar transistors with $\beta = 100$ are provided. $V_{CE(\text{sat})} = 0.5$ V and $V_{BE} = 0.7$ V. The load driven by the Schmitt trigger requires a current of 1 mA.

Solution. The transistors are made of silicon; therefore, $V_{BE} = 0.7$ V. For the circuit to trigger at 3.4 V, V_E must be designed to be equal to $(3.4 - 0.7) =$ 2.7 V. Since the Schmitt trigger is to drive a load that requires 1.0 mA, it is good design practice to allow approximately 10 times this current to flow in the collector of Q_2. Collector current of Q_2 is then 10 mA; therefore, the emitter current is also 10 mA.

$$R_E = 2.7/10 \text{ k}\Omega = 270 \ \Omega$$

Q_2 is in saturation; therefore, the collector voltage must be equal to $(V_E + 0.5) = (2.7 + 0.5) = 3.2$ V. The voltage drop across R_{c2} is $(10.0 - 3.2) = 6.8$ V. Because 10 mA flows through R_{c2},

$$R_{c2} = 6.8/10 \text{ k}\Omega = 680 \ \Omega$$

The base voltage of Q_2 must be at the voltage $(V_E + 0.7) = (2.7 + 0.7) =$ 3.4 V. The base current of Q_2 is

$$I_B = I_c/\beta = 10/100 \text{ mA} = 100 \ \mu\text{A}$$

To maintain a reasonably stable trigger point, it is necessary to allow approximately 10 times the base current to flow in the resistive chain R_{c1}, R_1, and R_2. The current in the resistive chain is therefore 1 mA and its total resistance is 10 V/1 mA = 10 kΩ. Let R_1' be equal to $(R_{c1} + R_1)$ so that

$$\frac{R_1'}{R_2} = \frac{10.0 - 3.4}{3.4} = 1.94$$

But

$$R_1' + R_2 = 10\ \text{k}\Omega$$

Therefore

$$R_1' = 6.6\ \text{k}\Omega \qquad \text{and} \qquad R_2 = 3.4\ \text{k}\Omega$$

When Q_1 is in saturation, it must draw the same current as when Q_2 was in saturation,

$$R_{c1} = R_{c2} = 680\ \Omega$$

Therefore

$$R_1 = (6.6 - 0.68) = 5.92\ \text{k}\Omega$$

The capacitor C is a speed-up capacitor, which helps the transition of Q_2 from the on state to the off state. A reasonable value is 50 to 100 pF.

9.3.3b Sample-and-Hold Circuit. The circuit diagram of the ideal S/H circuit is shown in Figure 9.16. The switch S closes and the ideal voltage source charges capacitor C instantaneously. When S opens, the capacitor retains its charge indefinitely. In practice, S is not an ideal short-circuit when closed and the voltage source has an internal resistance R_s. This means that C charges up with a time constant $\tau = CR_s$. So long as the pulse width of the driving source is significantly longer than τ, the error can be regarded as being small. When S is opened, C loses its charge due to leakage in the dielectric of the capacitor and the finite impedance of the load driven by the S/H.

A simple but practical circuit for the S/H is shown in Figure 9.17. The operational amplifiers are connected as voltage followers with a gain of unity.

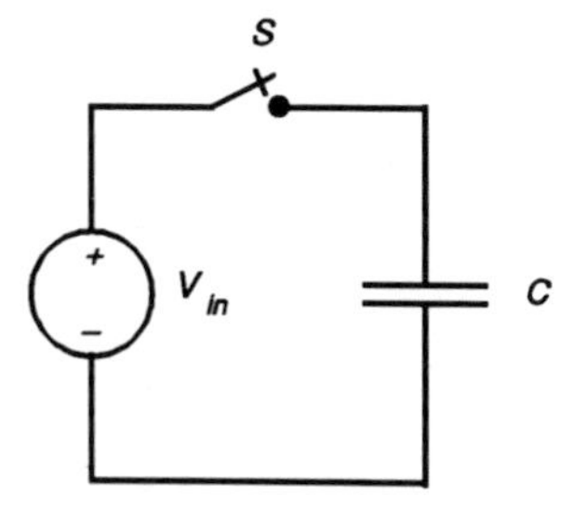

Figure 9.16. The ideal sample-and-hold circuit.

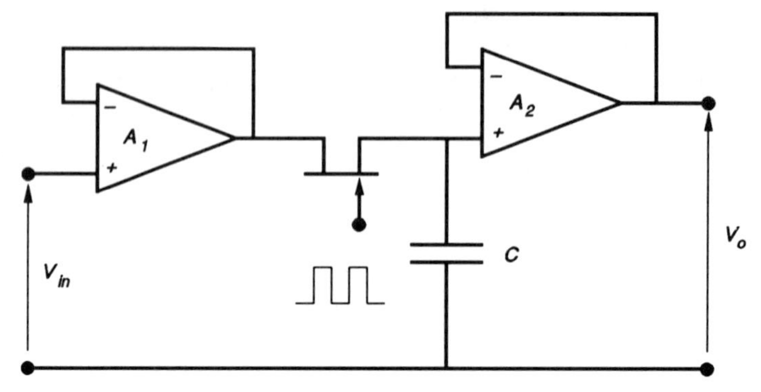

Figure 9.17. A practical sample-and-hold circuit.

The output impedance of A_1 is low enough for it to drive the required charge into the capacitor. The N-channel JFET is switched on by the pulse applied to the gate and the capacitor charges up to the value of the input voltage. When the JFET switch is turned off, the high input impedance of A_2 drains minimal current from C. The design of the S/H circuit is best illustrated by an example.

Example 9.3.2 Sample-and-Hold Circuit. The S/H circuit shown in Figure 9.17 uses a JFET as a series sampling gate. The voltage follower A_2 takes a current of 500 nA. The JFET has a 25-Ω source-to-drain resistance when it is in the on state and may be considered to be an open circuit when it is in the off state. The signal amplitude is 2.0 V, the sample time is 5 μsec, and the hold time is 500 μsec. The capacitor has a value 0.4 μF. Calculate the error in the output at the end of the hold time. Assume that the signal source has negligible resistance.

Solution. The equivalent circuit is shown in Figure 9.18. When the sampling gate is closed, assume that the current flowing into the capacitor is $i \gg 500$ nA (the leakage current taken by A_2). Then

$$\tau = CR_s = 25 \times 0.4 \times 10^{-6} \text{ sec}$$

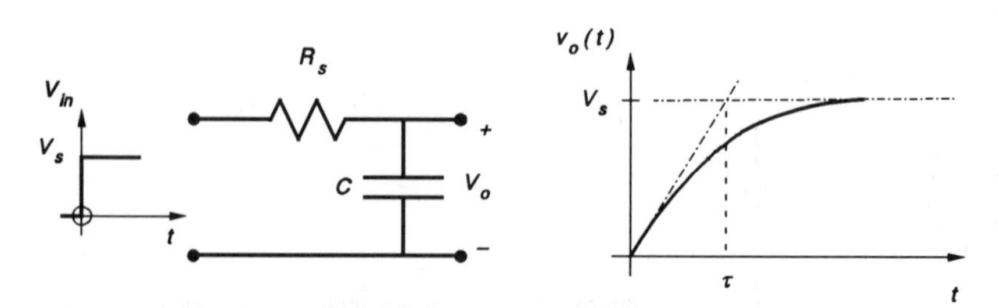

Figure 9.18. The RC circuit for the example showing the exponential response.

For the RC circuit,

$$v_c = V_s(1 + e^{-t/\tau}) \tag{9.3.1}$$

When $t = 5$ μsec and $V_s = 2.0$ V, $v_c = 1.987$ V. The charge on C at $t_1 = 5$ μsec is

$$\begin{aligned} Q &= Cv_c = 0.4 \times 10^{-6} \times 1.987 \\ &= 0.795 \times 10^{-6} \text{ Coulombs} \end{aligned}$$

The charge lost by C in time $t_2 = 500$ μsec is

$$\begin{aligned} \Delta Q &= I\Delta t = 500 \times 10^{-9} \times 500 \times 10^{-6} \text{ Coulombs} \\ &= 25 \times 10^{-11} \text{ Coulombs} \end{aligned}$$

and the charge on C at time t_2 is

$$\begin{aligned} Q - \Delta Q &= (0.795 \times 10^{-6} - 25 \times 10^{-11}) \\ &= 0.794 \times 10^{-6} \text{ Coulombs} \end{aligned}$$

Voltage across C at time t_2 is

$$v_{c2} = (Q - \Delta Q)/C = 1.986 \text{ V}$$

Percentage error in the output is 1.4%.

9.3.3c Other Circuit Blocks in the Decoder. The synchronized astable multivibrator is discussed in Section 7.2.7c and the ring counter in Section 9.3.2a. The design of the filters is beyond the scope of this book.

9.3.4 Pulse-Code Modulation Encoder / Multiplexer

As mentioned earlier, signals coded in PAM tend to be susceptible to corruption by noise and circuit nonlinearities because the information is contained in the amplitude of the pulses. Greater noise immunity could be obtained if the amplitude of the pulse could be coded in binary form. In the presence of noise and other forms of pulse degradation, it is much easier for the receiver to distinguish between the presence or absence of a pulse as opposed to the height of a pulse. PCM is the preferred technique in all modern telephone systems.

To illustrate the basic principles of the design, a four-channel PCM carrier system has been chosen; it is shown in Figure 9.19*a*. The four voice channels are sampled in sequence so the sampler output is a time-division multiplexed PAM signal. The PCM encoder converts the amplitude of each of the four samples into an eight-bit binary code. Figure 9.19*b* shows the sampled analog signals in a four-channel system, the interleaved PAM samples, and the PCM

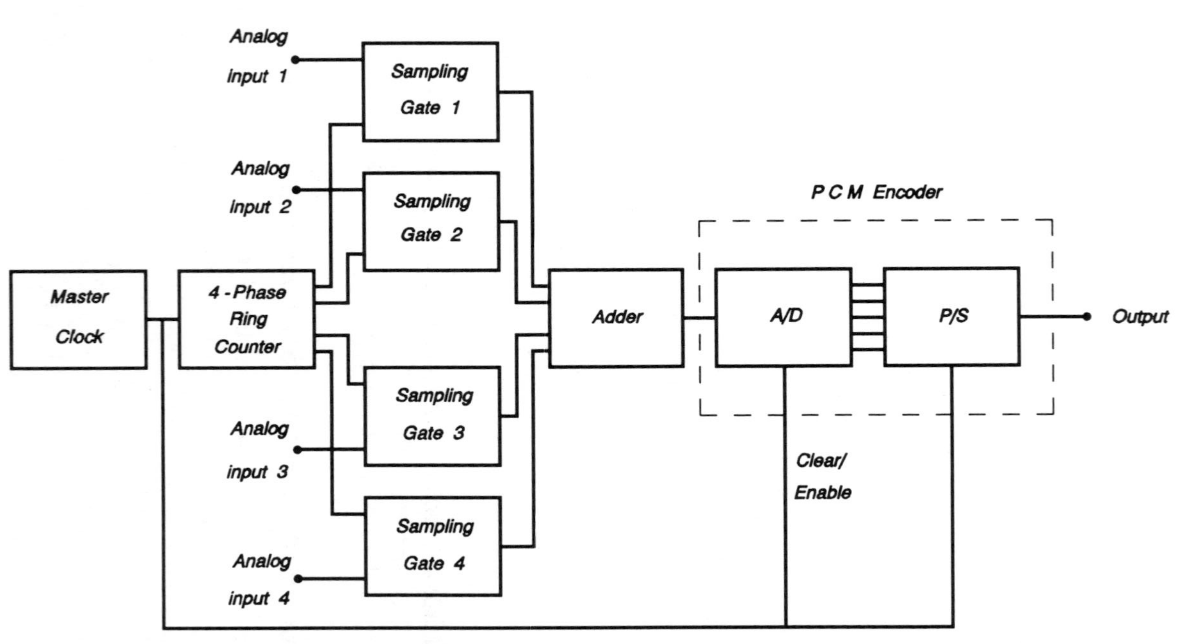

Figure 9.19. (*a*) The block diagram of a four-channel PCM system. Note that this is equivalent to a PAM circuit feeding into the coder of the codec. (*b*) An illustration of the signals in the four channels, the interleaved PAM and the 8-bit PCM equivalent. Reprinted with permission from *Transmission Systems for Communications*, 5th ed., AT&T Bell Labs, 1982.

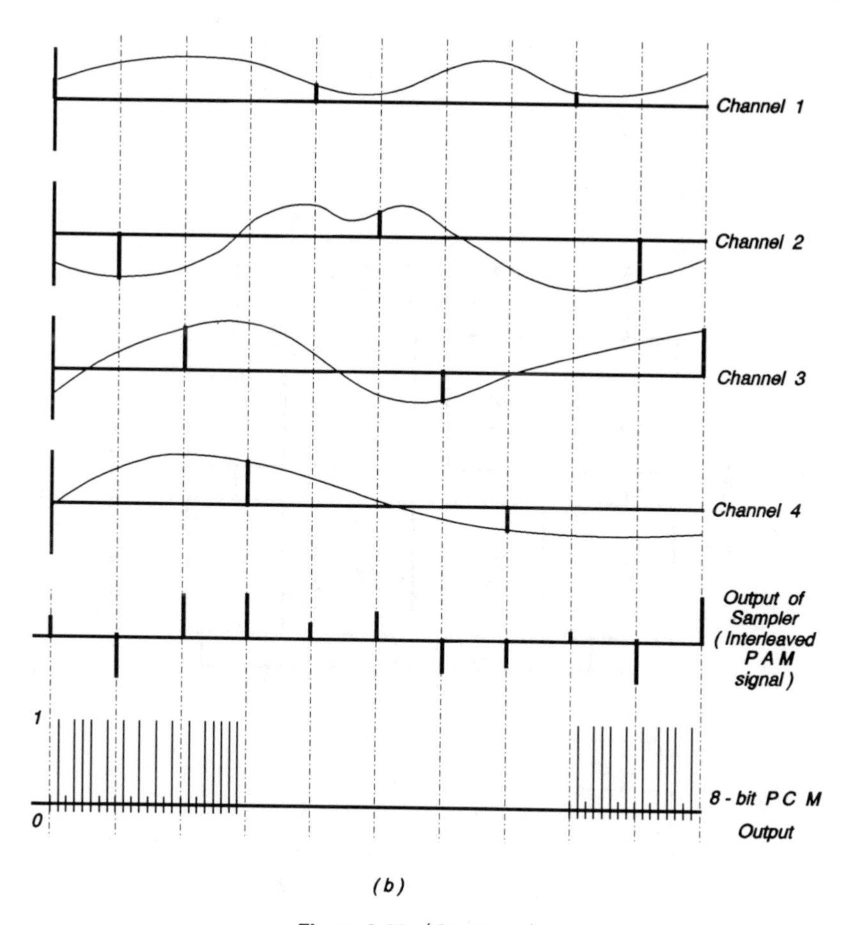

Figure 9.19. (*Continued*)

output. The 32 (4×8) bits are transmitted over a twisted-pair telephone wire. One bit (frame bit) is added for synchronization purposes; it may be wider than the other pulses, or it may have a specially coded sequence of 1s and 0s, which can be recognized by the receiver easily.

Comparing Figures 9.7 and 8.34 with Figure 9.19*a* shows that the PCM is a PAM system feeding into the coder of the codec (A/D followed by a P/S converter). The designs of the ring counter and the sampling gates were discussed in Section 9.3.2. The A/D and P/S converters were discussed in Section 8.5.1.

9.3.5 Pulse-Code Modulation Decoder/Demultiplexer

The system diagram of the decoder/demultiplexer is shown in Figure 9.20. The *frame bit extractor* is designed to recognize the frame bit and send out a pulse to synchronize the local oscillator to the master oscillator in the PCM

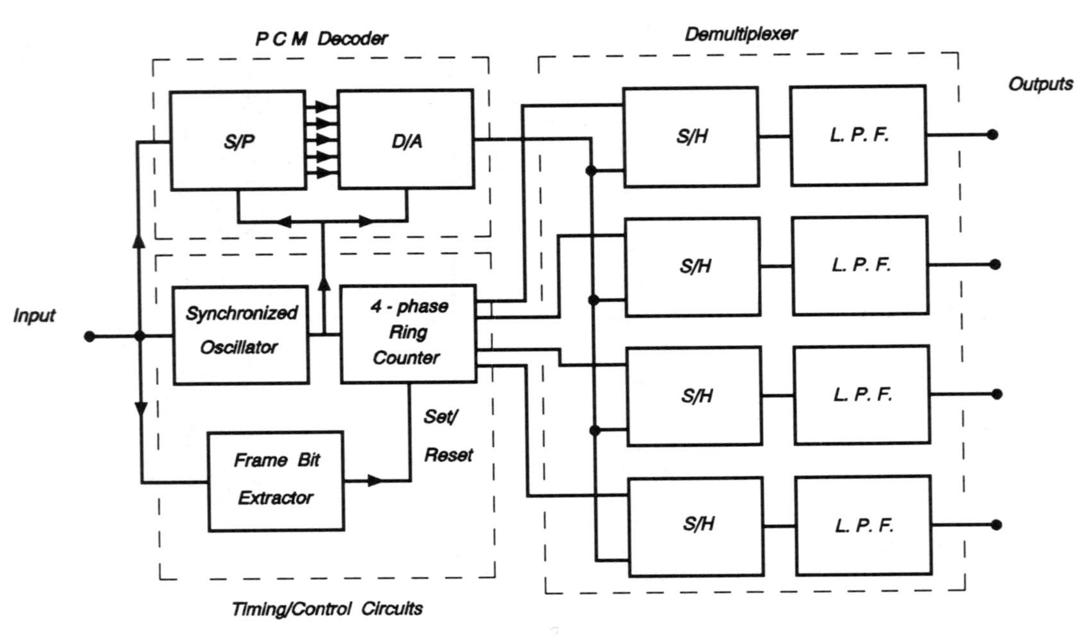

Figure 9.20. The block diagram of the PCM decoder / demultiplexer.

encoder. The output of the local oscillator is used to drive the S/P converter, the A/D converter, and the four-phase ring counter. The output from the D/A converter and the ring counter are fed to the S/H circuits. If the timing is correct, the decoder reconstructs the four samples from the eight-bit codes. The decommutator assigns each sample to its correct channel (demultiplex) and the outputs of the low-pass filters follow the envelope of the PAM signal producing a replica of the original audio frequency signal. Comparing Figures 8.34 and 9.14 and with Figure 9.20 shows that the decoder/demultiplexer is, in fact, the decoder of the codec followed by the PAM demultiplexer. The designs of the S/P and D/A converters were discussed in Sections 8.5.1e and 8.5.1a, respectively. The design of the circuit components for the PAM demultiplexer were discussed in Section 9.3.3.

9.3.6 Bell System T-1 PCM Carrier

The Bell T-1 carrier system uses an eight-bit PCM in 24 voice-channel banks. The sampling rate is 8 kHz. The number of bits generated for one scan of the channels (frame) is $24 \times 8 = 192$. One bit (frame bit) is required for synchronization, so the total number of bits per frame is 193. The analog signal is bandlimited to approximately 3.5 kHz, and, because the sampling rate is 8 kHz, the bit rate is $(193 \times 8000) = 1.544$ Mbits/sec. The minimum bandwidth required to transmit the signal is 1.5 MHz. The use of the eight-bit code means that the voice signals are quantized at 256 (2^8) levels. Some of the less significant bits may be robbed and used for signaling purposes, such as dialing and detection of on/off hook condition. The quantization error resulting from this is considered to be tolerable, although several compression schemes are used to minimize its effect [9].

When the signal is sent over the twisted-pair telephone cable, it suffers considerable degradation from noise, bandwidth limitation, and phase delay. It is therefore necessary to place repeaters and equalization circuits at intervals of approximately 6000 ft to restore the pulses. A 6000-ft ($\sim$ 2000 m), 22-gauge twisted-pair nonloaded cable has a 3-dB bandwidth of approximately 4 kHz.

9.3.7 Telecom Canada Digital Network

The Telecom Canada digital network DS-system is similar to the American Telephones and Telegraph (AT&T) T-system. In the Telecom Canada digital network [1], 24 voice channels each operating at 64 kbit/sec are multiplexed into a 1.544 Mbit/sec channel, designated DS-1. Four DS-1 channels are multiplexed into DS-2, which operates at 6.312 Mbit/sec. DS-3 has seven DS-2 feeding into it and it has a bit rate of 44.736 Mbit/sec. Six DS-3 channels are multiplexed to form DS-4. Its bit rate is 274.176 Mbit/sec, and it can handle 4032 voice channels. The signals from DS-1 and DS-2 may be transmitted on number 22 shielded twisted pairs. DS-3 and DS-4 use coaxial cable. All DS-system output pulses are put into the alternate mark inversion (AMI) format. This format eliminates the dc content of the signal, and it can be used to detect errors in transmission as well as provide substitution codes for time synchronization. The elimination of the dc content of the signal means that the dc power required to operate the repeater can be sent on the same line as the signal.

The output from these PCM switches can be used to modulate a suitable carrier for transmission over terrestrial microwave links, fiber optic cables or satellite links.

Newer and more elaborate digital systems keep coming on the market offering more channels, higher bit rates, greater flexibility, higher voice fidelity, and lower cost. The basic idea of a hierarchy of systems made up of modules that are interchangeable remains.

9.3.8 Synchronization Circuit

For large switches such as the DS-2 or DS-3, the simple timing circuit used in the example of the four-channel system is evidently inadequate. A centrally located cesium beam atomic clock (oscillator) generates the primary signal. The frequency stability of this oscillator is guaranteed to be less than one part in 10^{11} during its lifetime. The output signal is distributed to all nodes of the network where timing information is required according to a hierarchy determined by the importance of the node. At each node, the signal is used to synchronize a local crystal-controlled oscillator. The local oscillator (nodal clock) then has the accuracy of the primary cesium beam atomic clock. The nodal clock must be accurate on its own (one part in 10^{10} per day) so that at times when the primary source is lost, the system can function satisfactorily. All DS-1 signals are therefore synchronized indirectly to the cesium clock. The synchronization of the nodal clocks is carried out as shown in Figure 9.21. This is a phase-locked loop (see Section 5.2.7c) in which the frequency of the local voltage-controlled crystal oscillator (VCXO) is adjusted by a

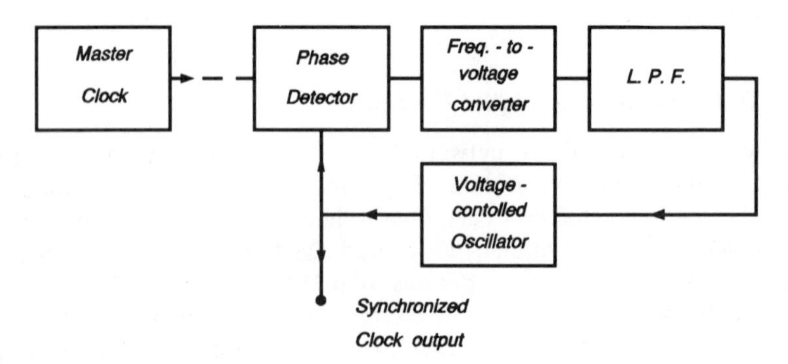

Figure 9.21. A block diagram of the system for the synchronization of the master clock to the nodal clock.

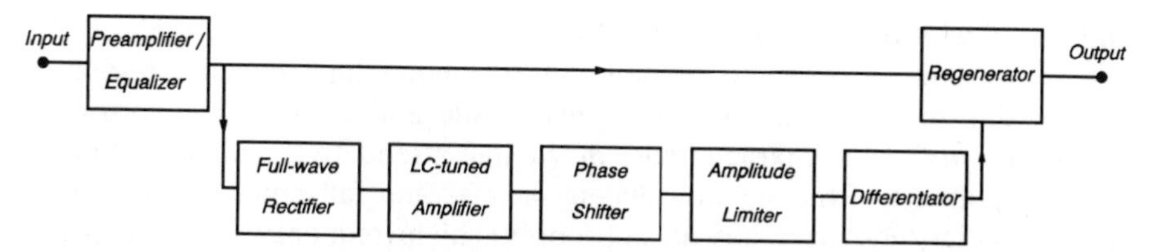

Figure 9.22. The block diagram of the regenerative repeater. The bit frame detector is not shown. Reprinted with permission from *Transmission Systems for Communications*, 5th ed., AT&T, Bell Labs, 1982.

voltage derived from the frequency difference between the two oscillators. When the difference is zero, no adjustment is applied and the VCXO stays in synchronism.

9.3.9 Regenerative Repeater

It was mentioned earlier that the output of the DS-1 switch was transmitted on the bandwidth-limited twisted-pair telephone line. The result of this was a rapid degradation of the signal with other factors such as noise contributing to it. It was therefore necessary to install regenerative repeaters about 1800 m apart to detect the degraded pulses and send new ones [3, 4]. The action of a regenerative repeater includes amplification, equalization, detection, timing, and pulse generation. The regenerative repeater does not accumulate noise the way a repeater (amplifier) on an analog line would; it generates a new clean pulse. A block diagram of the basic regenerative repeater is shown in Figure 9.22.

The incoming signal is fed into a preamplifier/equalizer to boost the level of all frequencies present in the signal to a suitable value. The output of the amplifier then goes to the input of the frame bit detector (not shown) and the full-wave clock rectifier. The frame detector identifies and the start/stop of each frame, and the rectifier changes the bipolar form of the signal to unipolar, thus producing a discrete frequency component at the signaling rate. A high Q factor LC-tuned amplifier selects the sinusoidal frequency component at the clock rate. The resulting sinusoid is amplified and fed to a phase shifter. The phase shifter output is further amplified and then goes to the amplitude limiter, which produces a squarewave. The squarewave is fed into a differentiator to generate positive- and negative-going clock pulses at the zero crossings. A phase shift network is used to adjust the phase of the clock pulses so that the positive clock pulses coincide with the maximum points (see Fig. 9.26) on the incoming signal (amplified and equalized). The resulting pulse and the output of the preamplifier/equalizer go to the regeneration repeater, where a decision in each time slot has to be made, whether the received bit was a 1 or a 0. If the decision is that the bit was a 1, the regenerator produces a pulse of appropriate dimensions.

9.3.9a Preamplifier / Equalizer. As mentioned in Section 9.3.6, the PCM signal transmitted over the twisted-pair telephone line, degenerates quite rapidly due to electrical noise, both man-made and natural. The restricted bandwidth and delay characteristics of the line attenuate the higher frequencies components of the signal reducing the rise and fall times of the pulses. The amplifier/equalizer has an appropriate high-frequency response boost built into it to compensate approximately for 1800 m of 22-gauge twisted-pair telephone line. This circuit may include an automatic gain control to keep the signal amplitude within specified limits.

9.3.9b Frame Bit Detector. The frame consists of 192 bits (8-bit code $\times$ 24 channels). Since the sampling frequency is 8 kHz, each frame must be scanned in 125 μsec. It is necessary, however, to have a marker to indicate where the frame starts or ends, so that each 8-bit word can be identified and steered to the correct channel at the receiving end [6]. To provide such a marker, one extra bit called the framing bit, is added at the end of each frame. By changing the framing bit from a 1 to a 0 and back again at the end of every frame, it is possible to identify it at the receiving end and to use it for synchronization of the channels. Positive identification is possible because the alternating value of the framing bit represents a frequency of 4 kHz, but all the signals in the other channels were prefiltered to remove all frequencies above 3.5 kHz. It follows that none of the other bits derived from signals in the channel can have a pattern of alternating bits from frame to frame.

The incoming pulses are scanned for the alternating bit in the 193rd position and, when it is found, the system latches onto it. When a specified number of errors occur in the 193rd time slot within a given time period, the system reactivates itself and starts the scanning process over again. While this is going on information is lost, and the telephone user might hear a click or two.

The circuit of the frame bit detector and the reframing process when an error occurs is quite extensive and will not be discussed in detail here.

9.3.9c Clock Rectifier. The full-wave rectifier restores the timing frequency to its original value and reverses the AMI process. This can be seen in Figure 9.23*a*. The rectification process is best achieved by a center-tapped transformer and two diodes as shown in Figure 9.23*b*.

9.3.9d The LC-Tuned Amplifier. The Q-factor of the amplifier is designed to be in the range 75–100. This ensures a narrow bandwidth for the precise recovery of the timing frequency and the so-called flywheel effect; that is, the circuit continues to operate even in the absence of a few pulses. The output is sinusoidal. Narrow bandwidth LC-tuned amplifiers were discussed in Section 2.5.1.

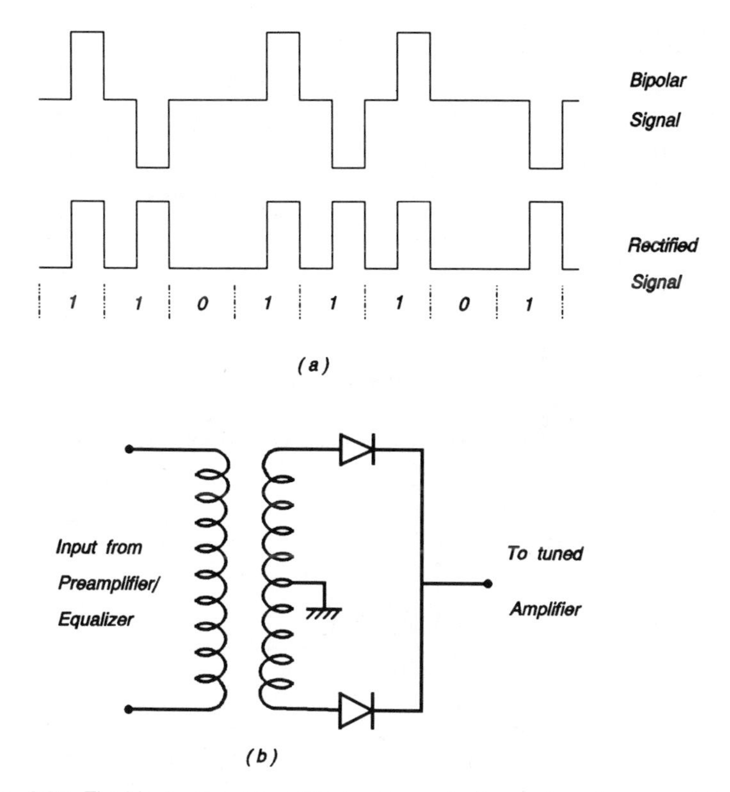

Figure 9.23. The bipolar signal is rectified by the center-tapped transformer and diodes.

9.3.9e Phase Shifter, Limiter, and Differentiator. The phase shifter, limiter, and differentiator are shown in Figure 9.24. The phase shifter shown in the diagram provides an output that leads the input signal; R_1 and C_1 can be interchanged if a phase lag is required. The time-constant R_1C_1 determines the phase shift.

The limiter is an operational amplifier with two back-to-back zener diodes. The output of the operational amplifier is clamped in the positive direction by D_2 acting as an ordinary diode and D_1 acting as a zener diode. Their roles are reversed when the output voltage goes negative. The resulting square-wave output goes to the differentiator.

The choice of the time-constant for the "differentiator" is governed by the condition $R_2C_2 \ll T$, the period of the input frequency.

9.3.9f Pulse Regenerator. The circuit of the pulse regenerator [4] is shown in Figure 9.25. The pulse regenerator consists of a bistable multivibrator whose operation is controlled by the clock pulse and the received signal pulse (amplified and equalized) through the transformer T_1 and diodes D_1 to D_6.

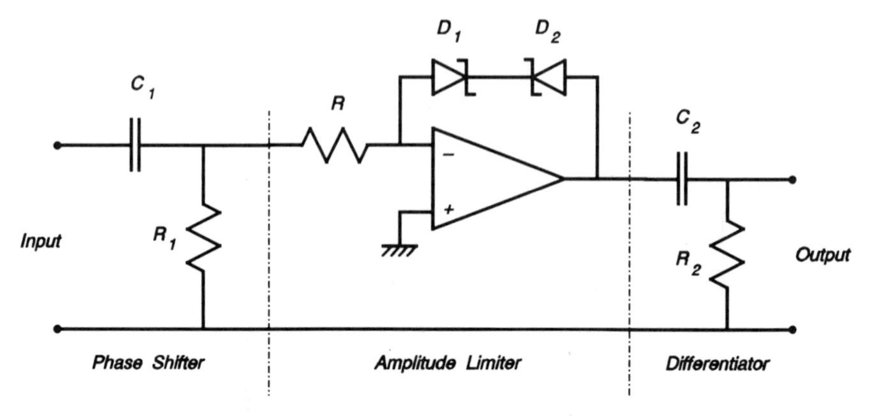

Figure 9.24. The phase shifter, amplitude limiter, and differentiator.

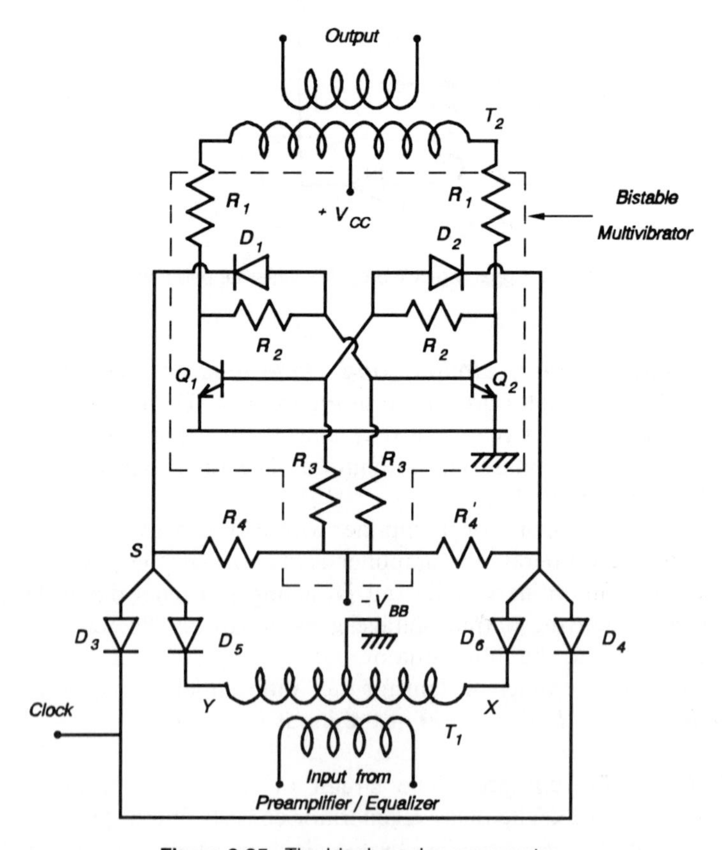

Figure 9.25. The bipolar pulse regenerator.

The received signal pulse is applied to the primary of the transformer T_1 whose secondary center-tap is grounded. Assuming that the voltage at node X is positive and also that this happens to coincide with a positive clock pulse, the diodes D_6 and D_4 are reverse biased. Assume further that Q_1 is on and in saturation and Q_2 is off; the base of Q_1 is then at $+0.7$ V. Diode D_2 will conduct current through R'_4 due to the dc supply V_{BB}. There will be a tendency for Q_1 to conduct a little less current than before. Meanwhile, the voltage at node Y is negative, D_5 is forward biased, and the voltage at node S is more positive than the voltage on the base of Q_2. Therefore, D_1 is reverse biased, and, consequently, the base voltage of Q_2 is free to change. Returning to Q_1, its collector voltage will tend to rise, which will cause the base voltage of Q_2 to rise too. The collector voltage of Q_2 will start to drop and this will cause the base voltage of Q_1 to fall even faster. This is the regenerative phase of the multivibrator action, which ends with Q_1 cut off and Q_2 in saturation.

The signal from the preamplifier/equalizer is still in the AMI format so the next signal pulse input to the regenerator will cause node Y to go positive and if this happens to coincide with a positive clock pulse. The multivibrator will change states with Q_1 back on and Q_2 cut off.

Since the collectors of Q_1 and Q_2 are coupled through the primary winding of transformer T_2 to the positive dc supply V_{CC}, whenever they change states, a pulse of current will flow in opposite directions in the primary winding of T_2 causing a pulse of opposite polarity to occur across the secondary. The output pulses are therefore in the AMI format (bipolar).

The time constant of the transformer inductance, its associated load, and the current that flows in the transformer when the multivibrator changes states determine the width of the pulse. The pulse shape is primarily controlled by the B-H characteristics of the transformer core material. A boxlike hysteresis loop produces a pulse with fast rise and fall times. This is because the core flux saturates early and any further increase in the current has no corresponding effect on the flux.

There are a number of different circuits that can be used in a regenerative repeater. Several versions use a blocking oscillator to restore the degraded PCM pulses.

To optimize the operation of the regenerative repeater, the phase shifter is adjusted so that the positive-going clock pulse coincides with the maximum value of the preamplifier output. This is shown in Figure 9.26.

9.4 DATA TRANSMISSION CIRCUITS

Data transmission over metallic loops predated voice traffic by at least 30 years. In 1837, Wheatstone and Cooke in England, and Morse in the United States established communication systems using binary information exclusively. Later, much improved forms of the telegraph (Teletype or Telex)

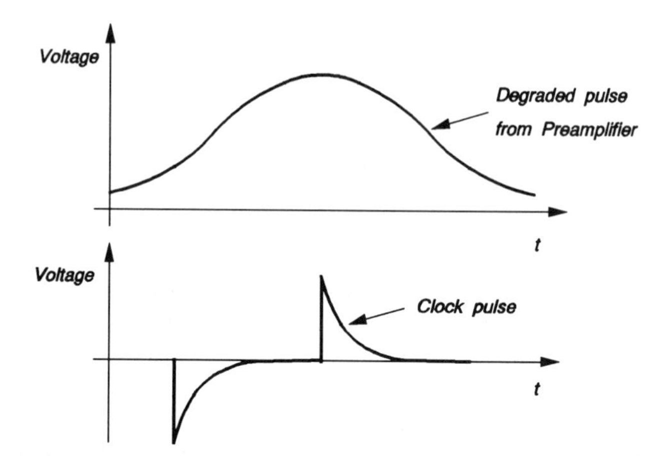

Figure 9.26. The clock pulse is adjusted to coincide with the peak of the degraded pulse. Reprinted with permission from J. S. Mayo, "Clock phase at various points of Figure 31," *Bell System Tech. Journal*, AT&T, 1962.

worked side by side with the telephone into the 1970s. One of the major reasons for its continued use was the need for the written word in business transactions. The advent of fast electronic computers capable of handling data at an ever faster rate hastened the development of transmission systems designed to cope with high-speed data processing. The evolution of the teletype into the integrated services digital network will now be traced briefly.

The printing telegraph was invented in 1855; it was the first attempt to transmit data automatically. In its practical form (Teletype), it had a keyboard based on the typewriter. In the transmit mode, the pressing of a key would cause a start code to be transmitted followed by a 5-bit message code followed in turn by a stop code. It was therefore not necessary to synchronize the transmitter to the receiver. This was all done by a set of mechanical cams, which opened and closed contacts to produce the necessary pulses. In the receive mode, the current pulses were decoded and used to operate a suitable electromechanical system to print the required character. The information was transmitted over telegraph lines. This system was capable of speeds of 50 to 150 words per minute. It was usual practice to use a punched paper tape to compose and edit the message before transmission.

The next significant improvement came with the introduction of the Teletypewriter Exchange [1]. This used a voice channel on the telephone system including the switching system and represented an early example of service integration. It ran at about 100 words per minute.

The addition of store-and-forward capability to the TWX provided a new service called the Message Switched Data Service. Store and forward, as the name suggests, is the ability to accept and store a message and to transmit it only when a channel is available and the receiver is ready. It also can send the message to more than one recipient terminal.

The next advance made data transmission compatible with voice telephony; it uses the analog telephone channel by interfacing through a modem (*mo*dulator-*dem*odulator). It was called Dataphone. The digital signal is used to modulate an analog carrier in a frequency-shift keying (FSK) scheme. The digit 1 or mark is coded as one audio frequency and the digit 0 or space is coded as a different audio frequency. A typical modem generates 980 Hz for a mark and 1180 Hz for a space. These frequencies were chosen to be compatible with the bandwidth limitations of the channel and the speed of the transmission. At the receiving end, the frequencies are decoded into a binary form. Data transmission rates up to 9600 bit/sec have been achieved using this technique.

The need to reduce the cost of data transmission led to the introduction of an essentially digital network called Dataroute. It operates at speeds up to 56 kbit/sec, and it can handle both synchronous and asynchronous data traffic at various speeds by using suitable conversion codes. The data may be inserted into the DS-1 PCM system and transmitted over the 1.544-Mbit/sec channel.

The advantages offered by the use of computers in the telephone system are exploited in the Datapac system. The system consists of nodes interconnected by 56-kbit/sec transmission trunks. The nodes have dedicated computers that receive, check for errors, package, address, and transmit data to other nodes in the network. This is an example of a packet switched network. It has facilities for retransmission of data when errors are detected.

The next significant improvement in data transmission was provided by Datalink, which transmits at rates of 2400, 4800, and 9600 bit/sec. It is independent of code sets and protocols and compatible with the digital telephone network.

Integrated Services Digital Network (ISDN) is the result of international efforts to standardize the format to be used in the existing and growing number of new services offered by the telephone system and other organizations. It uses packet-switching techniques, in which the datastream is separated into packets of modest size and stored in a buffer until a channel becomes available for its onward transmission. It is quite likely that different packets of a message reach their destination at different times and by different routes. It is therefore imperative that every packet be labeled so the message can be reconstructed at the receiving end. Current international agreements assume that ISDN would carry a whole range of telecommunication services including digital video signals. The suggested classes of service are as follows:

(1) Simple voice-channel service
(2) Complex telephone service (smart telephone)
(3) Message storage and delivery

(4) Telemetry
(5) Low bit rate data
(6) 8 to 64 kbit/sec data
(7) $N \times 64$ kbit/sec data services, where N is large

Discussions of these systems are ongoing.

9.4.1 Modem Circuits

The block diagram of a typical modem is shown in Figure 9.27. Typical frequencies are:

$$f_1 = 980 \text{ Hz} \qquad f_2 = 1180 \text{ Hz}$$

$$f_3 = 1650 \text{ Hz} \qquad f_4 = 1850 \text{ Hz}$$

Incoming data (1) is fed into the input port of modem (1), where the FSK transmitter converts it into f_1 and f_2 for mark and space, respectively. It is then coupled to the telephone line through a 4-to-2 wire hybrid (In older models, the carbon microphone and the telephone receiver were used in conjunction with an acoustic coupler; the 4-to-2 wire hybrid would then be the station set hybrid). At the receiving end, the modem (2) hybrid directs data (1) to the FSK decoder, which converts f_1 and f_2 into mark and space, respectively. Modem (2) communicates with modem (1) in the same way except for the difference in the frequencies used.

The 4-to-2 wire hybrid is discussed in Section 8.5.1f.

9.4.1a Frequency-Shift Keying Modulator (Transmitter). The FSK modulator of the modem has the block diagram shown in Figure 9.28. A crystal-controlled oscillator provides an accurate clock frequency, which is fed into two divide-by-N circuits. The outputs are passed through narrow bandpass filters tuned to the required frequencies f_1 and f_2. The outputs of the bandpass filters go to the inputs of the multipliers. The multiplicands are the binary signal to be FSK-coded and its complement. Note that the binary signal and its complement are used to control the flow of signal from the bandpass filters to the adder. When the binary input is a 1, a signal of frequency f_1 is fed to the adder; f_2 is cut off. When it is a 0, f_1 is cut off and f_2 reaches the adder.

The crystal-controlled oscillator was discussed in Section 8.4.5g. An inexpensive and readily available crystal designed to operate at 3.58 MHz

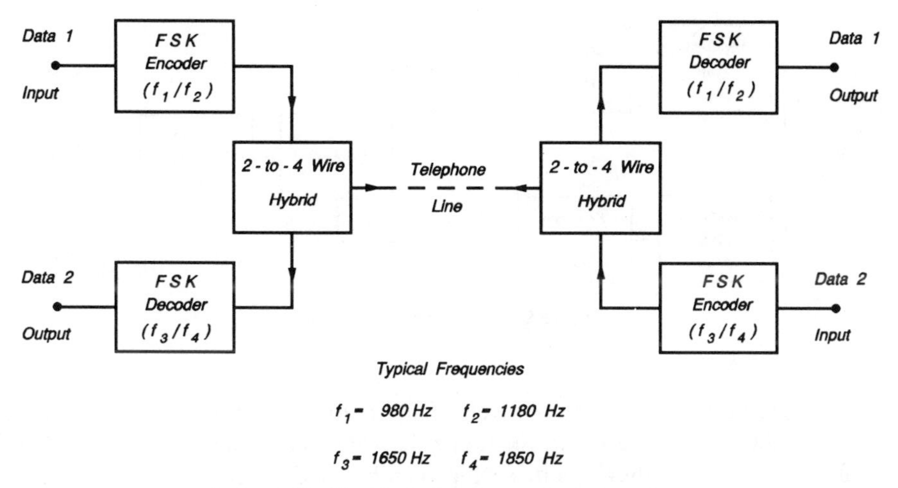

Figure 9.27. The block diagram of a typical modem.

(originally designed for the colour burst carrier of television sets; see Section 8.4.5b) may be used, in which case we have the following table:

Target Frequency, f	Required Divisor	Integer N	Actual Frequency, f'	Error, %
980	3653.06	3653	980.02	+0.002
1180	3033.90	3034	1179.96	−0.003

The divide-by-N circuits can be realized by using two 12-stage ripple counters ($2^{12} = 4096$) with suitable resets to get the required values of N. The bandpass filters can be simple LC-tank circuits with modest Q-factors tuned

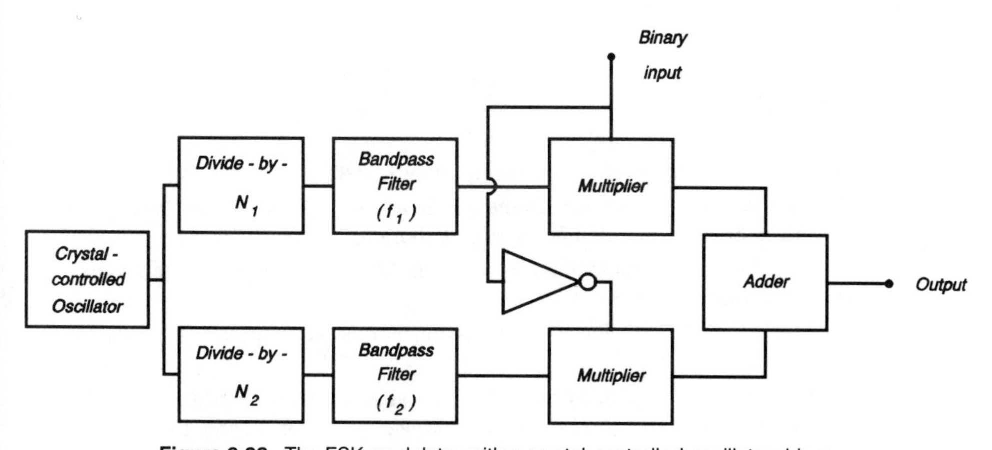

Figure 9.28. The FSK modulator with a crystal-controlled oscillator driver.

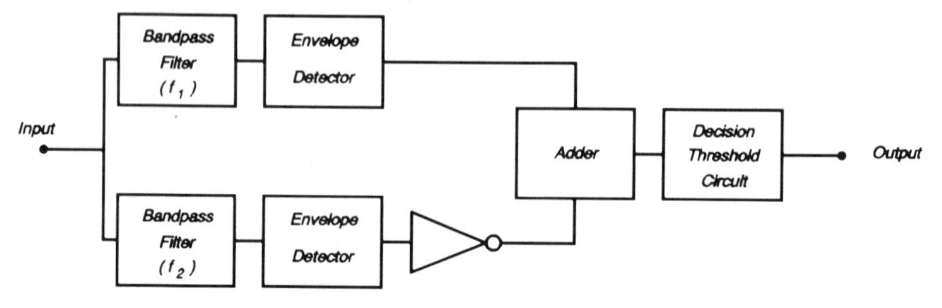

Figure 9.29. The FSK demodulator.

to 980 Hz and 1180 Hz, respectively. The analog multiplier was discussed in Section 2.6.3, the inverter in Section 8.4.5c and the adder in Section 4.4.2e. There are several other schemes which can be used to realize the FSK coder.

9.4.1b Frequency-Shift Keying Demodulator (Receiver). A simple FSK demodulator of the modem is shown in Figure 9.29. Two narrow bandpass filters are tuned to the two frequencies present in the FSK signal. Their outputs are used to drive the envelope detectors. Note that the envelopes of the two frequencies f_1 and f_2 are both squarewaves. The output of one of the envelope detectors is inverted and added to the other to give the required binary output.

The narrow bandpass filters can be LC parallel-tuned amplifiers with modest Q-factors. The envelope detector is discussed in Section 3.4.6, the inverter in Section 8.4.5c and the adder in Section 4.4.2e. The decision threshold circuit is a Schmitt trigger, which is discussed in Section 9.3.3a.

There are several modems on the market which use sophisticated phase-locked loop techniques for both the modulator and demodulator; most use integrated circuits.

REFERENCES

1. Telecom Canada, *Digital Network Notes*, Ottawa, 1983.
2. Staff, *Transmission Systems for Communications*, 4th ed., Western Electric Co. Inc. Winston-Salem, N.C. Bell Telephone Laboratories Inc., 1970.
3. J. S. Mayo, "Bipolar Repeater for Pulse Code Modulation Signals," *Bell System Technical Journal*, 41, 25–97, 1962.
4. F. T. Andrews, "Bipolar Pulse Transmission and Regeneration," U.S. Patent, 2,996,578, Aug. 15, 1961.
5. P. R. Gray and D. G. Messerschmitt, "Integrated Circuits for Local Digital Switching Lines Interfaces," *IEEE Comm. Mag.*, 18 (3), pp. 12–23, 1980.
6. A. J. Cirillo and D. K. Thovson, "D2 Channel Bank: Digital Functions," *Bell System Technical Journal*, 51 (8), 1972.

7. F. G. Stremler, *Introduction to Communication Systems*, 3rd ed. Addison-Wesley, Reading, MA, 1990.
8. B. E. Briley, *Introduction to Telephone Switching*, Addison-Wesley, Reading, MA, 1983.
9. C. L. Dammann, L. D. McDaniel and C. L. Maddox, "D2 Channel Bank: Multiplexing and Coding," *Bell System Technical Journal*, 51 (8), 1972.
10. Staff, *Transmission Systems for Communications*, 4th ed., Bell Telephone Labs. Inc., 1970. Publisher: Western Electric Co. Inc., Winston-Salem, North Carolina.
11. Staff, *Transmission Systems for Communications*, 5th ed., Bell Telephone Labs. Inc., 1982. Publisher: Western Electric Co. Inc., Winston-Salem, N.C.
12. B. P. Lathi, *Modern Digital and Analog Communication Systems*, CBS College Publishing, New York, 1983.

PROBLEMS

9.1 What are the advantages of single-sideband suppressed carrier transmission system? What are its disadvantages?

An AM radio transmitter delivers a maximum of 12 kW of radio frequency power to its antenna when the modulating signal is sinusoidal and the index of modulation is 0.5. Calculate

(1) The power in the carrier

(2) The power in the side frequencies

Calculate the maximum power that can be radiated if the transmitter were used in a single-sideband suppressed carrier (SSB-SC) system. What is the power gain offered by the SSB-SC system?

9.2 A telephone system uses frequency division multiplex for 12 channels in a basic group configuration (see Figure 9.2), two tones of frequencies $f_1 = 400$ Hz and $f_2 = 2.5$ kHz are applied to channel 5 (76 to 80 kHz). Unfortunately, the amplifier in channel 5 has distorted characteristics, which can be represented by

$$i_0 = av_i + bv_i^2 + cv_i^3$$

Derive an equation showing all the frequencies present in the output current. Which channels in the basic group are likely to be disturbed by the distortion caused by the channel 5 amplifier?

9.3 What are the advantages and disadvantages of a pulse code modulation system for the transmission of telephone signals? A telephone system uses frequency division multiplex for 600 channels in a basic mastergroup configuration (see Figure 9.4). The output is to be pulse code modulated, using 128 levels. Calculate the minimum sampling frequency allowing for one guard space between PCM samples.

9.4 Twelve audio channels in a telephone system are each bandlimited to 3.5 kHz and sampled at 8 kHz for transmission in a time division multiplex PAM system.

(1) Calculate the minimum clock frequency, allowing for a suitable synchronizing pulse. Where would you place the synchronizing pulse?

(2) Describe how the output of the PAM system can be used to amplitude modulate a carrier and calculate the bandwidth required to transmit the signal assuming that it is necessary to include up to the third harmonic of the pulse train.

9.5 A PAM wave has the waveform shown in Figure P9.1

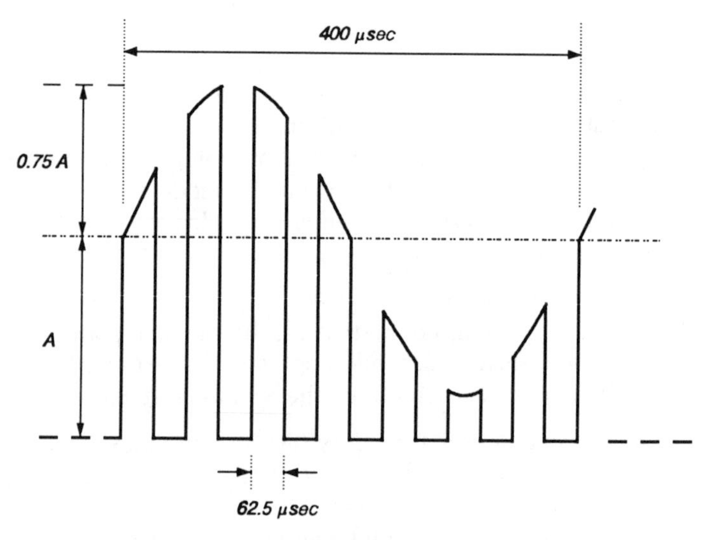

Figure P9.1.

where the mark-to-space ratio is 1. Derive an expression for the PAM waveform given that for a squarewave,

$$f(t) = \frac{1}{2} + \frac{2}{\pi} \sum_{n=1}^{\infty} \frac{1}{n} \sin n\omega t$$

(1) The PAM waveform is applied to the input of an ideal low-pass filter with a cut-off frequency equal to 10 kHz. Find an expression for the output of an ideal low-pass filter. Sketch the waveform.

(2) The PAM waveform is now applied to the input of an ideal bandpass filter of bandwidth 13 to 19 kHz. Find an expression for the output of the filter. Sketch the waveform.

(3) The bandpass filter in (2) is replaced by another ideal bandpass filter with bandwidth 45 to 51 kHz. Find an expression for the output of the filter. Sketch the waveform.

Which of the above outputs would you choose in order to recover the information in the original PAM waveform and why?

9.6 In the telephone system, for both FDM and TDM, channels are grouped to form hierarchies for transmission. Discuss the advantages of this method of organizing the channels.

9.7 What is a ring counter? Illustrate your answer with the design of an eight-phase ring counter. Suggest an application for such a ring counter.

9.8 Design a Schmitt trigger to trigger when a voltage in excess of 4.5 V is applied to its input. The dc supply voltage is 12 V and the load driven by the Schmitt trigger draws a current of 0.5 mA. Two enhancement-mode N-channel MOSFETs with threshold voltage equal to 3 V are provided.

9.9 Design an astable multivibrator to drive a sampling circuit in which the sample time is 1 μsec and the hold time is 15 μsec. The dc supply voltage is 6 V and the sampling gate draws a current of 1 mA when it samples the signal. What special precautions would you take in view of the large difference between the sample and hold times?

9.10 Design a series sampling gate using a bipolar transistor. The source of the signal to be sampled has a voltage of 5 V peak amplitude with an internal source resistance of 100 Ω. The load seen by the sampling gate is 10 kΩ. The transistor is NPN and has $\beta = 100$, $V_{CE(\text{sat})} = 0.5$ V and a base leakage current (when the transistor is off) $I_{E(\text{off})} = 50$ nA. Calculate the errors in the output.

10

TELECOMMUNICATION TRANSMISSION MEDIA

10.1 INTRODUCTION

In this chapter, the characteristics of the media in which the transmission of signals takes place will be discussed. Humans basically communicate through speech/hearing and by sight. We hear sounds from 20 Hz to 20 kHz, and we see only the portion of the radiation spectrum from approximately 4.3×10^{14} Hz (infra red; $\lambda = 7 \times 10^{-7}$ m) to approximately 7.5×10^{14} Hz (ultra violet; $\lambda = 4 \times 10^{-7}$ m). These communication channels occupy only a small portion of the detectable frequency spectrum, which has no lower boundary but has an upper boundary of 10^{22} Hz (gamma rays). Acoustic radiation in the frequency range 20 Hz to 20 kHz is attenuated quite severely in our environment, even when attempts are made to guide it along a conduit. It is therefore quite inefficient to transmit an acoustic signal over any distance that would qualify as *tele*communication. The same observation can be made about visible light. To communicate over distances greater than what we can bridge by shouting, or see reliably, it is necessary to convert the signal into another form that can be guided (by wire, waveguide, or optical fiber) or can be radiated efficiently in free space.

Wires, coaxial cables, waveguides, optical fibers, and free-space transmission have characteristics that vary as frequency changes. A medium may be efficient in one frequency range but quite unsuitable for another frequency range. However, efficiency is not the sole criterion for choosing the frequency that audio and video signals are translated for transmission. To help keep some order and to minimize interference among the various users of communication services, it is necessary to assign various frequency bands for specific uses; governments arrogate to themselves the right to demand a licensing fee for the use of these bands. For example, satellite communication has been assigned 4 to 6, 12 to 14, and 19 to 29 GHz, but there is no technical reason why they cannot operate at frequencies in between these frequencies or indeed outside them.

10.2 TWISTED-PAIR CABLE

Twisted-pairs consist of two insulated wires twisted together to form a pair. Several to many hundred pairs may be put together to form a cable. When this is done it is usual to use different pitches of twist to limit electromagnetic coupling between them and hence cross-talk. The conductor material is copper (usually numbers 19, 22, 24, and 26 American Wire Gauge), and the insulation is usually polyethylene. Wax-treated paper insulation was used in the past, but the ingress of moisture into the cable was a problem in most applications. It is still a problem even with polyethylene insulated cables, which are sometimes filled with greaselike substances to take up all the air spaces and thus discourage moisture from entering. Such cables may be suspended from poles, where they are easy and inexpensive to service but are aesthetically undesirable, or they are buried, which makes them expensive and difficult to repair.

The frequency characteristics of a BST 26-gauge nonloaded cable terminated in 900 Ω are shown in Figure 10.1. It can be seen that the twisted pair has a low-pass characteristic. It should be noted that contrary to expectation, the primary constants of the twisted pair (series resistance, shunt capacitance, series inductance, and shunt conductance, all per unit length) change with frequency. The bandwidth of the twisted pair can be extended to a higher frequency by inductive loading of the line. Lumped inductances are connected in series with the line at specified distances. The best results are obtained when the interval is kept short and the value of the lumped

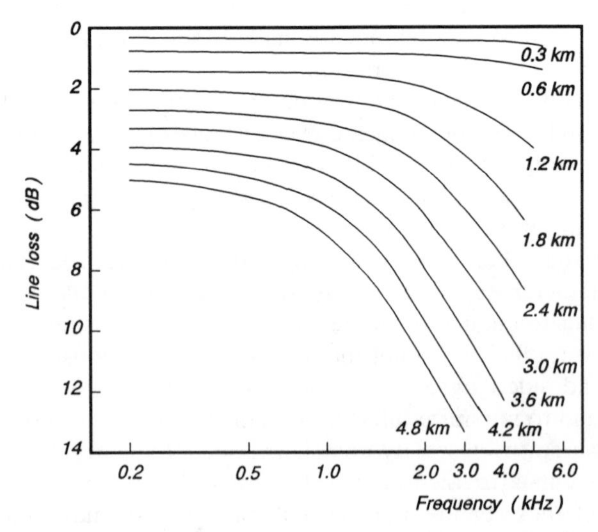

Figure 10.1. Frequency characteristics of 26-gauge BST nonloaded cable terminated in 900 Ω.

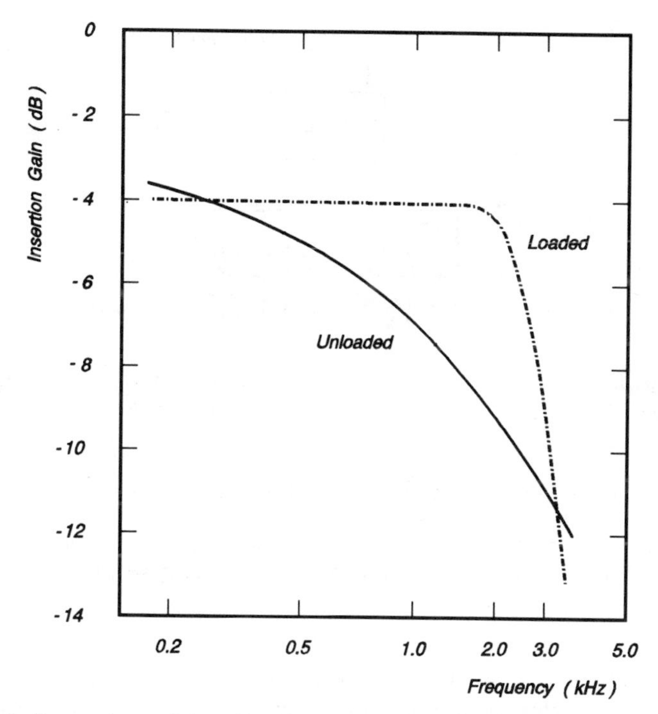

Figure 10.2. Comparison of loaded and unloaded 12,000 ft (3.7 km) 26-gauge cable terminated in 900 Ω and 2 mF. Reprinted with permission from *Transmission Systems for Communications*, 5th ed., AT&T, Bell Labs, 1982.

inductance is kept low, thus minimizing the discontinuities introduced by loading.

The frequency responses of a 12,000 ft (3.7 km) 26-gauge cable with 900-Ω terminations for the loaded and unloaded cases are shown in Figure 10.2. It can be seen that, although loading solves the problem of limited bandwidth for the typical subscriber loop voice channel, it is quite inadequate for the analog (the basic supergroup requires a 552-kHz bandwidth) and the digital (DS-1 requires a 1.5-MHz bandwidth) carrier applications for which it is used. In both these cases, the line must be equalized by placing amplifiers or repeaters at specific distances along its length that emphasize the high-frequency response or regenerate the pulses. Lines used for digital transmission require phase equalization as well, otherwise pulse dispersion degrades the pulses. Dispersion causes the rate of rise of the leading and trailing edges of the pulse to slow down and the base to spread out over a much longer time than the original pulse.

It can be seen from Figure 10.1 that there is a flat loss at lower frequencies, so it is usual to combine the equalizer with the amplifier. An amplifier used for this purpose is called a *repeater*. A repeater can take a number of

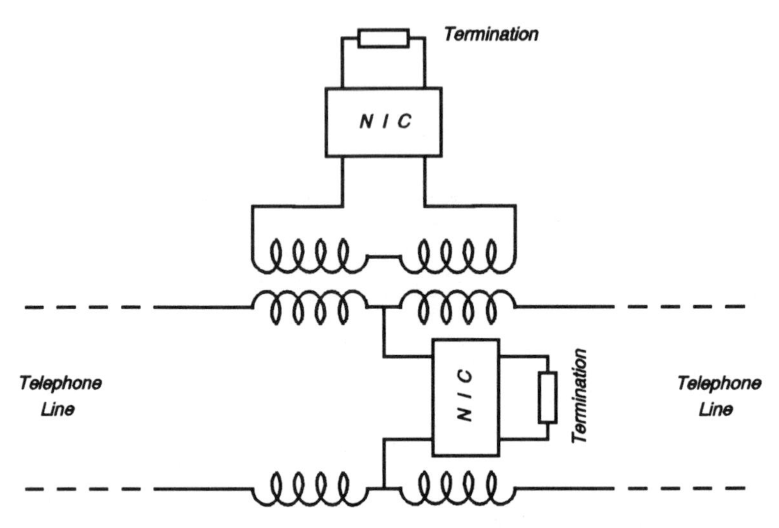

Figure 10.3. The connection of the negative impedance converter to the telephone line. Reprinted with permission from *Transmission Systems for Communications*, 4th ed., AT&T, Bell Labs, 1970.

forms. In a two-wire system where signals flow in both directions, a negative-impedance converter is coupled in series and/or in shunt with the line through a transformer. The configuration of the negative-impedance converter and its connection to the telephone line are shown in Figure 10.3.

Measures have to be taken to ensure that the negative impedance does not overwhelm the line impedance resulting in oscillation. The introduction of repeaters into the cable causes an impedance mismatch at the point of connection. This can cause problems with echo. Severe echo on the cable can impair the speech of most telephone users. There are circuits built into the cable or at the terminations to cancel the echo.

10.2.1 Negative-Impedance Converter

The negative-impedance converter is a two-port which converts an impedance connected to one port into the negative of the impedance at the other port. Consider the two-port shown in Figure 10.4, which is terminated at port 2 by an impedance Z_L. If the two-port is a negative-impedance converter then

$$Z_{in} = -kZ_L \tag{10.2.1}$$

where k is a constant.

Such a two-port is best described by a chain matrix equation. This gives

$$\begin{bmatrix} V_1 \\ I_1 \end{bmatrix} = \begin{bmatrix} A & B \\ C & D \end{bmatrix} \begin{bmatrix} V_2 \\ -I_2 \end{bmatrix} \tag{10.2.2}$$

$$V_1 = AV_2 - BI_2 \tag{10.2.3}$$

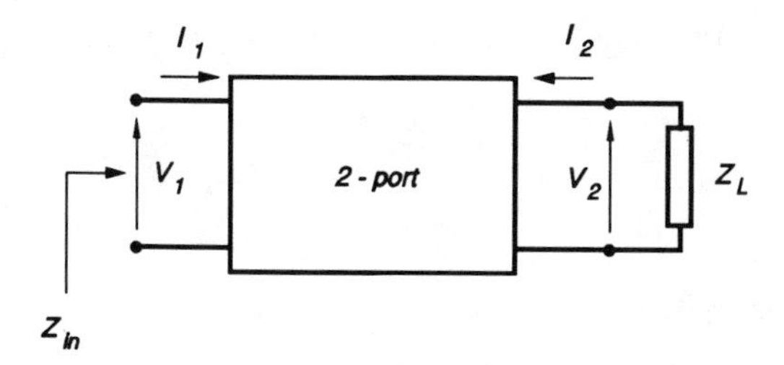

Figure 10.4. A two-port with its defining voltages and currents.

and

$$I_1 = CV_2 - DI_2 \tag{10.2.4}$$

From the termination, we have

$$-I_2 Z_L = V_2 \tag{10.2.5}$$

Substituting into Equations (10.2.3) and (10.2.4) gives

$$V_1 = AV_2 + \frac{BV_2}{Z_L} \tag{10.2.6}$$

and

$$I_1 = CV_2 + \frac{DV_2}{Z_L} \tag{10.2.7}$$

But for a two-port

$$Z_{in} = \frac{V_1}{I_1} = \frac{AZ_L + B}{CZ_L + D} \tag{10.2.8}$$

For Z_L to be equal to $-kZ_L$, $B = C = 0$; then

$$Z_{in} = \frac{AZ_L}{D} \qquad \text{so that} \qquad k = -\frac{A}{D} \tag{10.2.9}$$

There are two possibilities

$$\begin{bmatrix} A & B \\ C & D \end{bmatrix} = \begin{bmatrix} -k_1 & 0 \\ 0 & k_2 \end{bmatrix} \quad \text{or} \quad \begin{bmatrix} k_1 & 0 \\ 0 & -k_2 \end{bmatrix} \tag{10.2.10}$$

Both matrices satisfy the condition for a negative impedance converter, namely

$$Z_{in} = \frac{-k_1}{k_2} Z_L = -kZ_L \tag{10.2.11}$$

where $k = k_1/k_2$.

From the first matrix,

$$V_1 = -k_1 V_2 \tag{10.2.12}$$

This is called the voltage negative-impedance converter or VNIC [5]. From the second matrix,

$$I_1 = k_2(-I_2) \tag{10.2.13}$$

This is called the current negative-impedance converter, INIC or CNIC.

Without loss of generality, we can make $k_1 = k_2 = 1$, so that

$$\begin{bmatrix} A & B \\ C & D \end{bmatrix} = \begin{bmatrix} -1 & 0 \\ 0 & 1 \end{bmatrix} \quad \text{or} \quad \begin{bmatrix} 1 & 0 \\ 0 & -1 \end{bmatrix} \tag{10.2.14}$$

The negative-impedance converter is an example of what is described as a degenerate two-port; that is, it cannot be described by open-circuit impedance (Z) nor short-circuit admittance (Y) parameters. However, it has chain and hybrid parameters (both h and k).

The VNIC may be described in terms of its hybrid k parameters as follows:

$$\begin{bmatrix} k_{11} & k_{12} \\ k_{21} & k_{22} \end{bmatrix} = \begin{bmatrix} 0 & -1 \\ 1 & 0 \end{bmatrix} \tag{10.2.15}$$

The basic form of the VNIC is shown in Figure 10.5.

The transistors Q_1 and Q_2 may be represented by the low-frequency T-equivalent model shown in Figure 10.6. Assuming that C_1 and C_2 are

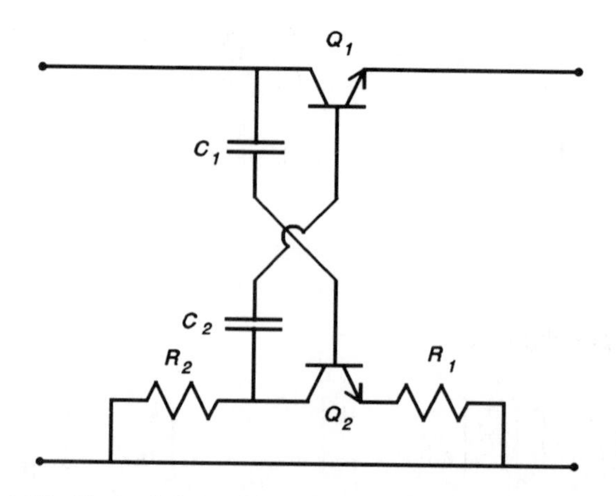

Figure 10.5. The basic form of the voltage negative-impedance converter.

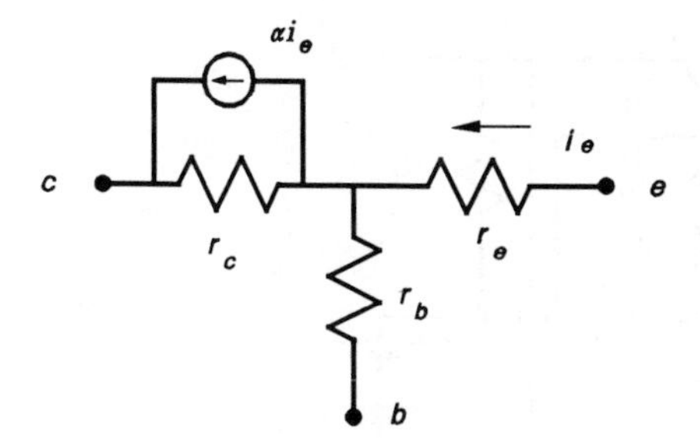

Figure 10.6. The *T*-equivalent model of the bipolar transistor.

short-circuits at the frequency of operation, the equivalent circuit of the VNIC is shown in Figure 10.7. The hybrid k parameters of the circuit in Figure 10.7 are

$$\begin{bmatrix} k_{11} & k_{12} \\ k_{21} & k_{22} \end{bmatrix} = \begin{bmatrix} \dfrac{1-\alpha_2}{R_1 + r_{e2} + r_{b2}(1-\alpha_2)} & -\alpha_1 \\ \dfrac{-\alpha_2 R_2}{R_1 + r_{e2} + r_{b2}(1-\alpha_2)} & r_{e1} + (r_{b1} + R_2)(1-\alpha_1) \end{bmatrix} \tag{10.2.16}$$

When $R_1 = R_2 = R$, r_e is small compared to R, and $\alpha_1 = \alpha_2 \approx 1$, the circuit behaves like a VNIC. The practical version of the VNIC circuit is shown in Figure 10.8.

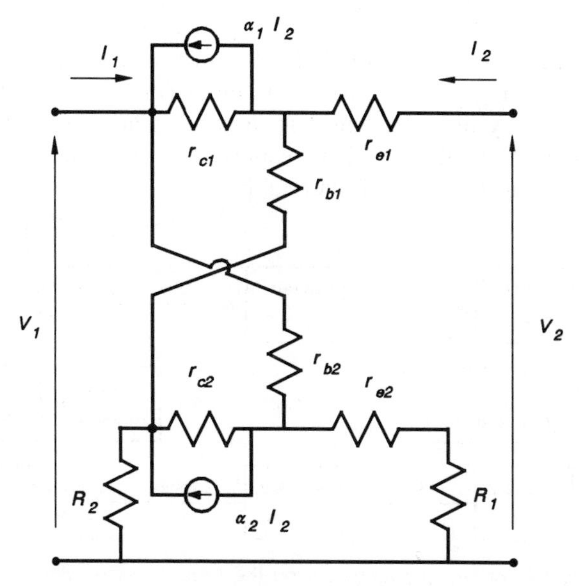

Figure 10.7. The basic VNIC when the transistors have been replaced with the *T*-equivalent circuit.

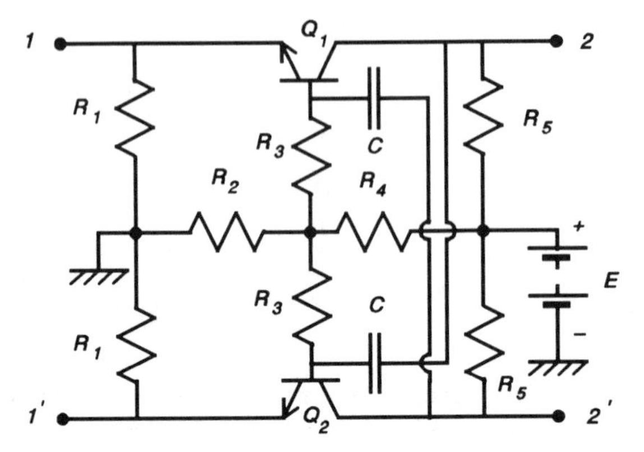

Figure 10.8. The circuit of the voltage negative-impedance converter.

10.2.2 Four-Wire Repeater

In a four-wire system, the forward and return paths are different and ordinary amplifiers may be used. This is shown in Figure 10.9. Again precautions have to be taken to counteract the possibility of instability through the hybrid-to-hybrid feedback path.

As frequency increases, the twisted pair has the tendency to lose signal power through radiation. Ultimately, its usefulness is limited by cross-talk between pairs.

10.3 COAXIAL CABLE

In a coaxial cable, one conductor is in the form of a tube with the second running concentrically along the axis. The inner conductor is supported by a solid dielectric or by discs of dielectric material placed at regular intervals along its length. A number of these cables are usually combined together with twisted pairs to form a multipair cable.

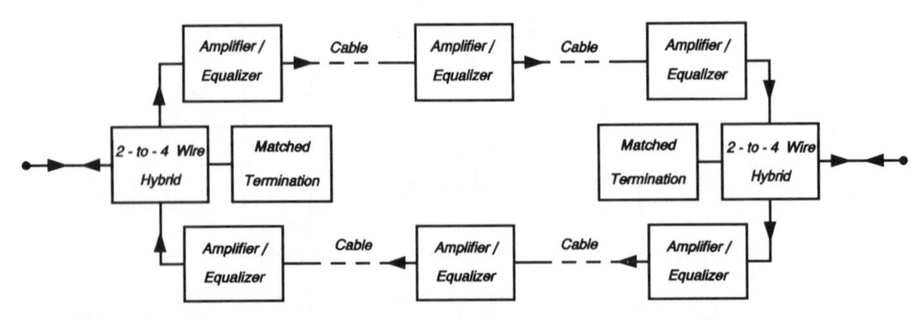

Figure 10.9. The use of ordinary amplifiers on the telephone line with 2-to-4 wire hybrid. Reprinted with permission from *Transmission Systems for Communications*, 4th ed., AT&T, Bell Labs., 1970.

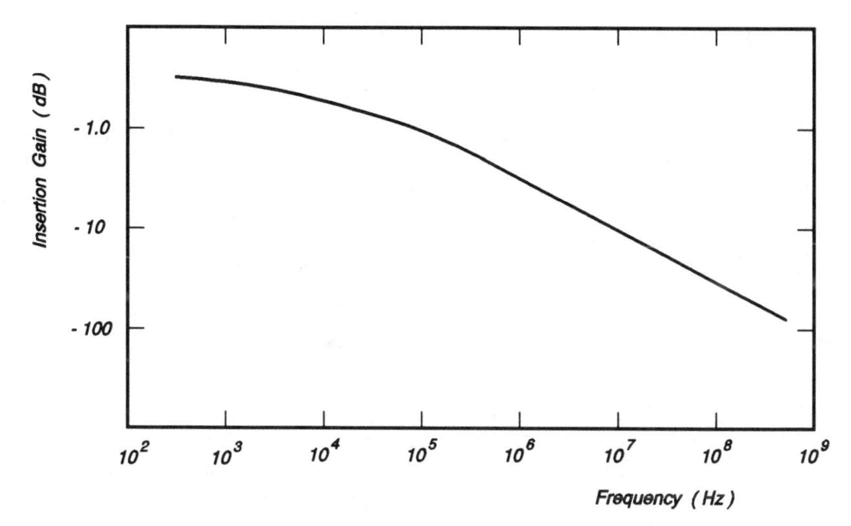

Figure 10.10. The insertion characteristics of terminated 0.375-in. coaxial cable. Reprinted with permission from *Transmission Systems for Communications*, 4th ed., AT&T, Bell Labs., 1970.

The structure of the coaxial cable ensures that at normal operating frequencies, the electromagnetic field generated by the current flowing in it is confined to the dielectric. Radiation is therefore severely limited. At the same time, the outer conductor protects the cable from extraneous signals such as noise and cross-talk.

The primary constants of the coaxial cable are much better behaved than those of the twisted pair. The inductance L, capacitance C, and conductance G per unit length are, in general, independent of frequency. The resistance R per unit length is a function of frequency due to skin effect; it varies as a function of $\sqrt{f}$.

The frequency characteristics of a 0.375-in. (9.5 mm) coaxial cable are shown in Figure 10.10. As expected, the coaxial cable has a much larger bandwidth than the twisted-pair. However, it still requires repeaters and frequency equalizers for analog lines and phase equalization for digital signal transmission. The characteristics of the repeaters are usually adaptively controlled to correct for changes in temperature and other operating conditions.

Coaxial cable is used for transmitting data at 274.176 Mbit/sec in the LD-4 (Bell-Canada) and T4M (Bell System in the U.S.A.) systems. They have 4032 voice channels or the equivalent video or digital data traffic. Their regenerators are spaced at 1.8-km intervals and the total length of the line can be 6500 km [1].

Specially constructed coaxial cables with repeaters of very high reliability are used for submarine cable systems. Because of the very high cost of these cables, they are used to transmit messages in both directions by assigning separate frequency bands to each direction. In spite of the development of satellite communication channels, submarine cables are still viable for trans-Atlantic and trans-Pacific traffic. Because of the propagation delay involved in traveling to the satellite and back, most trans-Atlantic telephone conversation use the satellite link in one direction only; cable is used in the opposite direction. In 1976, the TAT-6 (SG) trans-Atlantic cable system was installed. It used a 43-mm diameter coaxial cable with a 4200 voice channel capacity over a distance of 4000 km [2].

10.4 WAVEGUIDES

A waveguide may be viewed as a coaxial cable with the central conductor removed. The outer conductor guides the propagation of the electromagnetic wave. In its most common form it has a rectangular cross section with an aspect ratio of 2 : 1. The wider dimension must be about one-half the wavelength of the wave that it will transmit. Therefore the waveguide has a low-frequency cut off. There are a number of modes in which the wave can propagate, but in every case the electric and magnetic fields are orthogonal. When the electric field is at right angles to the axis of the waveguide, it is described as *transverse electric* mode. When the magnetic field is at right angles to the axis it is called *transverse magnetic* mode.

The mechanical structure of waveguide disqualifies it from being used for long-range transmission. Irregularities on the walls, such as projections, holes, lack of a perfect match at joints, bends, twists, and imperfect impedance matching at the terminations, can cause reflection and spurious modes to be generated, all of which result in signal loss.

Waveguides are used mainly as feedlines to antennas in terrestrial microwave relay systems. For frequencies above 18 GHz, they are superior to all other media in terms of loss, noise, and power handling.

10.5 OPTICAL FIBER

The use of the optical fiber as a medium for telecommunication was made possible by several developments.

(1) The *laser*, which is a coherent frequency source of the order of 10^{14} Hz and can be modulated. A *light emitting diode* (LED), which produces noncoherent light can also be used.
(2) A low-loss glass fiber that can be used as a waveguide for the light.
(3) A detector for the signal at the receiving end.

The laser can be modulated at a rate in the range of 10^9-bits/sec; the LED can operate at 10^8-bits/sec. The information-bearing capacity of the system is enormous. Ongoing research continues to increase the bit rate limits.

The optical fiber is essentially a high-quality glass rod of about 50-μm diameter for multimode propagation and 8-μm diameter for single mode propagation. The mechanical properties of a glass rod that small will make the system impracticable. In practice, a second layer of glass concentric with the optical fiber proper is deposited on the outside bringing the overall diameter to 125 μm. The outer glass sheathing, referred to as cladding, has a different refractive index and the signal is therefore confined to the core. The core may be of uniform refractive index or it may be graded. These techniques have resulted in optical fibers that have attenuation lower than 0.2 dB/km. Various types of protective covering may be put on the fiber and several fibers put together to form a cable.

The optical receiver is a reverse-biased semiconductor junction diode, which is coupled to the fiber so that the incoming light falls on the junction. The energy in the light is transferred to the electrons in the semiconductor lattice causing them to break away and move into the conduction band. The high electric field sweeps the electrons out of the junction into the external circuit, where they can be detected as a current. Silicon diodes have been used for 1-μm wavelength detectors. For longer wavelengths, such as 1.3 and 1.5 μm, germanium, InAs, and InSb are used [3].

In 1988, a trans-Atlantic optical fiber communication system went into service (TAT-8). It spans a distance of 6500 km and provides the equivalent of 40,000 telephone channels. It operates on the 1.3 μm wavelength; repeater separation is 35 km and the bit rate is 274 Mbit/sec.

10.6 FREE-SPACE PROPAGATION

The transmission media discussed earlier had one thing in common; the propagation of the signal was guided by a twisted pair, a coaxial cable, a waveguide, or an optical fiber. We now consider transmission systems that rely on propagation through free space.

In 1873, Maxwell showed theoretically that electromagnetic waves can propagate through free space. It took three decades to demonstrate experimentally that this was possible, when Hertz constructed the first high-frequency oscillator—the famous spark-gap contraption.

A large number of factors have to be taken into account when designing a free-space propagation communication system, including the following:

(1) The distance between transmitter and receiver
(2) The carrier frequency of the transmission
(3) The physical size of the antenna
(4) The power to be radiated
(5) The effect of the transmission on other users of the same and adjacent channels

Figure 10.11 shows different paths by which a signal radiated from a transmitter can reach a receiver.

10.6.1 Direct Wave

As the name suggests, the signal travels directly from the transmitter antenna to the receiver antenna. This requires that the two antennas be in line of sight. The direct wave is the major mode of propagation for medium-wave AM radio (540 to 1600 kHz), commercial FM broadcasting (88 to 108 MHz), terrestrial microwave relay systems, used for long-distance telephone and television signals (2, 4, 6, 11, 18, and 30 GHz) and satellite transmission systems (4, 6, 8, 12, 14, 17 to 21, 27 to 31 GHz).

10.6.2 Earth-Reflected Wave

Part of the propagated wave is reflected off the surface of the Earth and may arrive at the antenna with a different phase from the direct wave. Depending on the magnitude of the reflected wave, it can cause signal fluctuation and sometimes even complete cancellation of the direct wave. This problem has the most noticeable effect on terrestrial microwave relay systems, where automatic gain control, diversity protection, and adaptive equalization may be used to counteract it.

10.6.3 Troposphere-Reflected Wave

At a distance of approximately 10 km from the Earth's surface, there is an abrupt change in the dielectric constant of the atmosphere. This portion of the upper atmosphere is called the troposphere. High-frequency signals, such as those used in terrestrial microwave relay systems, may be reflected from the troposphere and have the same effect as the Earth-reflected wave at the receiver. On the other hand, the troposphere may be used as part of the communication channel in cases where direct line-of-sight conditions are not possible. This is called tropospheric scatter propagation and the best frequencies for this are 1, 2, and 5 GHz. It is most commonly used by the military for communication over difficult terrain typically over a distance of 300 to 500 km.

10.6.4 Sky-Reflected Wave

Surrounding the Earth at an elevation of approximately 70 km to approximately 400 km is a layer of ionized air caused by constant bombardment of ultraviolet, α, β, and γ radiation from the sun, as well as cosmic rays. It is called the ionosphere and it consists of several layers that have different reflective, refractive, and absorptive effects on radio waves.

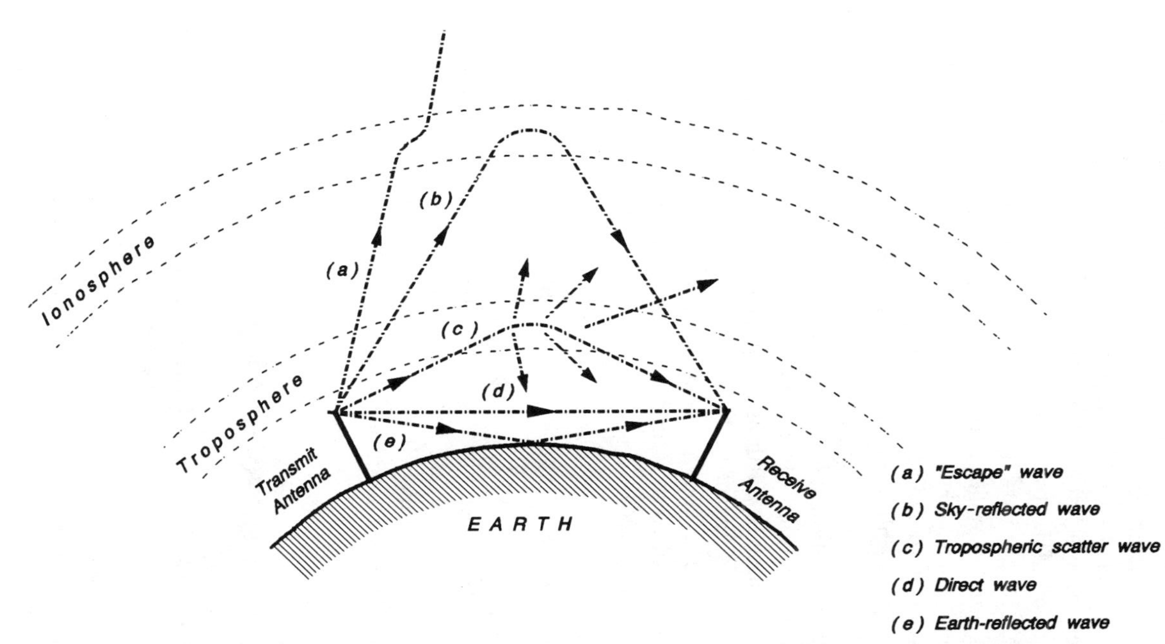

Figure 10.11. Free-space transmission showing various paths. Reprinted with permission from B. P. Lathi, *Modern Digital and Analog Communication Systems*, CBS College Publishing, New York, 1983.

When a radio signal reaches the ionosphere, a number of things can happen depending on the frequency and the angle of incidence. The wave may be reflected back to Earth, it may undergo refraction and eventually be returned to Earth or it may pass through the ionosphere and escape into outer space. With the correct choice of transmission frequency and angle of incidence it is possible to establish communication between two points on the Earth's surface where line-of-sight does not exist. This is the basis of short-wave (3 to 30 MHz) transmission.

The ionosphere changes its nature from day to night and with such phenomena as sun spot cycles and other cosmic events. The turbulent nature of the ionosphere makes communication by shortwave rather unreliable since the signal is subject to fading—both long and short term. This is due to cancellation and/or reinforcement of the different parts of the signal arriving at the receiver by diverse routes. To improve the reliability of communication via the ionosphere, automatic gain control and diversity protection techniques, such as utilization of two or more carrier frequencies, are used.

The signal returning to Earth after reflection from the ionosphere may be reflected from Earth's surface up to the ionosphere once more. On its second return to Earth it may be received by a suitably placed receiver. This phenomenon is described as multiple-hop and can be used to reach places beyond the single-hop distance.

10.6.5 Surface Wave

At very low frequencies (VLF = 3 to 30 kHz), the ionosphere and Earth's surface form two parallel conducting planes and can act as a waveguide. Very low frequencies signals are used for worldwide communication and navigation aids. At higher frequencies, temperature inversions and other local phenomena can generate a surface wave, which can add to the problem of fading.

10.7 TERRESTRIAL MICROWAVE RADIO

Terrestrial microwave radio is a relatively inexpensive medium for long-haul telephone and television signals. Its assigned frequencies are 2, 4, 6, 11, and 18 GHz, which make the system a line-of-sight operation. The distance between repeater stations is approximately 40 km at the lowest frequency and 3 km at the highest frequency, where rain can cause severe attenuation. It is well suited to difficult terrain, where the cost of burying or stringing up on posts any form of cable would be prohibitively expensive. The repeater stations can be placed in strategic positions, such as hill tops with easy access for maintenance personnel.

The block diagrams of a microwave transmitter and receiver are identical to that used in any other radio system. There are intermediate frequency amplifiers, oscillators, modulators (up-converters), demodulators (down-converters), filters, equalizers, and automatic gain control (AGC). However, because of the higher frequencies involved, the hardware used to realize the various functions look very different. What appears to the untrained eye to be a piece of printed circuit board may be, in fact, tuned circuits, transmission lines, open and short circuits, and so on.

At microwave frequencies, the electromagnetic wave behaves increasingly like light. Therefore antennas for microwave radio take the form of parabolic dishes or the horn reflector (hog-horn); these structures focus the electromagnetic wave into a narrow beam for optimal transmission. Parabolic dishes are usually used for one frequency band only and the signals may be polarized in the vertical or horizontal direction. The horn reflector type are multiband and may also be polarized.

A detailed discussion of the design of microwave, antennas, oscillators, amplifiers, modulators, frequency changers, and so on is outside the scope of this book. A limited list of books on the subject are given in the bibliography.

10.7.1 Analog Radio

In analog radio, a signal made up of a large number of voice frequency telephone channels or its equivalent is formed into a basic group, supergroup etc. The formation of the signal to be transmitted was discussed in Section 9.2. The appropriate subcarrier frequencies were given. The next step is to use the composite signal to modulate an intermediate frequency carrier of 70 or 140 MHz before the final up-conversion to microwave frequencies. Frequency modulation is the preferred method.

10.7.1a Terminal Transmitter and Receiver. The block diagram of the transmitter is shown in Figure 10.12*a*. The modulating signal, 70 MHz, itself modulated by a 3600-voice-frequency telephone channels (jumbogroup) or its equivalent, is amplified and used to drive the modulator, mixer, or up-converter. The other input to the up-converter is from the microwave oscillator. The output of the up-converter has to be bandpass filtered to remove undesirable products of the modulation process. After amplification by the radio frequency amplifier, the signal goes to the transmit antenna for radiation.

The receiver is shown in Figure 10.12*b*. The received signal is bandpass filtered to eliminate all but the required signal. It is then mixed with the output of the local oscillator to produce an intermediate frequency signal. After amplification, equalization, and the application of AGC, it goes to a demodulator where the original modulating signal (70 MHz) is recovered. It should be noted that in the case of the 3600-voice-frequency telephone channels, several levels of demodulation have to be carried out before the voice frequency signals are obtained.

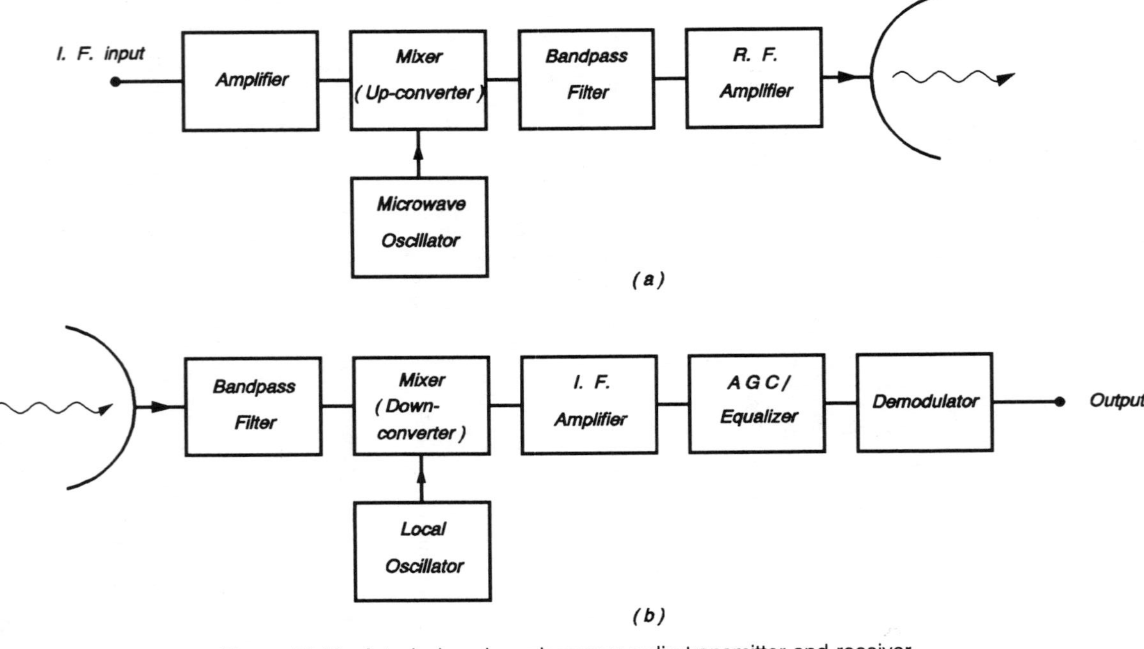

Figure 10.12. A typical analog microwave radio transmitter and receiver.

10.7.1b Repeater. A typical analog microwave radio repeater is shown in Figure 10.13. The antenna is highly directive and it must be secured in a position where it faces the transmitter antenna as directly as possible; small deviation can cause significant loss of signal power. Antennas 1 and 2 are used simultaneously for transmission and reception. The received signal, RF1 goes to the receive circulator, which directs it to the bandpass filter where all signals other than the desired signals are attenuated. The local oscillator and the mixer down-convert the signal to an intermediate frequency. After amplification, equalization, and the application of AGC, the signal is up-converted by the microwave oscillator, and the up-convert mixer to a different frequency RF2 (changing the frequency reduces the possibility of instability in the system). The radio frequency amplifier boosts the signal and feeds it to the transmit circulator, which passes it on to the antenna for onward transmission.

The signal traveling in the opposite direction, RF3 (different from both RF1 and RF2), follows an identical path in the opposite direction. The intermediate frequency used in this path is, in general, different from the one used earlier. The signal leaves the system at a frequency RF4, different from all the others.

10.7.2 Digital Radio

All the functional blocks of the analog radio shown in Figure 10.13 are present in some form or another in the digital radio. The baseband signal to be transmitted may be the output of a digital switch operating at a rate of 1.544, 6.312, 44.736, or 274.176 Mbit/sec. The binary output of the switch is used in a modulation scheme similar to that of the modem discussed in Section 9.4.1. The digit 1 is assigned a frequency f_1 and 0 is assigned a different frequency f_2. Usually f_1 and f_2 are equally displaced from the carrier frequency f_0; that is $(f_0 - f_1) = (f_2 - f_0)$. At the receiving end, the demodulation process converts f_1 and f_2 to 1s and 0s.

10.7.2a Regenerative Repeater. The block diagram of the regenerative repeater is shown in Figure 10.14. The system is identical to that shown in Figure 10.13 except that the signal is demodulated to the baseband and then regenerated, modulated, filtered, up-converted, and amplified for onward transmission. The signal traveling in the opposite direction is subjected to the same processing steps with the exception that the carrier radio frequencies are selected to minimize the possibility of instability. The regenerative repeater is described in Section 9.3.9f.

10.8 SATELLITE TRANSMISSION SYSTEM

A satellite transmission system is just another microwave relay system with a single repeater located in outer space. Because of its height, it can cover a large area of the globe, making it possible to cover the entire surface with

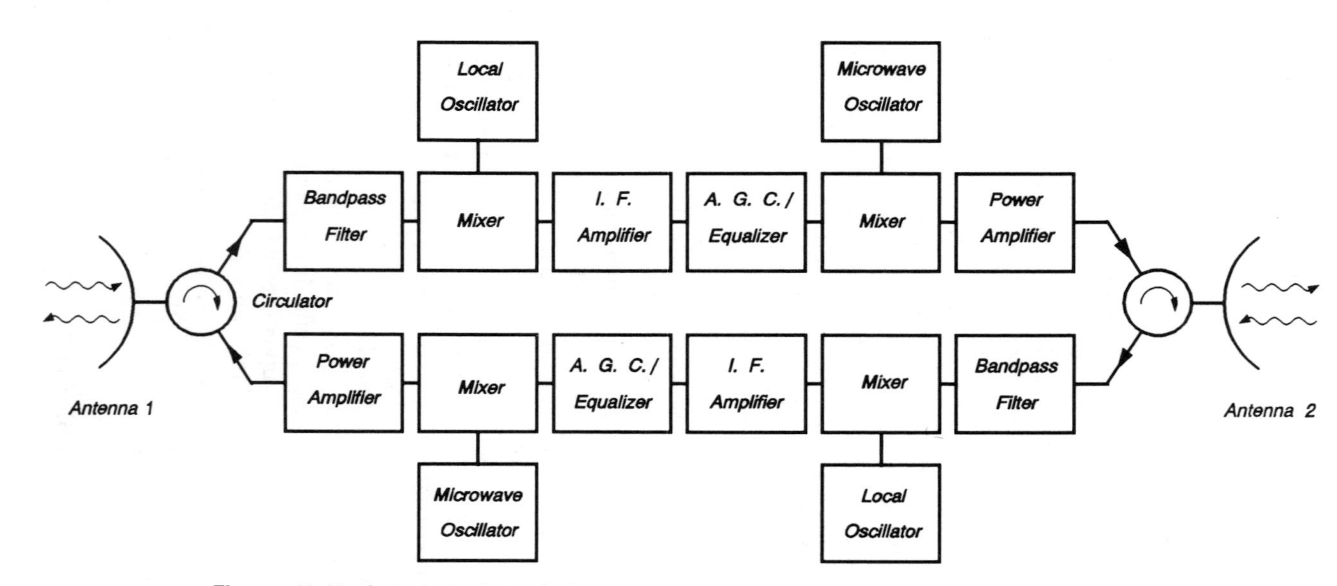

Figure 10.13. A typical analog microwave repeater. Note that the antennas are used for transmission and reception simultaneously.

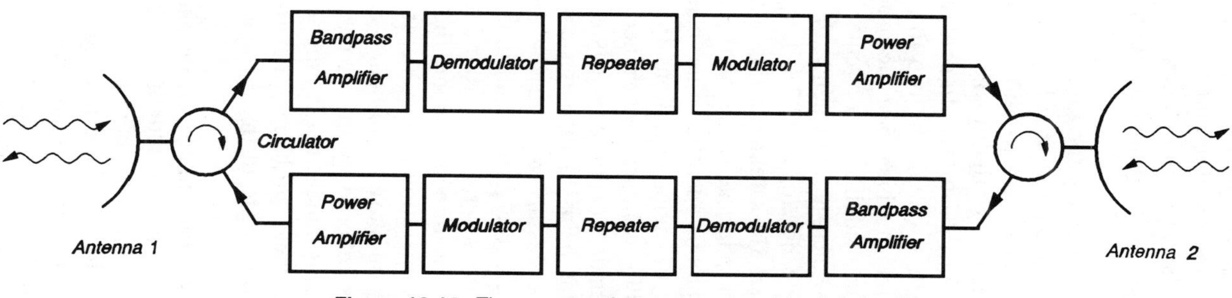

Figure 10.14. The regenerative repeater used in digital radio.

three geostationary satellites. It has applications in point-to-point communications as well as broadcasting, and it can reach remote parts of the earth where other systems cannot reach without large expenditures of money and effort. All that is required is the satellite and the terminal equipment in the earth stations. Satellites are however, not cheap. The cost of launching them and the difficulty of making repairs should anything go wrong, dictate very high levels of reliability.

To avoid the complexity of tracking the satellite from rising to setting and to maintain continuous communication, it is best to "park" the satellite at a point directly above the equator and to choose the correct speed in the direction of rotation of the Earth, so that it stays in a relatively fixed position above a reference point on Earth's surface. In general, satellites rotate in an elliptical orbit with the Earth at one of the two foci of the ellipse. The two foci can be at the same point, in which case the orbit is circular. There are distinct advantages to having a satellite in a circular orbit, because this fixes the distance traveled by the message and hence the delay. To satisfy these conditions, an Earth satellite has to be 42,230 km from the center of the Earth. It is not enough to park the satellite in this position, as it tends to drift away slowly due to the nonspherical shape of the Earth, the gravitational influence of the sun, the moon, and the other planets. A small rocket is provided for the correction of minor deviations from the nominal position. The useful life of the satellite is hence determined by how long the rocket fuel lasts. The satellite is made to spin, or it has a wheel in it which spins. The gyroscopic effect of this helps to further stabilize the satellite. If the satellite itself spins, it is necessary to spin the antenna in the opposite direction to keep it pointing at the Earth at all times.

For about 30 min a day for several days around the equinoxes (March 21 and September 21) the sun is directly behind the satellite and the noise generated by the sun makes operation impossible. It is necessary to switch to an alternate satellite or rely on more Earth-bound means of communication. Once a day, the satellite is eclipsed by the earth, and it not only loses its source of power from the solar panels, but it experiences a drastic change in temperature. To overcome these problems, a battery is provided and all components are designed to operate in the extreme temperature conditions of outer space.

Placing the satellite above the equator means that Earth stations in the extreme north and south are not well illuminated by the satellite. Furthermore, the low angle of elevation of the Earth station antenna makes it vulnerable to atmospheric fading and interference from terrestrial manmade noise. One solution to this problem is to use several satellites in nonequatorial, nonstationary orbits, so that as one satellite sets, the next one is rising. This means that tracking is necessary. The Russian domestic communication satellite system (*Molniya*) was designed on this basis, because most of the territory it is designed to serve is in the Northern Hemisphere.

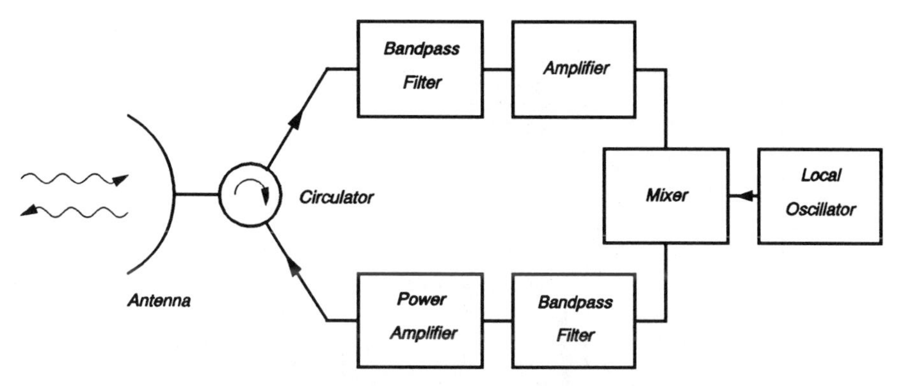

Figure 10.15. The communication satellite transponder. Reprinted with permission from V. I. Johannes, "Transmission system for telecommunications," in D. G. Fink and D. Christiansen (eds.) *Electronic Engineers Handbook*, McGraw-Hill, New York, 1982.

The frequency bands assigned for satellite communication are 4 to 6, 12 to 14 and 19 to 29 GHz. In the higher frequency bands, there is increasing attenuation due to signal power dissipation in rain droplets and water vapor in the atmosphere. Frequency reuse is possible so long as the satellites are sufficiently far apart in orbit for an Earth station to focus on only one of them at a time.

The repeater on a satellite is usually referred to as a *transponder*. The basic structure of a typical transponder is shown in Figure 10.15. The antenna is used for both transmission and reception. The path for a signal arriving from Earth is through the circulator to the bandpass filter where all but the desired band of frequencies containing the carrier are eliminated. Some gain is provided by an amplifier; the mixer, and the local oscillator change the carrier frequency to a different value. After further bandpass filtering to remove the unwanted products of the mixing process, the power amplifier boosts the signal power to the appropriate level for the return trip to Earth through the circulator and antenna. System instability is prevented by the correct choice of circulator characteristics and the change of the carrier frequency.

The loss of signal power between an Earth station and the satellite is approximately 200 dB [10]. The transponder power output is usually 10 W; higher power is not desirable because of the possibility of interference with other Earth-based communication systems. Besides, it is expensive to provide the dc power to run a high-power amplifier in outer space. To overcome these limitations, Earth stations use large antennas and high-power transmitters and very-low-noise amplifiers in the receiver. Sometimes the amplifiers are cooled with liquid nitrogen to achieve the necessary low-noise performance. Figure 10.16 shows the gains and losses of power in a television signal as it makes its way up to the satellite and back down to Earth.

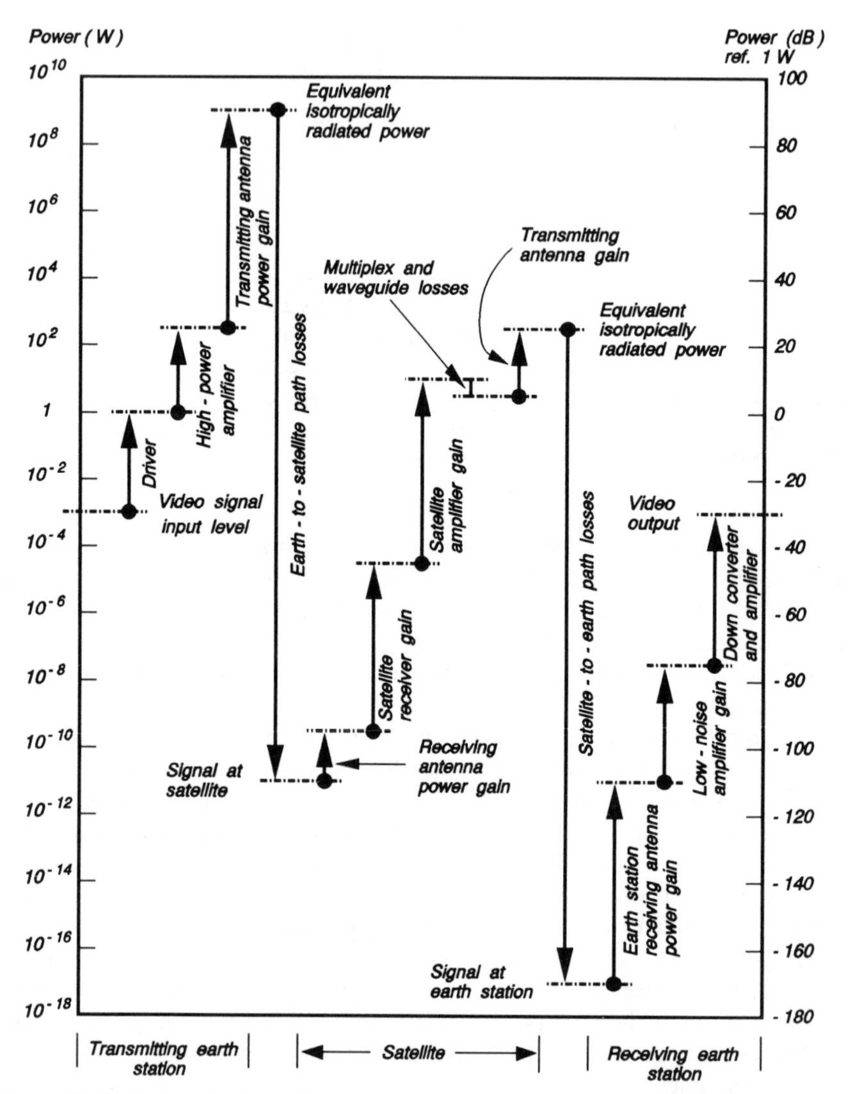

Figure 10.16. Power levels in the transmission of a TV signal via satellite. Reprinted with permission from B. I. Edelson, "Global satellite communications," *Scientific American*, February 1977.

The great distances traversed by the signal from one station to the next causes enough delay (240 msec) to make a telephone conversation over a satellite channel somewhat difficult; one person starts to talk when the other is already talking, without realizing it. This leads to a series of interruptions in the conversation. In general this is not intolerable, but the situation certainly gets much worse when the delay is doubled for a two-way conversation. To get around this problem, the normal practice in a trans-Atlantic telephone conversation is that when one link is via satellite, the other must be via cable. The effect of echo is psychological and can cause people with no speech impediment to stutter. The circuits used in satellite transmission usually have an echo cancellation feature.

To give some idea of how quickly satellite communication systems are changing, it is worth noting that in early 1990 there were some 94 communication satellites in operation with another 194 launchings planned in the near future. The first of a series of the International Telecommunication Satellites (INTELSAT I) was launched in 1965 and it had 240 voice channels. In 1980 INTELSAT V went into service with 12,000 voice and 2 television channels.

REFERENCES

1. Telecom Canada, *Digital Network Notes*, Ottawa, Canada, 1983.
2. R. L. Easton, "Undersea Cable Systems—A Survey," *IEEE Commun. Mag.*, 13(5), 12–15, 1975.
3. W. B. Jones, *Introduction to Optical Fiber Communication Systems*, Holt, Rinehart and Winston, New York, 1988.
4. Staff, *Transmission Systems for Communications*, 5th ed., Bell Telephone Laboratories Inc., 1982. Publisher: Western Electric Co. Inc., Winston-Salem, North Carolina.
5. J. G. Linvill, "Transistor Negative Impedance Converter," *Proc. IRE*, 41, 725–729, 1953.
6. T. Li, "Advances in Optical Fibre Communication: an Historical Perspective," *IEEE J. Selected Areas Commun.*, SAC-1(3), 356–372, 1983.
7. M. I. Schwartz, "Optical Fiber Transmission—From Conception to Prominence in 20 Years," *IEEE Comm. Mag.*, 22(5), 38–48, 1984.
8. W. L. Pritchard, "The History and Future of Commercial Satellite Communications," *IEEE Commun. Mag.*, 22(5), 22–37, 1984.
9. V. I. Johannes, in *Electronics Engineers' Handbook*, D. G. Fink and D. Christiansen (Eds.), McGraw-Hill, New York, 1982.
10. B. I. Edelson, "Global Satellite Communications," *Sci. Am.*, Feb., 1977.
11. B. P. Lathi, *Modern Digital and Analog Communication Systems*, CBS College Pub. New York, NY, p. 334, 1983.

BIBLIOGRAPHY

G. R. Basawapatna and R. B. Stancliff, "A Unified Approach to the Design of Wideband Microwave Solid-State Oscillators," *IEEE Trans.* MTT-27 (5), 379–385, 1979.

T. C. Edwards, *Foundations for Microstrip Circuit Design*, Wiley, New York, 1981.

O. P. Ghandi, *Microwave Engineering Applications*, Pergamon Press, New York, 1981.

G. Gonzalez, *Microwave Transistor Amplifiers: Analysis and Design*, Prentice-Hall, Englewood Cliffs, NJ, 1984.

T. T. Ha, *Solid-State Microwave Amplifier Design*, Wiley, New York, 1981.

T. K. Ishii, *Microwave Engineering*, Harcourt Brace Jovanovich, San Diego, 1989.

S. Y. Liao, *Microwave Circuit Analysis and Amplifier Design*, Prentice-Hall, Englewood Cliffs, NJ, 1987.

T. A. Midford and R. L. Bernick, "Millimeter-Wave CW-IMPATT Diodes and Oscillators," *IEEE Trans.* MTT-27 (5), 483–492, 1979.

M. N. O. Sadiku, *Elements of Electromagnetism*, Saunders Publishing, New York, 1989.

J. White, *Microwave Semiconductor Engineering*, Van Nostrand Reinhold, New York, 1982.

PROBLEMS

10.1 Describe how you would transmit a video signal of approximate bandwidth 1.5 MHz over a twisted-pair cable.

10.2 Derive an expression for the input impedance of the circuit shown in Figure P10.1, assuming that the operational amplifier is ideal. How can this circuit be used to improve the transmission characteristics of the twisted-pair or coaxial cable?

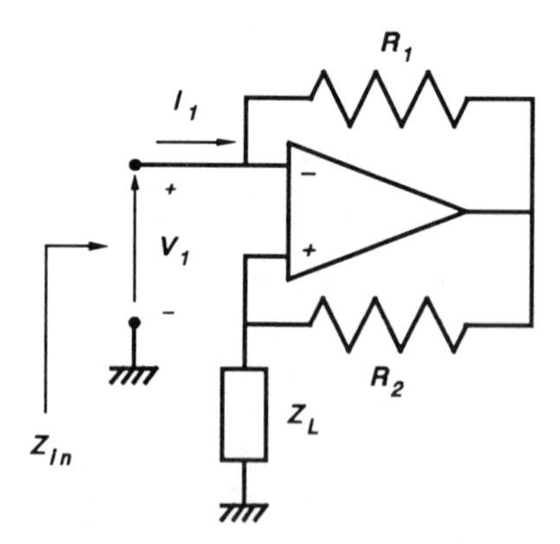

Figure P10.1.

10.3 Describe two techniques that can be used for the compensation of gain loss on a twisted-pair and/or coaxial cable. Discuss how these techniques can be modified to compensate for the relative loss of gain as a function of frequency.

10.4 Discuss the merits of optical fiber transmission relative to satellite communication.

10.5 Discuss the role of the ionosphere in the free-space propagation of very low frequencies (VLF) and shortwave transmission between two points on the Earth's surface.

10.6 Terrestrial microwave radio transmission is subject to severe attenuation in the atmosphere. What elements in the atmosphere are responsible for this; can you suggest a simple explanation of this phenomenon?

10.7 Describe the processes that take place at a typical terrestrial microwave analog relay station using a suitable block diagram. What are "up-conversion" and "down-conversion" and why are they necessary?

10.8 Using suitable block diagrams, discuss the major differences between analog and digital microwave relay stations.

10.9 Compare and contrast the satellite transponder to the microwave relay station paying particular attention to signal processes that occur in them.

APPENDIX A

THE TRANSFORMER

A.1 INTRODUCTION

A transformer is a device that is used for coupling a signal or power from a source to a load in an efficient manner. It does this by changing the impedance of the source to match that of the load. A transformer can also be used for phase inversion and dc isolation. The transformer consists of two or more coils of wire wound on a common core, so that the magnetic field of one coil links with the other coils. When a changing current is applied to one winding, it causes a changing magnetic field in the common core. The changing magnetic field induces a voltage in all the other coils, and current will flow if a path exists. The coils are said to be mutually coupled.

Consider the two coils connected as shown in Figure A.1. The application of KVL to loop 1 gives

$$j\omega L_{11} I_1 + j\omega M_{21} I_2 - E = 0 \tag{A.1}$$

where the voltage $j\omega M_{21} I_2$ is induced in loop 1 as a result of the current I_2 flowing in loop 2; the voltage opposes the applied voltage E. The application of KVL to loop 2 gives

$$j\omega M_{12} I_1 - (j\omega L_{22} + Z_L) I_2 = 0 \tag{A.2}$$

where the voltage $j\omega M_{12} I_1$ is induced in loop 2 as a result of the current I_1 flowing in loop 1.

It is evident that

$$M_{12} = M_{21} = M \tag{A.3}$$

where M is the mutual inductance between the two coils. Two coils may be closely or loosely coupled depending on how much of the magnetic field generated by one links with the other. The coupling coefficient k is related to

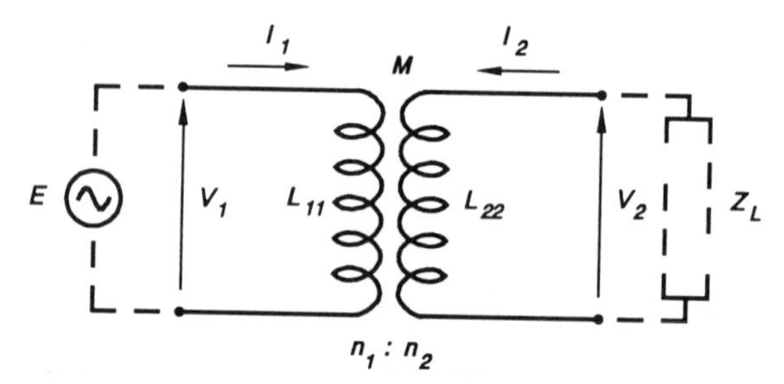

Figure A.1. Two mutually coupled coils with voltages and currents indicated.

the mutual inductance M and the self-inductances of the two coils by

$$k = \frac{M}{\sqrt{L_{11}L_{22}}} \tag{A.4}$$

A.2 THE IDEAL TRANSFORMER

A transformer may be considered to be ideal if it satisfies all the conditions below:

(1) $$a = \frac{n_1}{n_2} = \frac{V_1}{V_2} = \frac{I_2}{I_1} = \sqrt{\frac{L_{11}}{L_{22}}} \tag{A.5}$$

(2) The primary inductance L_{11} and the secondary inductance L_{22} both approach infinity while the ratio L_{11}/L_{22} remains finite
(3) The coupling coefficient $k = 1$
(4) The windings have no resistance and the magnetic core has no losses
(5) Stray capacitances of the coils are zero and stray capacitance between the coils is zero

To investigate the impedance transformation property of the transformer, consider the circuit shown in Figure A.2.

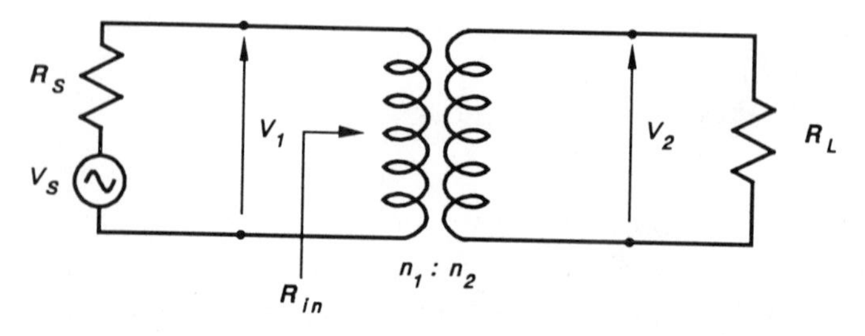

Figure A.2. An ideal transformer with source and load connected.

Assuming that the resistance seen across the primary of the transformer is R_{in} then power flowing into the transformer

$$P_{in} = V_1^2/R_{in} \tag{A.6}$$

Power flowing into the load R_L

$$P_o = V_2^2/R_L \tag{A.7}$$

Since the transformer is ideal, $P_{in} = P_o$

$$R_{in} = (V_1/V_2)^2 R_L \tag{A.8}$$

But

$$a = V_1/V_2 \tag{A.9}$$

Therefore

$$R_{in} = a^2 R_L \tag{A.10}$$

The source then sees a load equal to $a^2 R_L$ instead of R_L if the load had been connected directly to the source. Since the turns ratio, a, is a variable, it can be chosen to optimize the transfer of power to the load. This occurs when

$$a^2 R_L = R_s \tag{A.11}$$

The result can be generalized for the load impedance

$$Z_L = R_L + jX_L \tag{A.12}$$

and the source impedance

$$Z_s = R_s + jX_s \tag{A.13}$$

as

$$Z_s^* = a^2 Z_L \tag{A.14}$$

A.3 THE PRACTICAL TRANSFORMER

A transformer can be constructed in which the conditions in Equation (A.5) apply within a reasonable approximation. However, it is impossible to make the self-inductances L_{11} and L_{22} infinite. The coupling coefficient k can be made approximately equal to unity by using bifilar windings on a high permeability core. Winding resistance and core losses can be reduced consid-

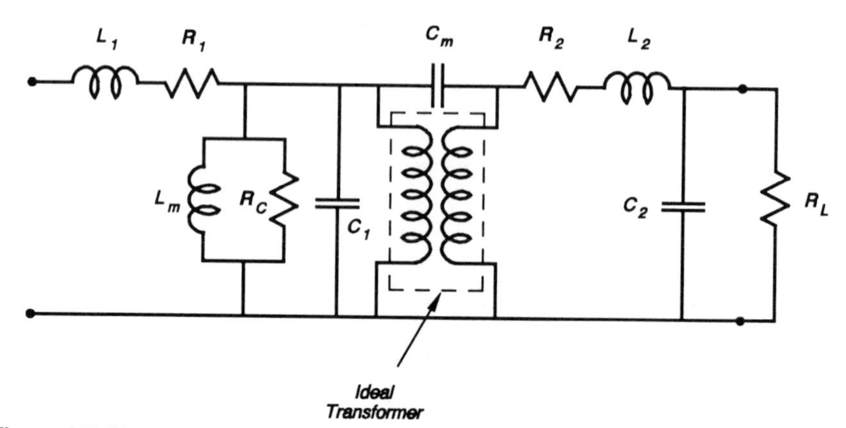

Figure A.3. The practical loaded transformer consisting of the ideal transformer with all sources of error shown outside it.

erably by a judicious choice of materials. Stray capacitance can be minimized increasing the thickness of insulation between layers of the windings. By taking into account all the imperfections, an equivalent circuit of the practical transformer is shown in Figure A.3, where

R_1 is the winding resistance of the primary
R_2 is the winding resistance of the secondary
L_1 is the leakage inductance of the primary
L_2 is the leaking inductance of the secondary
C_1 is the total stray capacitance of the primary
C_2 is the total stray capacitance of the secondary
R_c represents the core losses
L_m is the primary inductance in which the magnetizing current flows
C_m is the total stray capacitance between primary and secondary.

A number of the components shown in Figure A.3 can be made negligible by taking the following precautions:

(1) By using low resistance, wire for the windings will make R_1 and R_2 negligibly small compared to the source and load resistances.
(2) By using bifilar windings in which the primary and secondary coils are wound side by side, turn by turn, on a highly permeable core, the leakage inductances L_1 and L_2 can be considered to have been eliminated. It is evident that the use of bifilar windings will increase the value of C_m.

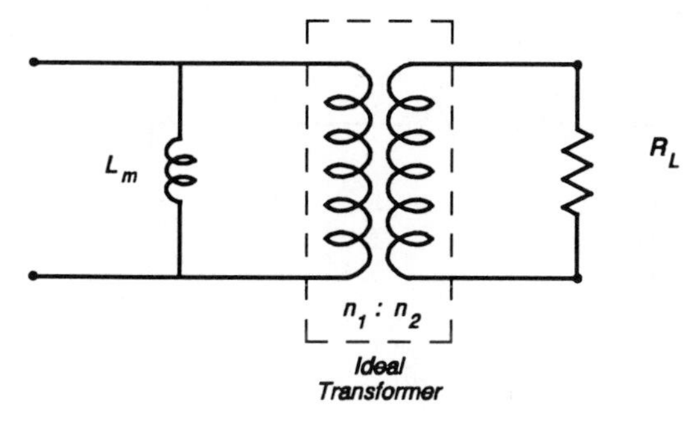

Figure A.4. A first-order approximation of the practical transformer.

(3) The use of a thicker insulation between layers of winding will reduce C_1 and C_2 but will increase the leakage inductances.
(4) By laminating the core, thus increasing the resistance of the path for eddy currents to flow, the core losses can be reduced and R_c can be left out of the equivalent circuit.

It should be noted that some of the steps described above are contradictory and suitable compromises have to be made to get the simplified equivalent circuit shown in Figure A.4.

The ideal transformer and its load R_L can be replaced by a resistance equal to a^2R_L as shown in Figure A.5. By applying Thevenin's theorem to the left of $X - X'$, the circuit can be simplified as shown in Figure A.6

$$V_{th} = \frac{a^2R_L}{a^2R_L + R_s}V_s \tag{A.15}$$

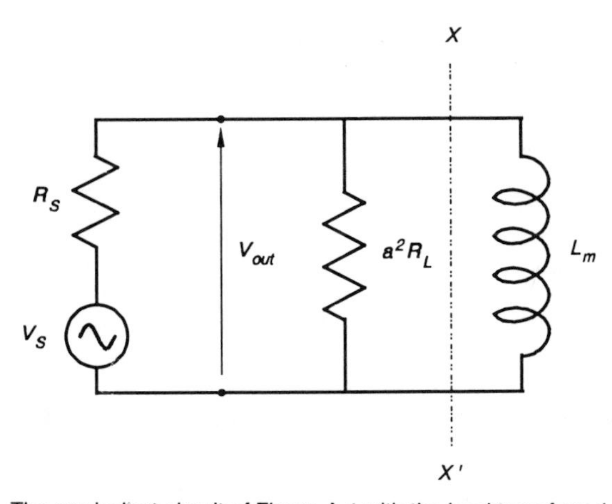

Figure A.5. The equivalent circuit of Figure A.4 with the load transferred to the primary.

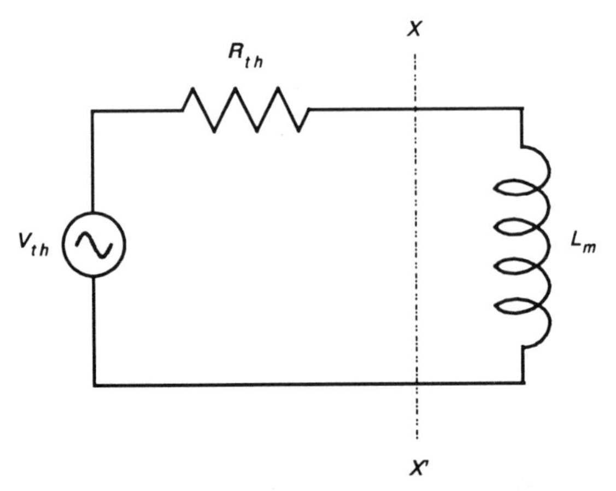

Figure A.6. The equivalent circuit of Figure A.5 when Thevenin's theorem has been applied to the left of XX'.

It can be seen that this is a high-pass filter with a slope of 6 dB/octave and a corner frequency

$$\omega_c = \frac{a^2 R_L R_s}{L_m(a^2 R_L + R_s)} \tag{A.16}$$

It is advantageous to make L_m (the primary inductance L_{11}) as large as possible in order to keep the corner frequency as low as possible.

The condition for maximum power transfer exists, if

$$a^2 R_L = R_s \tag{A.17}$$

Then

$$\omega_c = \frac{R_s}{2L_m} \tag{A.18}$$

APPENDIX B

DESIGNATION OF FREQUENCIES

30 – 300	Hz	Extremely Low Frequency (ELF)
300– 3	kHz	Voice Frequency (VF)
3 – 30	kHz	Very Low Frequency (VLF)
30 – 300	kHz	Low Frequency (LF)
300– 3	MHz	Medium Wave Frequency (MW)
3 – 30	MHz	Short Wave Frequency (SW)
30 – 300	MHz	Very High Frequency (VHF)
300– 3000	MHz	Ultra High Frequency (UHF)
3 – 30	GHz	Super High Frequency (SHF)
30 – 300	GHz	Extremely High Frequency (EHF)

Channel	Frequency Band (MHz)	Video Carrier (MHz)
	VHF Television Frequencies	
1	(Not normally used)	
2	54–60	55.25
3	60–66	61.25
4	66–72	67.25
5	76–82	77.25
6	82–88	83.25
	Commercial FM Broadcast Band (88–108 MHz)	
7	174–180	175.25
8	180–186	181.25
9	186–192	187.25
10	192–198	193.25
11	198–204	199.25
12	204–210	205.25
13	210–216	211.25
	UHF Television Frequencies	
14	470–476	471.25
15	476–482	477.25
16	482–488	483.25
17	488–494	489.25
18	494–500	495.25
19	500–506	501.25
20	506–512	507.25
21	512–518	513.25
22	518–524	519.25
23	524–530	525.25
24	530–536	531.25
25	536–542	537.25 etc.

The channels follow in order, each one being 6 MHz in bandwidth with the carrier at 1.25 MHz above the lower cut-off frequency. The total number of channels is 69 and channel 69 has a bandwidth from 800 to 806 with the carrier at 801.25 MHz. It should be noted that in every case the FM carrier for the audio is 4.5 MHz above the video carrier.

APPENDIX C

THE ELECTROMAGNETIC SPECTRUM

Figure C.1 shows a simplified electromagnetic spectrum starting from sub-audio frequency to gamma rays spanning nearly 18 decades of frequency. The corresponding wavelengths are given. The bandwidths of some common telecommunication media are indicated.

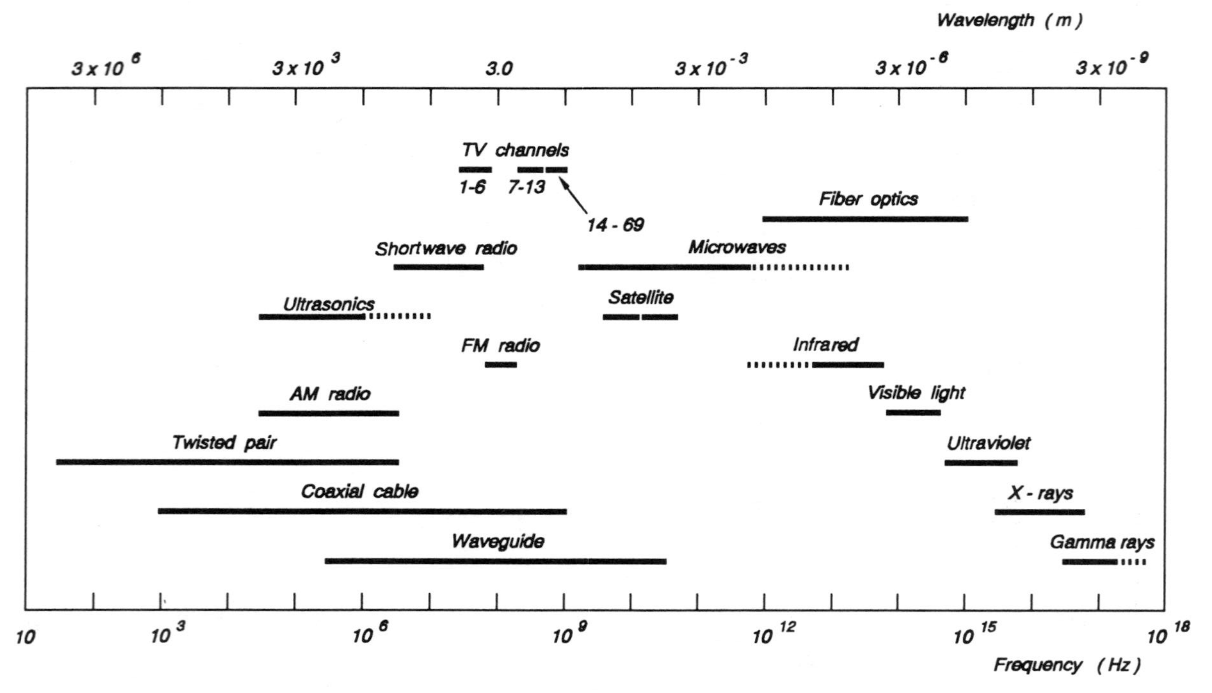

Figure C.1. A simplified electromagnetic spectrum.

INDEX